These two volumes are the proceedings of a major International Symposium on General Relativity held at the University of Maryland 27 to 29 May 1993 to celebrate the sixtieth birthdays of Professor Charles Misner and Professor Dieter Brill. Colleagues, friends, collaborators and former students, including many of the leading figures in relativity, have contributed to the volumes, which cover classical general relativity, quantum gravity and quantum cosmology, canonical formulation and the initial value problem, topology and geometry of spacetime and fields, mathematical and physical cosmology, and black hole physics and astrophysics. As invited articles, the papers in these volumes have an aim which goes beyond that of a standard conference proceedings. Not only do the authors discuss the most recent research results in their fields, but many also provide historical perspectives on how their subjects developed and offer individual insights in their search for new directions. The result is a collection of novel and refreshing discussions. Together they provide an authoritative and dynamic overview of the directions of current research in general relativity and gravitational theory, and will be essential reading for researchers and students in these and related fields.

Directions in General Relativity

Proceedings of the 1993 International Symposium, Maryland

VOLUME 1

Professor Charles W. Misner

Directions in General Relativity

Proceedings of the 1993 International Symposium, Maryland

VOLUME 1

Papers in honor of
Charles Misner

B. L. Hu
University of Maryland

M. P. Ryan Jr.
Universidad Nacional Autónoma de México

C. V. Vishveshwara
Indian Institute of Astrophysics

CAMBRIDGE
UNIVERSITY PRESS

CAMBRIDGE UNIVERSITY PRESS
Cambridge, New York, Melbourne, Madrid, Cape Town, Singapore, São Paulo

Cambridge University Press
The Edinburgh Building, Cambridge CB2 2RU, UK

Published in the United States of America by Cambridge University Press, New York

www.cambridge.org
Information on this title: www.cambridge.org/9780521452663

First published 1993
This digitally printed first paperback version 2005

A catalogue record for this publication is available from the British Library

ISBN-13 978-0-521-45266-3 hardback
ISBN-10 0-521-45266-X hardback

ISBN-13 978-0-521-02139-5 paperback
ISBN-10 0-521-02139-1 paperback

CONTENTS

viii

CONTENTS

CONTENTS

CONTENTS

CONTRIBUTORS

Jeeva S. Anandan
Department of Physics and Astronomy, University of South Carolina, Columbia, SC 29208, USA

Richard L. Arnowitt
Center for Theoretical Physics, Department of Physics, Texas A & M University College Station, TX 77843-4242, USA

Abhay Ashtekar
Department of Physics, Syracuse University, Syracuse, NY 13244-1130, USA

Beverly K. Berger
Physics Department, Oakland University, Rochester, MI 48309, USA

Dieter Brill
Department of Physics, University of Maryland, College Park, MD 20742-4111, USA

Stanley Deser
Martin Fisher School of Physics, Brandeis University, Waltham, MA 02254, USA

George F.R. Ellis
Department of Applied Mathematics, University of Cape Town, Cape Town 7700, South Africa

David Finkelstein
Department of Physics, Georgia Institute of Technology, Atlanta, Georgia 30332-0430

James B. Hartle
Department of Physics, University of California, Santa Barbara, CA 93106-9530, USA

CONTRIBUTORS

William A. Hiscock
Department of Physics, Montana State University, Bozeman, MT 59717, USA

Sergio Hojman
Department de Física, Facultad de Ciencias, Universidad de Chile, Casilla 653, Santiago, Chile

Bei-Lok Hu
Department of Physics, University of Maryland, College Park, MD 20742-4110, USA

James Isenberg
Department of Mathematics and Institute of Theoretical Science, University of Oregon, Eugene, OR 97403, USA

Werner Israel
Department of Physics, University of Alberta, Edmonton, Alberta T6G 2J1 Canada

Martin Jackson
Department of Mathematics and Computer Sciences, University of Puget Sound, Tacoma, WA 98416, USA

Karel V. Kuchař
Department of Physics, University of Utah, Salt Lake City, Utah 84112, USA

D. R. Matravers
School of Mathematical Studies, Portsmouth Polytechnic, Portsmouth P01 2EG, England

Richard A. Matzner
Center for Relativity, University of Texas, Austin, TX 78712, USA

Vincent Moncrief
Department of Physics, Yale University, 217 Prospect Street, New Haven, CT 06511, USA

CONTRIBUTORS

Pran Nath
Department of Physics, Northeastern University, Boston, MA 02115

James M. Nester
Department of Physics, National Central University, Chung-Li, Taiwan 32054, Republic of China

Yavuz Nutku
Department of Mathematics, Bilkent University, 06533 Ankara, Turkey

Denjoe O'Connor
Dublin Institute for Advanced Studies, 10 Burlington Road, Dublin 4, Ireland

Juan Pablo Paz
T6 Theoretical Astrophysics, MS 288, Los Alamos National Laboratory, Los Alamos, NM 87545, USA

Roger Penrose
Mathematical Institute, The University of Oxford, 24-29 St. Giles, Oxford OX1 3LB, UK

Kay-Thomas Pirk
Max Planck Institut für Astrophysik, Karl Schwarzschild Strasse 1 D-8046 Garching bei Müchen, Germany

A. K. Raychaudhuri
Relativity and Cosmology Centre, Physics Department, Jadavpur University, Jadavpur, Calcutta, 700 032, India

Michael P. Ryan, Jr.
Instituto de Ciencias Nucleares, Universidad Nacional Autónoma de México, Apartado Postal 70-543, 04510 DF, Mexico

David Samuel
Department of Physics, Montana State University, Bozeman, MT 59717, USA

Dennis Sciama
SISSA, Strada Costierall, 34014 Trieste, Italy

CONTRIBUTORS

and Department of Astrophysics, Oxford University, 1 Keble Road, Oxford OX1 3NP, UK

Stuart L. Shapiro
Center for Radiophysics and Space Research, Cornell University, Ithaca, NY 14853, USA

Sukanya Sinha
Instituto de Ciencias Nucleares, Universidad Nacional Autónoma de México, Apartado Postal 70-543, 04510 DF, Mexico

Chris R. Stephens
Instituut voor Theor. Fisica, Rijksuniveriteit Utrecht, Princetonplein 5, P.O. Box 80.006, 3508TA Utrecht, The Netherlands

Alan R. Steif
DAMTP, Cambridge University, Silver Street, Cambridge, CB3 9EW, UK

Ranjeet S. Tate
Physics Department, University of California, Santa Barbara, CA 93106-9530

Saul A. Teukolsky
Center for Radiophysics and Space Research, Cornell University, Ithaca, NY 14853, USA

Kip S. Thorne
Department of Theoretical Astrophysics, California Institute of Technology, Pasadena, CA 91125, USA

Claes Uggla
Department of Physics, Syracuse University Syracuse, NY 13244-2230, USA

C. V. Vishveshwara
Indian Institute of Astrophysics, Bangalore 560 034, India

Robert M. Wald
Enrico Fermi Institute and Department of Physics, University of Chicago, 5640 S. Ellis Avenue, Chicago, IL 60637, USA

CONTRIBUTORS

Joseph Weber
Department of Physics, University of Maryland, College Park, MD 20742-4110, USA

John A. Wheeler
Department of Physics, Princeton University, Princeton, NJ 08544, USA

G. Wilmot
Department of Physics, University of Maryland, College Park, MD 20742-4110, USA

James W. York, Jr.
Department of Physics and Astronomy, University of North Carolina, Chapel Hill, NC 27599, USA

an International Symposium on

DIRECTIONS IN GENERAL RELATIVITY

in celebration of the sixtieth birthdays of
Professors Dieter Brill and Charles Misner

University of Maryland, College Park, May 27-29, 1993

Time	Thursday, May 27	Friday, May 28	Saturday, May 29
9:00-9:45	Deser	B. DeWitt	Bennett
9:45-10:30	Choquet	Horowitz	Sciama
10:30-11:00	*break*	*break*	*break*
11:00-11:45	Moncrief	Ashtekar	Matzner
11:45-12:30	C. DeWitt/ Cartier	Kuchar	Thorne
12:30-2:00	*lunch*	*lunch*	*lunch*
2:00-2:45	Wald	Penrose*	Hawking*
2:45-3:30	York	Sorkin	*Directions in GR*
3:30-4:00	*break*	*break*	*Disc. leader*: Hartle
4:00-4:45	*Classical Relativity*	*Quantum Gravity*	Wheeler
4:45-5:30	*Disc. leader*: Isenberg	*Disc. leader*: Smolin	

* Unconfirmed as of Mar. 1, 1993.

Organization

Advisory Committee : Deser, C. DeWitt, Choquet, Hartle, Sciama,
Toll, Wheeler
Scientific Committee: Hu, Isenberg, Jacobson, Matzner, Moncrief, Wald
Local Committee: Hu, Jacobson, Romano, Simon
Festschrift Editors: Hu, Jacobson, Ryan, Vishveshwara

Sponsors

National Science Foundation
University of Maryland

Papers from Both Volumes Classified by Subject

GEOMETRY, GAUGE THEORIES, AND MATHEMATICAL METHODS

QUANTUM THEORY

QUANTUM GRAVITY

BLACK HOLES

RELATIVISTIC ASTROPHYSICS AND COSMOLOGY

Preface

Charlie Misner's contributions, characterized by profound physical insight and brilliant mathematical skill, have left an indelible mark on general relativity during its course of development for more than the past three decades. Equally important has been his influence on his colleagues and coworkers. To his students he has been a gentle guide, a model mentor and a source of inspiration. Charlie's curriculum vitae included at the end of this volume offers a glimpse of his scholarship and achievements. At the same time, the excerpts from the messages gathered for him on the occasion of his sixtieth birthday, June 13, 1992, are an eloquent testimony to the affection, respect and gratitude of his friends, colleagues and students.

The articles that follow have been written by experts in their respective fields. The areas covered range over a wide spectrum of topics in classical relativity, quantum mechanics, quantum gravity, cosmology and black hole physics. The latest developments in these subjects have been presented, often with reference to the perspective of the past and with indications of future directions. One can discern in most of these articles the influence of Charlie Misner in one form or another.

A novel feature of this Festschrift is that it represents the time-reversed version of the proceedings of an international symposium on *Directions in General Relativity* organized at the University of Maryland, College Park, May 27–29, 1993, at which the contents of some of these articles and related topics will be discussed in detail. The symposium is in honour of Charles Misner as well as Dieter Brill whose sixtieth birthday falls on August 9, 1993. The second volume of the proceedings edited by Ted Jacobson and one of us (BLH) is a Festschrift for Dieter.

Ever since we announced a symposium and festschrift for these two esteemed scientists in the Fall of 1991, we have been blessed with enthusiastic responses from friends, colleagues and former students of Charlie and Dieter all around the globe. Without their encouragement and participation this celebration could not have been realized.

The advisors and organizers of this symposium have shared with us their wisdom and effort. For this volume we would like to thank Professors Beverly Berger, Dieter Brill, James Isenberg and Richard Matzner for their contributions to the article on Professor Misner's research. One of us (BLH) would like to offer a special word of appreciation to Professor Stanley Deser, who made valuable suggestions from local problems to global issues; and to Professor Theodore Jacobson, for his thoughtful and unfailing help throughout the planning and execution of these events. He also wishes to thank Prof. Jordan Goodman for letting us use the VAX facilities, Drs. Joseph Romano and Jonathan Simon for their generous help with the organization of the symposium, Mrs. Betty Alexander for her patient typing and corrections, and Mr. Andrew Matacz, Drs. Joseph Romano and Jonathan Simon for their skillful TEXnical assistance in the preparation of manuscripts in both volumes.

We gratefully acknowledge support from the National Science Foundation and the University of Maryland.

We wish both Charlie and Dieter many more years of joyous creativity and inspiration.

Bei-Lok Hu
Michael Ryan
C. V. Vishveshwara

Charles W. Misner:
Insight and Discovery

This summary of Charles Misner's publications only hints at the richness of his research. Misner's work is characterized by a fascination with geometry in its broadest sense and by a desire to probe the physical manifestations of gravitation. Many of his papers initiated new areas of study in general relativity. These areas either provide continuing research interest, have experienced one or more revivals, or have developed from an essential ingredient he provided. This review will emphasize those aspects of his research which have become part of the essential background of our subject. References [n] are to Misner's list of publications near the end of this volume.

To appreciate Misner's impact on general relativity one need only recall the state of this field when he began his research at Princeton in the 1950's. Major activities in the previous decades included the then-ignored work by Oppenheimer and Snyder on gravitational collapse and by Alpher and Herman predicting a 5 degree cosmic background radiation. Apart from cosmology, the appreciated work included the Einstein-Infeld-Hoffman equations of motion results from the late 1930's, Bergmann's studies of quantum gravity from the early 1950's, and the studies of the initial value problem by Lichnerowicz and Fourès (Choquet-Bruhat). Active centers with an interest in general relativity as Wheeler started his group at Princeton included those led by Bergmann at Syracuse, Lichnerowicz and Fourès-Bruhat in France, Bondi in London, Klein and Møller in Scandinavia, Synge and Pirani in Dublin (one of Schild's sojourns also), Jordan and Ehlers in Hamburg, Infeld in Warsaw, and a few others. However, the state of the subject was in great flux. For example, arguments were raging over whether pure gravitational waves exist and carry energy. All this changed as a result of several developments tied to Wheeler's group (e.g., Misner, Weber, Brill) , to the concurrent Bondi-Sachs analysis of metric behavior along outgoing null hypersurfaces, and especially with the analysis by Misner (with Arnowitt and Deser) of the dynamical properties of the gravitational field. At about the same time that Dirac constructed a Hamiltonian formulation of general relativity, Arnowitt, Deser, and Misner (ADM) [5–9, 11–17, 4^B, 5^B, 7^B] independently began their study of the dynamical structure of the theory. An excellent summary of this effort is found in [4^B]. In close analogy with the parameterized formulation of the non-relativistic particle and the gauge degree of freedom of the electromagnetic field, they showed how to extract the true dynamical degrees of freedom of the gravitational field from the original "already parameterized" Hamiltonian formulation. In addition, they constructed physically meaningful measures of gravitational energy and momentum and demonstrated the distinction between physical and coordinate waves. The Einstein-Hilbert action was expressed in 3+1 form to yield a spacetime as the evolution of spaces. The metric and connection varied independently (the Palatini form of the action) generated a first order formalism in terms of the induced metric h_{ij} and extrinsic curvature K_{ij} of the spacelike hypersurface. The structure of Einstein's equations was then seen as evolution equations for h and K (or the related π conjugate to h) off the hypersurface, supplemented by four constraints (preserved by the Bianchi identities during the evolution) that were functionals of the hypersurface initial data. To identify the dynamical degrees of freedom, four additional functionals of the initial data were imposed as coordinate conditions and the constraints solved for their conjugate momenta. The two remaining degrees of freedom are dynamical while the negative of the momentum conjugate to the intrinsic time becomes the true ADM Hamiltonian. As an example, ADM proposed a set of coordinates that allowed them to define the energy and momentum of gravitational fields in asymptotically flat spacetimes and to explore the properties of gravitational waves in analogy with linearized theory. The ADM analysis is a cornerstone of modern general relativity. Its many subsequent applications include positive energy theorems, canonical quantization, the initial value problem, quantum cosmology, and numerical relativity. The ADM formalism also provided a useful tool for Misner's subsequent research.

Misner had learned an initial value formulation from Lichnerowicz's text (Paris 1955), and claimed [1^B] at the Chapel Hill conference that Wheeler's sketches of wormholes could be turned into rigorous initial conditions, which he later produced [10]. His interest in the initial value problem, and in finding physically interesting examples, injected this technology not only into the ADM collaboration, but also into Wheeler's group where Brill soon used it in showing that large classes of pure gravitational wave initial conditions existed and had positive mass.

Prior to the mid-1960s cosmology as the study of the real universe was a quiet, almost moribund subject. (The exceptions were the primordial nucleosynthesis work of Gamow, Alpher, and Herman and the associated prediction of the cosmic microwave background, and the countervailing development of steady state cosmology by Bondi, Gold, and Hoyle which also stimulated much activity in radio astronomy.) Classical observational cosmology concentrated on the determination of the Hubble constant and the deceleration parameter. Theoretical cosmologists such as Behr, Heckmann and Schücking, Shepley, Melvin, Thorne, Taub, and others were beginning to investigate anisotropic (but mostly still homogeneous) cosmologies. At Princeton, Dicke was considering local effects of the universe's dynamics. Misner's interest in cosmology, encouraged by Sciama, was stimulated by the work of Peebles on nucleosynthesis and galaxy formation and by the Dicke and Peebles interpretation of the Penzias and Wilson observations of microwave background radiation. A spurious anisotropy in the distribution of known QSO's influenced Misner to begin looking at cosmologies more complicated than the isotropic homogeneous ones. Thus began his series on chaotic cosmologies. Could essentially generic cosmologies evolve by understandable physical processes to be like what we see today?

The first paper in this series [30] discussed the damping of inequalities within the horizon by neutrino viscosity. Misner noted that smoothing occurred by this mechanism only within the horizon (a few light seconds in size) at the time when the neutrinos became essentially collisionless. As yet no definite notice is taken of what we now call the "horizon problem"—how can opposite sides of the sky have the same temperature if they are not even now in each other's past light cone?

Then [31] Misner applied a similar analysis to homogeneous but anisotropic cosmologies. Here he did address the large-scale temperature anisotropy problem, more closely attacking the horizon problem. He pointed out that very large-scale anisotropies in the expansion of the universe would modify the distribution function describing essentially collisionless radiation (neutrinos). Neutrino interactions (which are dissipative due to equilibration of the very different energies of the anisotropically red or blue shifted neutrinos) tend to restore the isotropy of the distribution. Misner used a viscosity approximation, which is valid in the case that the neutrinos collide frequently (compared to the expansion rate of the universe). If this approximation were valid (as is assumed in [31]), the initial anisotropy can be totally dissipated. However, in realistic cosmologies, the long mean free path limit is correct, so that this mechanism does not operate. Then [32] Misner developed fully the long mean free path (neutrino) distribution function in homogeneous anisotropic Bianchi I cosmologies. The distribution function for the collisionless neutrinos is a function of the conserved anisotropic components of the neutrino momentum. This causes the initial thermal distribution to evolve anisotropically to yield a distribution with an angularly dependent temperature. The computed stresses for these strongly anisotropic neutrinos act to limit the excursion of the anisotropy and to cause it to oscillate with decreasing amplitude around the value zero as the universe expands. This yielded a prediction of the present microwave temperature anisotropy based on the assumption that the dissipation had removed most of the anisotropy before a redshift $\sim 10^{10}$. The photons' distribution function can be tracked when they become collisionless at $z \sim 10^3$. Although the underlying mechanism may be invalid, this type of argument led to upper limits on the microwave anisotropies that for years were much stronger than the observations.

Misner next [34] applied similar techniques to the (essentially vacuum) dynamics of the Bianchi IX (closed) anisotropic cosmology he called the Mixmaster universe. Here the curvature itself rather than the neutrino content provided restoring forces which limited the anisotropy during expansion. In this paper there was a full-fledged attack on the horizon problem. Misner showed here that there are repeated epochs in which the evolution acts to remove horizons to allow communication around the universe and thus solve the "horizon problem" (now popularly addressed by the inflationary scenario). However, the horizon-removing epochs are too rare to have happened since the Planck era for the mechanism to operate. The approach to the singularity through an infinite sequence of Kasner models (noted independently and elaborated by Belinskii, Lifshitz, and Khalatnikov) appeared to satisfy Misner's criteria for chaotic cosmology. (This inspired him to name the model after the Mixmaster brand electric food mixer.) The characterization of the amount of chaos in Mixmaster dynamics is once again the subject of active investigation. This model has also been the basis of a number of studies of quantum behavior, of chaos in Hamiltonian systems, and of rotating cosmologies.

The chaotic classical Mixmaster dynamics combined with Misner's studies in ADM quantization of homogeneous cosmologies [36] caused him to argue [37] that in fact the Planck epoch does not represent a dramatic change in the behavior of the cosmology. In his formulation, the independent degrees of freedom adiabatically increase their energy (looking at the universe evolving backward to the singularity) but don't change their quantum number. "Thus modes that are classical now (high quantum number) remain classical at the singularity." This means that in the approach to the singularity, the Mixmaster chaos need not be cut off at the Planck time. Thus all the time back to the absolute zero is available allowing an infinite number of (quasi-periodic in logarithmic time) horizon-removing epochs to have actually occurred. Since this result is factor-ordering dependent, relies on the validity of the minisuperspace approximation, and fails to answer the question of how the modes became classical in the first place, it has not been generally accepted. On the other hand, the inflationary scenario can be viewed as an example of the chaotic cosmology idea since it addresses the same issues. It predicts small current fluctuations—and suffers from the same problems as Misner's early dissipation work. The smooth state is an attractor but the solutions do not necessarily settle completely to the smooth state. Discussions of these points are frustrated by lack of a *measure* for initial parameters to characterize the probability for successful inflationary (or generally chaotic) models.

Misner's work in astrophysical cosmology contributed to an explosion of new work in this field. It introduced new words to the language (often misused)—neutrino viscosity, Mixmaster universe. The mathematical techniques he used became standard.

Misner's astrophysical and mathematical cosmology cannot be separated from the quantum studies. At the famous Chapel Hill conference in 1957, Misner presented results (based in part on his Ph. D. thesis) on Feynman quantization of general relativity [1]. Even then (influenced by Wheeler), he strongly believed that to quantize gravity required an approach different from that for other fields. "If gravity is to occupy a significant place in modern physics, it can do so only by being *qualitatively* different from other fields. As soon as we assume gravity behaves qualitatively like other fields, we find that it is quantitatively insignificant." The ingredients and issues—measure on superspace, treatment of general coordinate invariance, the role of time, identification of states and observables—were all present in this work and are still with us. Further progress, however, required a greater understanding of the dynamical structure of general relativity represented, for example, by the ADM collaboration [4^B]. One of the main objectives of the ADM program was to use their identification of the true dynamical degrees of freedom to construct the quantum theory. They were motivated in part by the contemporaneous ideas of Pauli, Klein, Deser, and others that quantum gravity might provide a universal cutoff for matter. (Many believe the unification of gravity and matter within a finite superstring framework to have realized this effect.)

Following DeWitt's application of Dirac quantization [i.e. solving the Wheeler-DeWitt equation] to the Friedmann-Robertson-Walker (FRW) universe, Misner began the program of research he christened "quantum cosmology" to apply the ADM formalism to spatially homogeneous universes. The procedure was to identify a convenient set of variables in minisuperspace (the reduction of Wheeler's superspace of all 3-geometries to those which are spatially homogeneous), to select as a time variable one of the directions in minisuperspace (usually taken to be the logarithmic volume), and to identify the negative of its canonically conjugate momentum expressed as a solution to the Hamiltonian constraint equation to be the ADM Hamiltonian. In general, this H_{ADM} was a function of the dynamical degrees of freedom and (usually) the intrinsic time. However, Misner recognized that a quantum theory based on H_{ADM} is ambiguous even for simple models. Since the single degree of freedom of the FRW models is used up as the time variable, the three (or more) degrees of freedom Bianchi I and IX (Mixmaster) cosmologies were used to test the formalism

[36, 16^B, 17^B]. The Bianchi I model is equivalent to a massless free particle in 2+1 dimensions. The quantization could proceed via a Schrödinger equation in H_{ADM} or a Klein-Gordon equation in H^2_{ADM}. (The Klein-Gordon equation is equivalent to the Wheeler-DeWitt equation with a different interpretation of the intrinsic time variable.) For Bianchi I, where square root operators can be defined by spectral representation, exact solutions may be found in both approaches. More interesting was the quantization of the Mixmaster model. Here the classical behavior suggested that the inability to choose preferred initial conditions would be irrelevant. The Klein-Gordon formalism was used to study approximate eigenfunctions for the model in the standard minisuperspace representation as a particle in an expanding potential well and, asymptotically, in the time independent potential well of the Misner-Chitre coordinates. As discussed above, Misner's formulation [37] led to the conclusion that the mode number was an adiabatic invariant. Recently, (single state) exact Mixmaster wave functions (for a special factor ordering) have been found independently by Kodama, Moncrief and Ryan, and Graham.

Although a trivial system for ADM quantization, the FRW models can be used to explore the possibility of physically relevant quantum effects in the early universe. In [17^B], Misner considers a quantized radiation filled FRW model. In a spirit similar to the myriad of calculations in the DeSitter model starting with Hartle and Hawking and Vilenkin in the early 1980's, he considers scattering off the minisuperspace potential that allows the existence of an empty, closed model near the singularity as a quantum tunneling event. Quantum cosmology results reported by Misner include collaboration with others (Fishbone, Hughston, Jacobs, Nutku, Ryan, and Zapolsky) in the Maryland group.

The ADM quantization procedure was then extended to the simplest spatially inhomogeneous model—the Gowdy T^3 universe—following Berger's thesis (itself based on Kuchař's treatment of quantized Einstein-Rosen waves). Here the issue at the time was quantum particle (graviton) creation as a possible mechanism for anisotropy dissipation and singularity avoidance. Misner emphasized [41] the importance of the Kasner-like behavior of the modes near the singularity, which caused the definition of graviton number valid at late times to yield spurious graviton creation at early times. This led to the more natural coherent state description of the quantized model.

Except for cosmology, the central focus of relativistic astrophysics is the strong gravitational fields in the vicinity of black holes, star collapse, and the cores of active galaxies. In the 1960's, Misner (and collaborators) conducted research which has formed a basis for study of the major targets of numerical relativity in these areas— black hole collisions and relativistic star collapse. Misner's wormhole initial conditions

[10], a solution to the time symmetric initial value problem, have been used as the starting point (e.g. by Smarr and Eppley) for the numerical study of the head on collision of two black holes. At the moment of time symmetry, the constraints of general relativity reduce (in a vacuum) to the requirement that the spatial scalar curvature vanish. The solution can be understood as two identical masses that are momentarily at rest. At this instant of time symmetry, the solution contains no gravitational radiation. In subsequent work [21], using an analog of the method of image charges, Misner was able to construct initial data for multiple Einstein-Rosen bridges. Very similar techniques have been used recently to construct general ADM initial data for interacting spinning black holes with arbitrary masses and momenta.

Misner and Sharp [23] derived the equations for spherical general relativistic gravitational collapse of matter with pressure and extended it to increasingly realistic matter [26]. (Previous work by Oppenheimer and Snyder had considered only dust.) This required the spherically symmetric Einstein equations with a fluid source coupled to the equations of motion for a relativistic fluid in a spherically symmetric gravitational field. Boundary conditions appropriate to the center and surface of the star were imposed. This formulation was designed for numerical implementation. (May and White found the initial set of equations independently, but remained in contact with Misner during their pioneering work.) Other studies by Misner included neutrino release during star collapse [24] and a simulation of energy loss to radiation using the Vaidya metric (with Lindquist and Schwartz [25]). The Misner and Sharp equations were a significant element of the "technology" necessary to study realistic star collapse.

Weber's report in 1970 of the possible detection of gravitational radiation from the galactic center stimulated furious theoretical activity. Misner proposed [39] to lower the energy requirements for a source at the galactic center by noting that the sun's location close to the galactic plane allows a natural restriction on emission angles. Since the resonant detectors also needed gravitational waves of frequency much higher than would be natural for orbits about black holes as massive as implied by the energy requirements, Misner suggested that a mechanism for gravitational synchrotron radiation would yield both high harmonics and beamed radiation. A large group of researchers gathered around Misner to exploit this idea, and showed in principle (e.g. [40, 42, 43]) that gravitational synchrotron radiation could occur in both the Schwarzschild and Kerr backgrounds. However, no astrophysically realistic mechanism could be found to explain the observations, either by the Maryland group or by the many others who studied the question. Although the observations are now discounted, the activity at that time was part of a synergy that generated a rapid increase in understanding of wave equations in black hole backgrounds, perturbations of black hole spacetimes, and mechanisms for the generation of gravitational radiation.

Misner also contributed to the understanding of solutions to Einstein's equations that were geometrically both fascinating and instructive, even if at the time they had no physical relevance. His pioneering studies were concerned with the causal behavior of solutions of Einstein's equations and with the nature of their singularities. During the 1960's, he and Taub [20, 33, 9^B] explored the recently discovered Newman, Unti, Tamburino (NUT) metric. They found that it serves as a non-globally-hyperbolic extension of the globally hyperbolic Taub spacetime. They showed that the resulting Taub-NUT spacetime is singular in the sense that it contains incomplete null geodesics. The spacetime curvature in Taub-NUT is, however, bounded. Rather, at the boundary of the Taub region, where the singularity occurs, one finds a Cauchy horizon. Past this Cauchy horizon (in the NUT region) one finds closed timelike lines which signal the breakdown of causality (although the unextended Taub spacetime is fully causal). The occurrence of this sort of behavior in a spacetime which can be developed from a seemingly reasonable set of initial data has worried relativists and has led to Penrose's conjecture of Strong Cosmic Censorship (that the spacetime development of generic initial data will not admit extensions of the sort seen in the Taub-NUT spacetime). This conjecture is one of the outstanding issues in classical mathematical general relativity. As a pedagogical tool to aid one's understanding of the behavior of the Taub-NUT spacetime, Misner invented a two-dimensional model spacetime in which similar behavior occurs. This model, which has come to be known as the "Misner spacetime," has been a particularly useful laboratory and prototype for the study of the question of the uniqueness of extensions in Taub-NUT-like spacetimes. These causality violating spacetimes are now receiving significant attention following Morris, Thorne, and Yurtsever's suggestion that the laws of physics may not forbid time machines. The Misner constructions also formed the basis for the recent work in three dimensional gravity inspired by Gott's causality violating spacetime. In fact, Deser, Jackiw, and 't Hooft showed this solution to be an example of the Misner identification. (As an historical footnote, Misner's well-known method for curvature computation using differential forms first appears in [20].)

The fundamental role of geometry and topology provides a common thread in Misner's work. An extensive discussion (with Wheeler) of physics as geometry appeared very early in his career [2]. Soon after, he and Finkelstein [3] showed that conserved integer quantities could be associated with nontrivial topology. These homotopic conservation laws are related to the "intrinsic nonlinearity" of the field theory that contains them. This structure could then be used to interpret a particle as a topological soliton. The modern flavor of these ideas and their relevance to, for example, cosmic strings and magnetic monopoles is striking.

In what might be regarded as a similar spirit, Misner much later proposed that harmonic maps be considered as models for field theories [44, 24^B, 28^B]. The harmonic maps take spacetime manifold points into points of a manifold of field values. The geometric structures of both manifolds can be interesting. Harmonic maps satisfy nonlinear field equations with different properties from those of the nonlinear Yang-Mills fields. As such, they might be expected to yield different classes (from those already known) of field theories in spacetime that might have physical application. (In particle physics harmonic maps are usually called, from an earlier example, nonlinear σ models.) The full ramifications of these ideas have yet to be explored.

In the mid 1980's, Misner turned his energies toward the pedagogy of introductory physics. His previous major pedagogical contribution, *Gravitation* [B], written in collaboration with Thorne and Wheeler (as an outgrowth of simultaneous attempts by at least two of the authors to write their own books), has become a classic among physics texts. Affectionately known as "MTW" or the "Telephone Book," it provides an almost complete discussion of the state of relativity (including significant portions of the authors' own research) up to its time. Recently, Misner (with Cooney) [C, 29^B] has developed a method of physics instruction based on the commercially available spreadsheet *Lotus 1-2-3*. The spreadsheet provides a numerical analysis and graphing engine with a relatively simple user interface. The basic motivation is to use the spreadsheet to construct and explore models of physical systems in a manner that can bypass the students' mathematical limitations. "Physics teachers used to imagine that conceptual models were being constructed inside students' heads, similar to the one good physicists have inside their heads. End of chapter exercises were debugging tools to test this hypothesis ... But students develop strategies to manufacture evidence instead of the model ... Now the model can be more external." [29^B] The associated computer program "Spreadsheet Physics Worksheets" was a winner of the second annual *Computers in Physics* contest for innovative software in physics education.

Beverly K. Berger
Dieter Brill
James Isenberg
Richard A. Matzner

REMARKS CONCERNING THE GEOMETRIES OF GRAVITY AND GAUGE FIELDS

JEEVA ANANDAN

Department of Physics and Astronomy, University of South Carolina, Columbia SC 29208, USA

Abstract

An important limitation is shown in the analogy between the Aharonov-Bohm effect and the parallel transport on a cone. It illustrates a basic difference between gravity and gauge fields due to the existence of the solder form for the space-time geometry. This difference is further shown by the observability of the gravitational phase for open paths. This reinforces a previous suggestion that the fundamental variables for quantizing the gravitational field are the solder form and the connection, and not the metric.

1. INTRODUCTION

I recall with great pleasure the discussions which I had with Charles Misner on fundamental aspects of physics, such as the geometry of gravity, gauge fields, and quantum theory. In particular, I remember the encouragement he gave to my somewhat unorthodox attempts to understand the similarities and differences between gauge fields and gravity from their effects on quantum interference, and their implications to physical geometry. It therefore seems appropriate to present here for his Festschrift some observations which came out of this investigation.

Geometry is a part of mathematics which can be visualized, and is intimately related to symmetries. This may explain the tremendous usefulness of geometry in physics. In section 2, I shall make some basic remarks about the similarities and differences between the geometries of gravity and gauge field. Then I shall illustrate, in section 3, an important difference between them that arises due to the existence of the solder form for gravity, using the Aharonov-Bohm (AB) effect and parallel transport on a cone. In section 4, I shall further illustrate this difference by the fact that the gravitational phase for a spinless particle is observable for an open path, unlike the AB effect. This implies that the translational gauge symmetry of the gravitational field is broken by the existence of the solder form. It is then argued that the solder form and the connection are the proper variables for quantizing the gravitational field.

2. LOCALITY OF GRAVITY AND GAUGE FIELDS

Something which Misner emphasized to me during a conversation was the fundamental role assigned to locality by the theory of relativity. Already in special relativity locality is incorporated in the fact that signals cannot travel faster than the speed of light. But in general relativity, locality plays an even more fundamental role: The principle of equivalence states that the laws of physics are locally Minkowskian. Also, because space-time is curved, there is no distant parallelism and vectors at two different points can only be compared by parallel transporting them to a common point with respect to the gravitational connection.

These three aspects of locality are also present in gauge fields[1] which are now being used to describe the three remaining fundamental interactions in physics. The principle of equivalence for gauge fields may be stated as follows: Given any point p in space-time, a gauge can be chosen so that the corresponding connection coefficients or vector potential vanishes at p for all fields interacting with the given gauge field. Also, there is no distant parallelism for vectors parallel transported using the gauge field connection if the curvature or the Yang-Mills field strength is non vanishing.

Contrary to what is sometimes said, gravity does not differ fundamentally from gauge fields simply because it is associated with a metric. Because if the gauge group is unitary then it leaves invariant a metric in the vector space at each space-time point that consists of all possible values of any matter field interacting with the gauge field at that point. The essential difference is that the gravitational metric can be used to measure distances along any curve in space-time, unlike the gauge field metric. But I shall show now, by means of physical arguments, that this fundmental difference between gravity and gauge fields exists even *prior* to introducing the metric.

3. AHARONOV-BOHM EFFECT AND PARALLEL TRANSPORT ON A CONE

It is an interesting fact that the phase shifts in quantum interference due to gravity and gauge fields are obtained in a simple manner from the distance due to the gravitational metric and parallel transport due to gravitational and gauge field connections along the interfering beams[2]. Conversely, the phase shifts in quantum interference can be used to *define* gauge fields and gravity[3]. This is most easily shown for the simplest gauge field, namely the electromagnetic field, by means of the AB effect[4].

We recall that the magnetic AB effect is the phase shift in the interference of two coherent electron beams which enclose a cylinder containing a magnetic flux. In the interference region, the wave function may be written as $\psi_1(\mathbf{r}, t) + \psi_2(\mathbf{r}, t)$, where ψ_1 and ψ_2 are the wave functions corresponding to the two beams. The introduction of the magnetic field inside the cylinder modifies this wave function to

$$\psi(\mathbf{r}, t) = \psi_1(\mathbf{r}, t) + F_\gamma \psi_2(\mathbf{r}, t),$$

in an appropriate gauge, where

$$F_\gamma = exp(-\frac{ie}{\hbar c} \oint_\gamma A_\mu dx^\mu). \tag{1}$$

Here the integral is along the curve γ going around the cylinder, A_μ is the electromagnetic 4-vector potential and e is the charge of the electron. Therefore the intensity distribution $|\psi(\mathbf{r}, t)|^2$ in the interference region is modified in an apparently non local way by the magnetic flux via F_γ even though the magnetic and electric field strengths vanish everywhere along the beams.

But this phenomenon is not surprising when we realize the analogy with the geometry of a cone[5]. The cone may be formed by taking a flat sheet of paper bounded by two straight lines making an angle θ and identifying the two straight lines (Fig. 1a); we denote this cone by C_θ. For $0 \leq \theta \leq 2\pi$, this is what we do when we make a cone by rolling this flat sheet so that these two lines coincide to form one of the generators of the cone. Since the paper is not stretched or compressed during this process, a cone has no intrinsic curvature except at the apex, which can be smoothed out so that the curvature is finite there. In the multiply connected geometry around the apex, the intrinsic curvature is zero everywhere, same as the flat geometry of the sheet which was rolled up to be the cone. In particular, a vector is parallel transported like on the flat sheet. But a vector V parallel transported around a closed curve drawn on the curvature free region of the cone so as to enclose the apex undergoes a rotation by the angle θ, which is the holonomy transformation associated with this curve.

If the curvature at the apex is regarded as analogous to the magnetic field in the cylinder then the zero intrinsic curvature everywhere else corresponds to the vanishing of magnetic field strength outside the cylinder. Then V moving in a curvature free region is analogous to beams traveling in a field free region. The rotation by the angle θ which relates V and V' (Fig. 1a) is analogous to the phase difference between the two beams due to F_γ. This suggests that the electromagnetic field may be a connection for parallel transporting the value of the wave function and the AB effect arises because a wave function when parallel transported around the closed curve γ, gets multiplied by F_γ. The electric and magnetic field strengths at each space-time point then constitutes the curvature of this connection at this point. Thus the phase factor (1) is the holonomy transformation associated with γ for this connection. The statement that the electromagnetic field is a gauge field is the same as saying that it is a connection as described above.

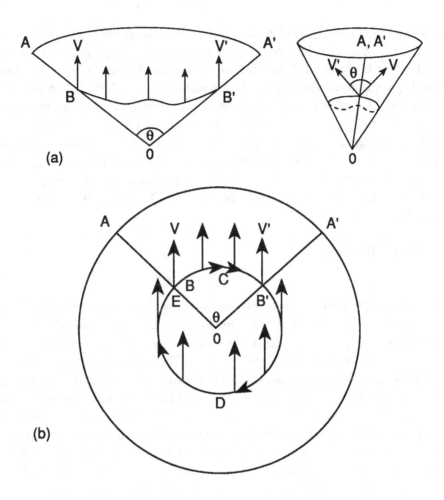

Figure 1

a) Analogy between the Aharonov-Bohm effect and parallel transport on a cone. The cone may be obtained by identifying the lines OA and OA' on a flat sheet. Therefore, the vector V parallel transported from B around the cone would come back to B (identified with B') as V' rotated by the angle θ. This is analogous to the AB phase shift with the magnetic field corresponding to the curvature at the apex of the cone. b) The limitation of this analogy when θ is changed to θ+2π by adding an extra sheet of paper. The vector parallel transported along the closed curve BCDEB' rotates by θ+2π with respect to the tangent vector to the curve. This enables one to distinguish this cone from the earlier one. This is unlike the AB effect which cannot distinguish between two enclosed magnetic fluxes that differ by one quantum of flux.

The above mentioned conical geometry describes the gravitational field in each section normal to a long straight string[6], such as a cosmic string. A gravitational analog of the AB effect is obtained if we interfere two coherent beams of identical particles with intrinsic spin around the string. The resulting phase shift due to the cosmic string is a special case of the phase shift due to an arbitrary gravitational field obtained before[2]. Basically, this phase shift consists of two parts, one due to the change in path lengths of the interfering beams, and the other due to the holonomy transformation, which in this case is a rotation undergone by the wave function when it is parallel transported around the interfering beams. This change in path length and holonomy transformation, and consequently the phase shifts, occur even though the space is locally flat.

Now if the AB phase

$$\phi_\gamma = \frac{e}{\hbar c} \oint_\gamma A_\mu dx^\mu$$

is changed by 2π, which corresponds to changing the magnetic flux inside the cylinder by a "quantum of flux", then (1) is unchanged. Therefore the AB experiment or for that matter any other experiment outside the cylinder cannot detect the difference between these two magnetic fluxes. Hence, Wu and Yang[7] stated that, because of the AB effect, the electromagnetic field strength $F_{\mu\nu}$ has too little information, ϕ_γ has too much information, and it is the phase factor or the holonomy transformation F_γ, for arbitrary closed curves γ, which has the right amount of information of the electromagnetic field. This has been generalised to an arbitrary connection by the theorem[8] which states that from the holonomy transformations the connection can be reconstructed and it is then unique up to gauge transformations. A simple physical system to illustrate the Wu-Yang statement is a superconducting ring enclosing a magnetic flux. No experiment performed in the interior of the ring using Cooper pairs can distinguish between a given enclosed flux Φ and $\Phi + n\Phi_0$, where Φ_0 is the quantum of flux for the Cooper pair and n is an integer. For example, if we measure the flux by inserting a Josephson junction in the ring and observe the Josephson current, we would obtain the same current for both fluxes. Because the AB phases for the two fluxes differ by $2\pi n$ and therefore (1) is the same for both fluxes, with e now being the charge of the Cooper pair.

An important and interesting limitation of the analogy of the AB effect with the cone emerges when we consider the meaning of increasing the flux of the curvature in the apex region of the cone by "one quantum". The new flux may be regarded as corresponding to the cone $C_{\theta+2\pi}$ which has one extra sheet of paper compared to C_θ. (To embed $C_{\theta+2\pi}$ into a three dimensional Euclidean space it needs to be twisted in some way but it is well defined by the identification stated above.) The holonomy transformations are the same for C_θ and $C_{\theta+2\pi}$ (Fig. 1b). Therefore the above mentioned theorem[8] implies that the cones C_θ and $C_{\theta+2\pi}$ are the same as far as their connections are concerned. Here a connection is regarded simply as a rule for parallel transporting abstract vectors attached to points on the cone and not regarded as tangent vectors. Physically, the phase shift arising from spin in an interference experiment which is determined by the holonomy transformation will be the same for both cones for a bosonic particle. For a fermionic particle, there is a difference of π between the phase shifts because this phase is acquired by a fermion when it is rotated by 2π radians. Therefore for fermions, C_θ is not equivalent to $C_{\theta+2\pi}$, but is equivalent to $C_{\theta+4\pi}$, because of the nature of the spinor connection. A straightforward application of the Gauss-Bonnet theorem shows that the flux or integral of the curvature at the smoothed out apex of the cone C_α is $2\pi - \alpha$. Therefore this flux is negative when $\alpha > 2\pi$. In the latter case, it follows via Einstein's field equations that if C_α represents the geometry around a cosmic string then the string has negative mass. In particular, $C_{\theta+2\pi}$ and $C_{\theta+4\pi}$ represent geomtries around cosmic strings with negative mass.

The two cones, which are the same as far as their linear connections are concerned, are of course, different when we take into account their metrics. This gives rise to the phase shift due to changes in the path lengths of the interfering beams[2]. But even if we forget their metrics, there is a subtle difference between the two cones. To see this, for each cone, parallel transport a vector around a closed smooth curve that encloses the apex and does not intersect itself. This vector rotates with respect to the tangent vector to the curve by the angle θ for C_θ and by the angle $\theta + 2\pi$ for $C_{\theta+2\pi}$. This difference, which can be observed by means of local measurements, arises ultimately because we identify the vectors being parallel transported with the tangent vectors to the cone. The mathematical concept used to make this identification in an arbitrary manifold is called the *solder form*, or the canonical 1- form, or the canonical form[9]. For the electromagnetic field the vector $\psi(\mathbf{x}, t)$ belongs to an internal space and cannot be compared with a tangent vector. Therefore the Wu-Yang statement[7] is valid. But the gravitational field connection is for parallel transporting tangent vectors. Hence there is such an identification. This is the most fundamental difference between gravity and gauge fields[10].

If C_θ and $C_{\theta+2\pi}$ have only the connections or only the solder forms then they are identical. Since the two connections in the the frame bundles over C_θ and $C_{\theta+2\pi}$ have the same holonomy transformations, there exists a fiber bundle isomorphism $\tilde{f}$ between the frame bundles which maps one connection into the other[8]. This $\tilde{f}$ induces a unique diffeomorphism f between the base manifolds C_θ and $C_{\theta+2\pi}$ in the obvious way. The differential f_* is a map between tangent vectors. It determines a fiber bundle isomorphism $\bar{f}$ between the two frame bundles that maps one solder form into the other. But $\tilde{f}$ and $\bar{f}$ are topologically different in the sense that one cannot be continuously deformed into the other. This is why we were able to distinguish between C_θ and $C_{\theta+2\pi}$ when the connections and the solder forms are both present, even when the metric is absent.

4. GRAVITATIONAL PHASE FACTOR

The phase shift in quantum interference due to an arbitrary gauge field, which generalizes the AB effect, is determined by the "phase factor"[2,11]

$$F_\gamma = P exp(-\frac{ig}{\hbar c} \oint_\gamma A_\mu^k T_k dx^\mu), \tag{2}$$

where T_k generate the Lie algebra of the gauge group, A_μ^k is the Yang-Mills gauge potential, P denotes path ordering, and γ is a closed curve through the interfering beams. Here, F_γ is an element of the gauge group. Its eigenvalues can be determined by interference experiments[3]. This shows the real significance of (1) as an element of the $U(1)$ group, which is a special case of the gauge group. When γ is an infinitesimal closed curve spanning a surface element represented by $d\sigma^{\mu\nu}$,

$$F_\gamma = 1 + \frac{ig}{2\hbar c} F_{\mu\nu}^k T_k d\sigma^{\mu\nu} \tag{3}$$

where $F^k = dA^k - gC^k{}_{ij} A^i \wedge A^j$ is the Yang-Mills field strength.

The phase shift in quantum interference of a particle due to the gravitational field is determined by[12]

$$F_\gamma = P exp[-\frac{i}{\hbar} \int_\gamma (e_\mu^a P_a + \frac{1}{2\hbar} \Gamma_\mu^{ab} M_{ab}) dx^\mu], \tag{4}$$

which is an element of the Poincare group that may be associated with any path γ in space-time. Here, P_a and $M_{ab}, a, b = 0, 1, 2, 3$ are respectively the energy-momentum and angular momentum operators which generate the representation of the Poincare group corresponding to the given particle, e_μ^a is dual to the frame e_a^μ used by local observers:

$$e_\mu^a e_b^\mu = \delta_b^a, \tag{5}$$

and Γ_μ^{ab} are the connection coefficients with respect to this frame field. If the local observers use orthonormal frames then

$$e_a^\mu e_b^\nu g_{\mu\nu} = \eta_{ab}, \tag{6}$$

where $g_{\mu\nu}$ is the space-time metric and η_{ab} are the coefficients of the Minkowski metric. When the particle has non zero intrinsic spin then the values of the wave function are what are observed by observers using the frame field e_b^μ. Then (4) implies that the spinor field is parallel transported in addition to a phase that it acquires due to its energy-momentum. This is obtained in the WKB approximation, disregarding here for simplicity a real factor which does not contribute to the phase[2].

For the special case of a spinless particle, $M_{ab} = 0$, the gravitational phase acquired by a locally plane wave is, to a good approximation,

$$\phi = \frac{1}{\hbar c} \int_{\gamma} e_{\mu}^{a} p_{a}, \qquad (7)$$

where p_a are the eigenvalues of the energy-momentum operators P_a and the integral is along the classical trajectory[13]. A remarkable feature of (7) is that it is observable for an open curve γ unlike the phase shifts for gauge fields which can be observed only for closed curves. For example, (7) may be observed by the Josephson effect for a path across the Josephson junction[3], or by the oscillation of strangeness in the Kaon system for an open time-like path γ along the Kaon beam[14]. Both these phases depend on the geometry of space- time as determined by the gravitational field.

To understand this difference between gravity and gauge fields note that the field e_{μ}^{a} plays three roles here: First, comparing (4) with (2) suggests that e_{μ}^{a} is like a connection or gauge potential associated with the translation group. Indeed, e_{μ}^{a} and Γ_{μ}^{ab} may be regarded as constituting the connection in the affine bundle. The curvature of this connection is obtained by evaluating (4) for an infinitesimal closed curve γ:

$$F_{\gamma} = 1 + \frac{i}{2\hbar}(Q_{\mu\nu}^{a}P_{a} + \frac{1}{2}R^{ab}{}_{\mu\nu}M_{ab})d\sigma^{\mu\nu}, \qquad (8)$$

using the Poincare Lie algebra, where $Q^{a} = de^{a} + \Gamma^{a}{}_{b} \wedge e^{b}$ is the torsion and $R^{ab} = d\Gamma^{ab} + \Gamma^{a}_{c} \wedge \Gamma^{cb}$ is the curvature. Eq. (8) is the analog of (3) for gravity. This is the most physical way that I know to regard gravity as the gauge field of the Poincare group. Second, e_{μ}^{a} represents the solder form referred to earlier. In geometrical language, it is the pullback of the solder form with respect to the local section e_{b}^{μ} in the bundle of frames, which follows from (5). The (Lie-algebra valued) 1-form $e_{\mu}^{a}P_{a}$ acts on the tangent vector to γ to give an element in the Lie algebra of the translational group, which is also an observable in the Hilbert space. When this observable acts on a WKB wave function it gives as an approximate eigenvalue the rate of change of phase along γ. When this is integrated along γ the phase (7) is obtained. Third, (5) and (6) imply that e_{μ}^{a} is like the square root of the metric:

$$e_{\mu}^{a}e_{\nu}^{b}\eta_{ab} = g_{\mu\nu}. \qquad (9)$$

In the first role, there is no restriction on the values of e_{μ}^{a} at any given point in space-time. Indeed e_{μ}^{a} may be made to vanish by an appropriate choice of gauge along any differentiable curve that does not intersect itself. The "gravitational field" then has the full gauge symmetry of the affine group $A(4, R)$, i. e. the group of inhomogeneous linear transformations on a four dimensional real vector space. The holonomy group is a subgroup of the Poincare group which enables only the generators of the Poincare Lie algebra to occur in (4). The corresponding "gravitational phase", like the AB phase, would then be meaningful only for a closed curve γ.

However, in the second role, the matrix e_μ^a is restricted to be non singular. The gauge symmetry group is reduced to the general linear group $GL(4, R) \subset A(4, R)$, with Γ_μ^{ab} being the connection or gauge field. It is the breaking of the translational gauge symmetry that enables the phase (7) to be observable. The solder form is the canonical 1-form defined on the frame bundle whose structure group is $GL(4, R)$. The discussion of the parallel transport of a vector around a cone in section 3 shows the important role played by the solder form which makes this theory richer than a gauge theory with $GL(4, R)$ as the internal symmetry. The e_μ^a now transforms as a tensor, instead of a connection, under local gauge transformations which corresponds to space-time dependent transformations of the frame field. Therefore the phase (7) is invariant under these gauge transformations for an open curve γ, as it should be because it is observable.

Despite the breaking of the translational gauge invariance, the torsion which appears in (8) as the curvature corresponding to this group nevertheless arises naturally from a physical point of view. Because the motion of the amplitude of a spinor wave function provides an operational definition of the connection which is independent of the Christoffel connection that comes from the metric[2,15]. Therefore, the connection, in a coordinate basis, can be non symmetric and the torsion is then twice the antisymmetric part of this connection. Hence the burden of proof is on gravitational theories with zero torsion to justify this constraint and not on torsion theories to justify introducing torsion, because kinematically torsion arises naturally whenever there are fields with intrinsic spin as seen above. But it is not necessary to introduce a metric in the first two roles of e_μ^a discussed here.

In the third role, the specification of the metric, which is the same as specifying the orthonormal frame field e_b^μ, breaks the gauge symmetry further to the Lorentz group $O(3, 1, R) \subset GL(4, R)$ that leaves this metric invariant. But from the observed phases (7), the metric may be constructed[3,15]. Therefore it does not appear to be as fundamental a physical variable as the solder form or connection. From an operational point of view, the motion of a quantum system in a gravitational field is influenced directly by the solder form and connection, and the metric seems to arise only as a secondary construct. Therefore in the reaction of the quantum system on the gravitational field, which needs to be described by quantum gravity, the solder form and the connection would be the fundamental dynamical variables that are affected.

It was therefore proposed that in quantizing the gravitational field the variables e_μ^a and Γ_μ^{ab} should be quantized and not the metric[15]. The arguments in this paper which show further the important role played by the solder form reinforce this view. The important role assigned to the vector potential by the AB effect finds its counterpart in quantum electrodynamics in which it is the vector potential which is quantized. Similarly, the quantum effects discussed above which depend on the gravitational phase factor (4) suggest that the variables e_μ^a and Γ_μ^{ab} should be quantized in order to obtain quantum gravitodynamics. It is noteworthy that (4) is an element of the Poincare group, even though the curvature of space-time classically breaks Poincare invariance. This is analogous to the phase factor (2) for gauge fields being an element of the corresponding gauge group. This role of groups in the fundamental interactions, together with the general role of symmetry in quantum physics, which is much more fundamental, substantive and determinative than in classical physics, suggest that the way forward in physics at the present time should perhaps be guided by the precept 'symmetry is destiny'.

Acknowledgements

I thank P. O. Mazur, R. Penrose and R. Howard for useful remarks. This work was partially supported by NSF grant no. PHY-8807812.

REFERENCES

1. C. N. Yang and R. L. Mills, Phys. Rev. 96 (1954) 191.
2. J. Anandan, Nuov. Cim. 53A (1979) 221.
3. J. Anandan, Phys. Rev. D 33 (1986) 2280; J. Anandan in *Topological Properties and Global Structure of Space-Time*, eds. P. G. Bergmann and V. De Sabbata (Plenum Press, NY 1985), p. 1-14.
4. Y. Aharonov and D. Bohm, Phys. Rev. 115 (1959) 485.
5. J. S. Dowker, Nuov. Cim. 52B (1967) 129.
6. L. Marder, Proc. Roy. Soc. A 252 (1959) 45; in *Recent Devolopments in General Relativity* (Pergamon, New York 1962).
7. T. T. Wu and C. N. Yang, Phys. Rev. D, 33 (1986) 2280.
8. J. Anandan in *Conference on Differential Geometric Methods in Physics*, edited by G. Denardo and H. D. Doebner (World Scientific, Singapore, 1983) p. 211.
9. S. Kobayashi and K. Nomizu, *Foundations of Differential Geometry* (John Wiley, New York 1963) p. 118.
10. A. Trautman in *The Physicist's Conception of Nature*, edited by J. Mehra (Reidel, Holland, 1973).
11. See also, I. Bialynicki-Birula, Bull. Acad. Pol. Sci., Ser. Sci. Math. Astron. Phys.

11 (1963) 135; D. Wisnivesky and Y. Aharonov, Ann, of Phys. 45 (1967) 479.

12. J. Anandan in *Quantum Theory and Gravitation*, edited by A. R. Marlow (Academic Press, New York 1980), p. 157.

13. J. Anandan, Phys. Rev. D, 15 (1977) 1448.

14. L. Stodolsky, J. Gen. Rel. and Grav.11 (1979) 391.

15. J. Anandan, Found. Phys. 10 (1980) 601.

Gravity and Unification of Fundamental Interactions
by R. ARNOWITT

Center for Theoretical Physics, Department of Physics
Texas A&M University, College Station, TX 77843-4242

and PRAN NATH

Department of Physics, Northeastern University
Boston, MA 02115

It is a pleasure to contribute this paper in honor of Charlie Misner for his many contributions to gravitational theory and for his warm friendship.

I. INTRODUCTION

The discovery of general relativity by Einstein and its early experimental verification excited at that time both the scientific and lay public alike. However, during the 1930s and 1940s the hope that gravity would be a unifying principle of nature faded. The discovery of the self-energy infinities of Lorentz covariant quantum field theory indicated the insufficiency of the quantum theoretical framework. Subsequently, it was realized that these infinities were even more virilant in the non-renormalizable general relativity. Most significant was the experimental discovery during this period of the weak and strong interactions, implying that the original ideas of Einstein and Weyl to unify gravity with electromagnetism were premature. Perhaps the one idea from this era that has remained in present day efforts to unify interactions was the most radical: the suggestion by Kaluza and Klein that there might exist additional compactified dimensions in space-time. Most remarkable was the work of Oscar Klein who, using dimensional reduction, discovered non-abelian gauge theory and applied it to construct a precursor of present day electro-weak theory. This was a spectacular theoretical tour-de-force which unfortunately did not appear to stimulate further work at that time.

The development of the Glashow-Weinberg-Salam model of electroweak interactions[1] combined with the QCD theory of strong interactions to form the Standard Model, has led to the recent approaches to build models of unified interactions. The Standard Model, based on the gauge group $SU(3)_c \times SU(2)_L \times U(1)_Y$ is one of the most successful theories ever constructed in that is is consistent with all data involving the strong, electromagnetic and weak interactions. It is not, however, a unified theory (it is based on a product group) nor does it include gravity. Efforts to construct a unified theory, e.g. based on the group $SU(5)$[2], ran into the well-known "gauge hierarchy" problem. In such models, unification occurs at the grand unification scale $M_G \approx 10^{14}$ GeV. However, the mass self-energy of the scalar Higgs boson (needed to spontaneously break the $SU(2) \times U(1)$ symmetry) diverges quadratically. Since the Higgs mass, m_H, must be of order of the Z boson mass, it is necessary to fine tune parameters in the basic Lagrangian to 24 decimal places if the divergence is to be cut off at the scale M_G!

The most elegant resolution of the gauge hierarchy problem appears to lie in supersymmetry (SUSY). Here one doubles the particle spectrum by having as many fermi as bose states, with a bose-fermi symmetry holding. The quadratic divergence of the Higgs self-mass then precisely cancels (the bose and fermi loops having opposite sign). However, supersymmetry must be a spontaneously broken symmetry, since experimental bounds imply that the squark, the bose partners of the quarks, are not degenerate with the quarks and must have masses greater than 100 GeV. Global supersymmetry is exceedingly difficult to break spontaneously[3], and no phenomenologically acceptable way of doing this has ever been found. Thus the resolution of the gauge hierarchy problem lies in the use of *local supersymmetry* (i.e. supergravity[4]) which based on the SUSY massless spin $j = \frac{3}{2}, 2$ multiplet, automatically includes gravity into the dynamics.

The above discussion shows how one is led to include gravity into the analysis in order to solve specifically particle interaction problems. In this note we will discuss some examples of how gravity (more exactly supergravity) structures particle interactions. In particular we will examine the origin of the absence of flavor changing neutral interactions (FCNI), and the nature of the SUSY mass spectrum. There is belief among many researchers that all interactions, including gravity, unify at the Planck scale $M_{Pl} = (1/8\pi G_N)^{\frac{1}{2}} \cong 2.4 \times 10^{18}$ GeV. (G_N is the Newtonian constant). Superstring theory, which illustrates this phenomena, also has the remarkable feature of eliminating the infinities of quantum general relativity (and of other interactions as well). However, at present there is no concrete string model which gives an acceptable theory of particle interactions. Nonetheless, there has been considerable progress of bringing together gravity with other forces of nature via $N=1$ supergravity unification. In this note, therefore, we will restrict our discussion to $N=1$ supergravity unified models, and describe some of their properties and predictions.

II. $N=1$ SUPERGRAVITY MODELS

Supergravity theories are specified by an integer $N=1...8$ which represents the number of spin $\frac{3}{2}$ (gravitino) fields. However, only $N=1$ supergravity couples to chiral matter, and so we restrict ourselves to this case. As in general relativity, one must couple supergravity to matter, now in a fashion that preserves both general covariance and local supersymmetry. The great success of the Standard Model tells us what kind of matter one must consider: $j=1,\frac{1}{2}$ massless vector multiplets to describe gauge bosons, and $j=\frac{1}{2}, 0$ massless chiral multiplets to describe quarks and leptons. The full Lagrangian describing the coupling of supergravity to vector and chiral matter is quite complicated.[5,6,7] The most general coupling depends on the following functions: (i) The superpotential $W(z_i)$, where z_i are the $j=0$ complex scalar field components of the chiral multiplets, (ii) the Kahler potential $d(z_i, z_i^*)$ and (iii) the gauge kinetic function $f_{\alpha\beta}(z_i, z_i^*)$ where α, β are gauge indices. Actually, only the combination $G = -\kappa^2 d - \ln[\kappa^6 WW^*]$, where $\kappa = 1/M_{Pl}$ enters into the Lagrangian. We examine here only the following terms in the Lagrangian:

(1) The effective potential:

$$V = -\kappa^{-4} e^{-G}\left[\left(G^{-1}\right)^i_j G_i G^j +3\right]+\frac{g^2}{2}\mathrm{Re}\left(f^{-1}\right)^{\alpha\beta} D_\alpha D_\beta \tag{1}$$

where $G_i = \partial G/\partial z_i$, $G^j = \partial G/\partial z_j^*$, $D_\alpha = -\kappa^{-4} G_i (T^\alpha)_y z_j$, and g and T^α are the gauge coupling constants and group generators [$(G^{-1})^i_j$ and $(f^{-1})^{\alpha\beta}$ are matrix inverses].

(2) Scalar kinetic energy:

$$-d^i_j \partial_m z_i \partial^m z_j^* \tag{2}$$

where $d^i_j = \partial^2 d/\partial z_i \partial z_j^*$.

(3) Gauge kinetic energy:

$$-\tfrac{1}{4}(\mathrm{Re}\, f_{\alpha\beta}) F^\alpha_{\mu\nu} F^{\mu\nu\beta} \tag{3}$$

where $F^\alpha_{\mu\nu}$ are the gauge field strengths.

Unlike global supersymmetry, the effective potential of supergravity, Eq. (1), allows for spontaneous breaking. These are due to the additional gravitational terms present proportional to $\kappa = 1/M_{Pl}$. Thus a simple linear form, $W = m^2(z + B)$, produces a spontaneous breaking of supersymmetry with $\langle z \rangle = O(M_{Pl})$ upon minimizing $V(z)$. (Alternate methods of breaking supersymmetry via gaugino condensates also exist.[8]) The significant point here is that it is the (super) gravity interactions that give rise to the breaking of supersymmetry. However, Planck size VEVs give rise to a new "gauge hierarchy" problem, i.e. the danger that the physical particles (e.g. quarks and leptons) will also grow Planck size masses. This problem can be solved by having the fields $\{z\}$, which generate the spontaneous breaking of supersymmetry, live in a "hidden sector", i.e. we assume that the $\{z\}$ fields are gauge singlets with respect to the physical group G, and that the superpotential decomposes as

$$W = W_{\text{phys}}(z_A) + W_{\text{hidden}}(z) \tag{4}$$

where W_{phys} involves only the physical fields. From Eq. (1) one finds that the "super Higgs" fields $\{z\}$ then communicate with the physical fields $\{z_A\}$ only gravitationally, i.e. with factors $\kappa^n = (1/M_{pl})^n$, suppressing any contributions to the physical masses.

One can now proceed to construct supergravity GUT models. One assumes that there is a GUT sector in W_{phys} which at scale M_G breaks the physical grand unification gauge group G to the Standard Model group $SU(3)_C \times SU(2)_L \times U(1)_Y$. This occurs by some of the fields z_A growing VEVs (and masses) of $O(M_G)$. The gauge kinetic energy Eq. (3) contains the factor $f_{\alpha\beta}(z_i)$. We will consider here those theories where $f_{\alpha\beta}$ has a power series expansion: $f_{\alpha\beta} = c_{\alpha\beta} + \kappa c_{\alpha\beta i} z_i + \dots$. After spontaneous breaking of supersymmetry and G, one may write $f_{\alpha\beta} = c'_{\alpha\beta} + \kappa c'_{\alpha\beta i} z'_i + \dots$ where $\langle z'_i \rangle = 0$. To put the theory into canonical form, it is necessary to make an orthogonal and scale transformation on the $F^\alpha_{\mu\nu}$ to reduce $f_{\alpha\beta}$ to the form

$$f_{\alpha\beta} = \delta_{\alpha\beta} + \kappa c''_{\alpha\beta i} z'_i + \dots \tag{5}$$

The additional terms, $\kappa c'' z' + \dots$, lead to non-renormalizable interactions which are very small below M_G as they are scaled by $\kappa = 1/M_{pl}$. [In many models, in fact, the higher terms begin at $O(\kappa^2)$.] A similar analysis on the scalar kinetic energy Eq. (2) allows one to reduce the "Kahler metric" d^i_j to canonical form $d^i_j = \delta^i_j + \kappa c^i_{jk}{}'' z'_k$, by a bilinear transformation on scalar fields z_i.

One may now proceed to integrate out the superheavy fields with masses $O(M_G)$ [from the breaking of the gauge group G] and eliminate the super Higgs fields. For models of the type being considered here, and which possess three light generations of quarks and leptons and two light Higgs doublets (as suggested by the LEP data), the following basic result holds[9]: the dynamics at the GUT scale can be described by (i) SUSY gauge interactions; (ii) an effective superpotential with renormalizable quadratic and cubic terms $W^{\text{eff}} = W^{(2)} + W^{(3)}$,

$$W^{\text{eff}} = \mu_0 H_1 H_2 + \lambda^{(u)}_{ij} q_i H_2 u^C_j + \lambda^{(d)}_{ij} q_i H_1 d^C_j + \lambda^{(e)}_{ij} l_i H_1 e^C_j \tag{6}$$

where $H_{1,2}$, q_i, l_i are the Higgs, quark and lepton doublets, $\lambda_{ij}^{(u,d,e)}$ are the Yukawa coupling constants and μ_0 is the Higgs mixing parameter; (iii) an effective potential of the form

$$V^{eff} = \left\{ \sum_a \left| \partial W^{eff} / \partial z_a \right|^2 + V_D \right\} + \left[m_0^2 z_a z_a^* + \left(A_0 W^{(3)} + B_0 W^{(2)} + h.c. \right) \right] \tag{7}$$

and a gaugino Majorana mass term $-m_{\frac{1}{2}} \bar{\lambda}^\alpha \lambda^\alpha$. In Eq. (7), $\{z_a\}$ are the remaining light squarks, sleptons and Higgs fields. Thus the brace is the usual global supersymmetric effective potential (V_D is the usual D term). The remaining terms, involving m_0, A_0, B_0 and $m_{\frac{1}{2}}$, are terms that break supersymmetry softly (i.e. maintain the cancellation of the quadratic divergences and hence preserve the gauge hierarchy). Thus the entire content of the hidden sector (where supersymmetry is broken) is characterized by these four parameters. Further, the soft-breaking parameters are a consequence of the supergravity interactions.

III. SUSY MASS SPECTRUM AND FLAVOR CHANGING NEUTRAL INTERACTIONS

One of the remarkable features of supergravity GUT models is that they offer a natural explanation (not found in the Standard Model) of the breaking of $SU(2) \times U(1)$ at the electroweak scale via radiative corrections[10]. In the Standard Model, one assumes the existence of a negative Higgs (mass)^2 to achieve electroweak breaking. In supergravity grand unification, spontaneous supersymmetry breaking gives each scalar a positive universal (mass)^2, m_0^2, at scale $Q = M_G$, as seen from Eq. (7). Using the renormalization group equations (RGE), one finds that at the electroweak scale, $Q = O(M_Z)$, the $m_{H_2}^2$ turns negative (for a wide range of parameters) due to the Yukawa interactions with the heavy t-quark. This causes the VEV growth, $\langle H_{1,2} \rangle \neq 0$, and hence the breaking of $SU(2) \times U(1)$. Thus the spontaneous breaking of $SU(2) \times U(1)$ is the joint consequence of the supergravity interactions which give rise to the soft breaking terms and the heavy t-quark Yukawa interactions. The condition that spontaneous breaking of $SU(2) \times U(1)$ occurs allows one to eliminate one GUT scale parameter, μ_0, and express B_0 in terms of $\tan \beta \equiv \langle H_2 \rangle / \langle H_1 \rangle$ and the other parameters. The theory then depends on only four new parameters, m_0, $m_{\frac{1}{2}}$, A_0, and $\tan_\beta$, and the as yet undetermined t-quark mass m_t.

The bose-fermi symmetry of supersymmetry implies the existence of 32 SUSY particles in addition to the usual quarks and leptons. These are twelve squarks, nine sleptons, two charginos (Winos), four neutralinos (Zinos), one gluino and four Higgs bosons. The masses of all these particles can be evaluated, using the RGE, in terms of the 4+1 parameters m_0, $m_{\frac{1}{2}}$, A_0, $\tan \beta$ and m_t. Thus there is a great deal of correlations between the 32 SUSY masses, and experimental predictions to test the theory can be made[11]. The nature of the predictions depends on the GUT group G [e.g. $G=SU(5)$, $O(10)$, flipped $SU(5)$] but generally one expects some of the SUSY particles to be observable at LEP2 and possibly the Tevatron, while some will require the SSC or LHC to be seen.

One remarkable feature of particle interactions is the apparent experimental absence of flavor changing neutral interactions. Thus the branching ratio for $K_s^0 \rightarrow \mu^+ + \mu^-$ is $< 3.2 \times 10^{-7}$ (90% CL). In the Standard Model, this is accounted for by the famous GIM mechanism with loop corrections being suppressed in the decay amplitude by a factor $(m_c^2 - m_u^2)/M_W^2 \approx 10^{-4}$. ($m_{u,c}$ are the u,c quark masses.) The smallness of the quark masses ($m_c \cong 1.4$ GeV, $m_u \cong 5$ MeV) then guarantees the smallness of such flavor changing processes. In supersymmetry, a companion diagram occurs with quarks replaced by squarks and $m_c^2 - m_u^2 \rightarrow m_{\tilde{c}}^2 - m_{\tilde{u}}^2$. However, here the squark masses $m_{\tilde{c}}, m_{\tilde{u}}$ are large, i.e. greater than 100 GeV, and one needs a remarkable cancellation to occur to one part in 10^4, if flavor changing interactions are still to be suppressed. As discussed above, the squark masses at the electroweak scale can be calculated in terms of the basic GUT scale parameters using the RGE to link the two energy scales. One finds for e.g. the left squarks $\tilde{u}_L$ and $\tilde{c}_L$ the result[10]:

$$m_{\tilde{u}_L}^2 = m_0^2 + \left(\sum_i f_i \alpha_i \right) m_{\frac{1}{2}} + \left(\tfrac{1}{2} - \tfrac{2}{3} \sin^2 \theta_w \right) M_Z^2 + m_u^2$$

$$m_{\tilde{c}_L}^2 = m_0^2 + \left(\sum_i f_i \alpha_i \right) m_{\frac{1}{2}} + \left(\tfrac{1}{2} - \tfrac{2}{3} \sin^2 \theta_w \right) M_Z^2 + m_c^2$$

(8)

The first term of Eq. (8) is the universal supersymmetry soft breaking mass. The second term arises from the gauge interactions in running the RGE from M_G to M_Z (and depends on the $SU(3)$, $SU(2)$ and $U(1)$ gauge coupling constants α_3, α_2 and α_1), the third term is from V_D of Eq. (7) and is also a gauge interaction, while the last term is from the Yukawa couplings. One has then that $m_{\tilde{c}_L}^2 - m_{\tilde{u}_L}^2 = m_c^2 - m_u^2$ and the *required cancellation automatically occurs*. (A similar result also holds for the right squarks: $m_{\tilde{c}_R}^2 - m_{\tilde{u}_R}^2 = m_c^2 - m_u^2$.)

It is of interest to investigate the origin of this enormous cancellation in the squark (mass)^2 differences. That m_0 is the same for each generation arises from supersymmetry being broken in the hidden sector. It thus communicates its effects to the physical sector gravitationally, and gravitational interactions are generation independent. The gauge interactions are also generation independents. Only the Yukawa couplings, leading to the quark masses of Eq. (8), distinguish the generations. Other additional corrections [neglected in Eq. (8)] such as the non-renormalizable additions to $f_{\alpha\beta}$ in Eq. (5) are negligible as they are scaled by powers of $\kappa = 1/M_{Pl}$. We see therefore that the gravitational nature of supersymmetry breaking is essential in maintaining the GIM mechanism in supergravity grand unification.

V. CONCLUDING REMARKS

In this note we have surveyed some of the aspects of supersymmetric grand unification. Gravity, in its supersymmeterized version (supergravity), plays a crucial role in the structure of particle interactions. Thus, it is local supersymmetry that gives rise to the spontaneous breaking of supersymmetry, the gauge hierarchy being maintained by the fact that the symmetry breaking is communicated to the physical sector gravitationally in the form of soft breaking terms. Using the renormalization group, these soft breaking terms then give rise to the spontaneous breaking of $SU(2) \times U(1)$ at the electroweak scale, leading to a number of experimental tests at LEP2, the Tevatron, the SSC and LHC. The supergravitational nature of the soft breaking terms was also seen to be crucial to account for the remarkable suppression of flavor changing neutral interactions found experimentally.

One of the reasons gravity plays such an important role for particle interactions in these models is that, in its supersymmeterized form, it is the square root of the Newtonian constant, $\kappa = 1/M_{Pl} = (8\pi G_N)^{1/2}$, that enters into phenomena, rather than G_N. The introduction of a fermion sector to gravity thus allows for larger effects. However, supergravity interactions are non-renormalizable, and one cannot think of these models as yet to be fundamental. Rather, supergravity (and ordinary gravity as well) represent effective field theories valid at best only below the Planck scale. A more fundamental theory (such as string theory) is needed for a true understanding of the role of gravity in the unification of fundamental interactions.

ACKNOWLEDGMENT

This work was supported in part by NSF Grant Numbers PHY-916593 and PHY-917809.

REFERENCES

1. S. L. Glashow, *Nucl. Phys.* **B22**, 579 (1962); S. Weinberg, *Phys. Rev. Lett.* **19**, 1264 (1967); A. Salam, *Proc. 8th Nobel Symposium*, Stockholm 1968, ed. N. Svartholm (Almquist and Wiksells, Stockholm, 1968).

2. H. Georgi and S. L. Glashow, *Phys. Rev. Lett.* **32**, 448 (1974).

3. S. Weinberg, *Phys. Lett.* **B62**, 111 (1976); E. Witten, *Nucl Phys.* **B202**, 253 (1982).

4. D. Z. Freedman, P. van Nieuwenhuizen and S. Ferrara, *Phys. Rev.* **D16**, 3214 (1976); S. Deser and B. Zumino, *Phys. Lett.* **B62**, 355 (1976).

5. E. Cremmer, S. Ferrara, L. Girardello and A. van Proeyen, *Phys. Lett.* **B116**, 231 (1982); *Nucl. Phys.* **B212**, 413 (1983).

6. A. H. Chamseddine, R. Arnowitt and P. Nath, *Phys. Rev. Lett.* **49**, 970 (1982); P. Nath, R. Arnowitt and A. H. Chamseddine, "Applied N=1 Supergravity," ICTP Lecture Series, Vol. 1, 1983 (World Scientific, Singapore, 1984).

7. E. Witten and J. Bagger, *Nucl. Phys.* **B222**, 125 (1983).

8. H. P. Nilles, *Phys. Lett.* **B115**, 193 (1981); S. Ferrara, L. Girardello and H. P. Nilles, *Phys. Lett.* **B125**, 457 (1983).

9. L. Hall, J. Lykken and S. Weinberg, *Phys. Rev.* **D27**, 2359 (1983); P. Nath, R. Arnowitt and A. H. Chamseddine, *Nucl. Phys.* **B227**, 121 (1983).

10. K. Inoue et al., *Prog. Theor. Phys.* **68**, 927 (1982); L. Alvarez-Gaumé, J. Polchinski and M. B. Wise, *Nucl. Phys.* **B250**, 495 (1983); J. Ellis et al., *Phys. Lett.* **B125**, 275 (1983); L. E. Ibañez and C. Lopez, *Nucl. Phys.* **B233**, 545 (1984); L. E. Ibañez, C. Lopez and C. Muñoz, *Nucl. Phys.* **B250**, 218 (1985).

11. R. Arnowitt and P. Nath, *Phys. Rev. Lett.* **69**, 725 (1992); P. Nath and R. Arnowitt, *Phys. Lett.* **B289**, 368 (1992); G. G. Ross and R. G. Roberts, *Nucl. Phys.* **B377**, 571 (1992); S. Kelley, J. Lopez, D. V. Nanopoulos, H. Pois and K. Yuan, *Phys. Lett.* **B273**, 423 (1991); B. Ananthanarayan, G. Lazarides and G. Shafi, B-92-29/PRL-TH-92/16.

Minisuperspaces: Symmetries and Quantization

Abhay Ashtekar [*]   *Ranjeet Tate* [†]   *Claes Uggla* [‡]

Abstract

In several of the class A Bianchi models, minisuperspaces admit symmetries. It is pointed out that they can be used effectively to complete the Dirac quantization program. The resulting quantum theory provides a useful platform to investigate a number of conceptual and technical problems of quantum gravity.

1. Introduction

Minisuperspaces are useful toy models for canonical quantum gravity because they capture many of the essential features of general relativity and are at the same time free of the technical difficulties associated with the presence of an infinite number of degrees of freedom. This fact was recognized by Charlie Misner quite early and, under his leadership, a number of insightful contributions were made by the Maryland group in the sixties and the seventies. Charlie's own papers are so thorough and deep that they have become classics; one can trace back to them so many of the significant ideas in this area. Indeed, it is a frequent occurrence that a beginner in the field gets excited by a new idea that seems beautiful and subtle only to find out later that Charlie was well aware of it. It is therefore a pleasure and a privilege to contribute an article on minisuperspaces to this Festschrift −of course, Charlie himself may already know all our results!

In this paper we shall use the minisuperspaces associated with Bianchi models to illustrate some techniques that can be used in the quantization of constrained systems

[*]Physics Department, Syracuse University, Syracuse, NY 13244-1130, USA and Inter-University Center for Astronomy and Astrophysics, Pune, 411017, India. This author was supported in part by NSF grant PHY90-16733, by a travel grant from the United Nations Development Office and by research funds provided by Syracuse University.

[†]Physics Department, Syracuse University, Syracuse, NY 13244-1130, USA. This author was supported in part by NSF grant PHY90-16733 and by research funds provided by Syracuse University.

[‡]Physics Department, Syracuse University, Syracuse, NY 13244-1130, USA and Department of Physics, University of Stockholm, Vanadisvägen 9, S-113 46 Stockholm, Sweden. This author was supported by a grant from the Swedish National Science Research Council and by research funds provided by Syracuse University.

–including general relativity– and to point out some of the pitfalls involved. We will carry out canonical quantization and impose quantum constraints to select the physical states *a la* Dirac. However, as is by now well-known, the Dirac program is incomplete; it provides no guidelines to introduce the inner product on the space of these physical states. For this, additional input is needed. A possible strategy [1] is to use appropriate symmetries of the classical theory.

Consider for example a "free" particle of mass μ moving in a stationary space-time (where the Killing field is everywhere timelike). The constraint $P^a P_a + \mu^2 = 0$ implies that the physical states in the quantum theory must satisfy the Klein-Gordon equation $\nabla^a \nabla_a \phi - \mu^2 \phi = 0$. To select the inner product on the space of physical states, one can use the fact [2, 3] that the space of solutions admits a unique Kähler structure which is left invariant by the action of the timelike Killing field. The isometry group of the underlying spacetime is unitarily implemented on the resulting Hilbert space; the classical symmetry is promoted to a symmetry of the quantum theory. In fact there is a precise sense [4] in which this last condition selects the Hilbert space structure of the quantum theory uniquely. We will see that a similar strategy is available for all class A Bianchi models except types VIII and IX. In these models, therefore, the Dirac program can be completed and the resulting mathematical framework can be used to test many of the ideas that have been proposed in quantum cosmology. For example, one can investigate if the resulting Hilbert space admits a preferred wave function which can be taken to be the ground state and hence a candidate for the "wave function of the universe". We shall find that, in general, the answer is in the negative.

The paper is organized as follows. In section 2, we recall some of the key features of the mathematical description of Bianchi models. In section 3, we point out that in type I, II, VI$_0$ and VII$_0$ models, the supermetric on each minisuperspace admits a null Killing vector whose action leaves the potential term in the scalar (or, Hamiltonian) constraint invariant. Such Killing fields are called *conditional symmetries* [5]. We then quantize these models by requiring that the symmetries be promoted to the quantum theory. We conclude in section 4 with a discussion of the ramifications of these results for quantum gravity in general and quantum cosmology in particular.

2. Hamiltonian cosmology

In this paper, we shall consider spatially homogeneous models which admit a three-dimensional isometry group which acts simply and transitively on the preferred spatial slices. (Thus, we exclude, e.g., the Kantowski-Sachs models.) In this case, one can choose, on the homogeneous slices, a basis of (group invariant) 1-forms ω^a ($a = 1 - 3$) such that

$$d\omega^a = -\tfrac{1}{2} C^a{}_{bc} \omega^b \wedge \omega^c , \tag{1}$$

where $C^a{}_{bc}$ are the components of the structure constant tensor of the Lie algebra of the associated Bianchi group. The vanishing or nonvanishing of the trace $C^b{}_{ab}$ divides these models into two classes called class A and class B respectively. This classification is important for quantization because while a satisfactory Hamiltonian formulation is available for all class A models, the standard procedure for obtaining this formulation fails in the case of general class B models (although the possibility that a modified procedure may eventually work is not ruled out. See, e.g., the suggestion made in the last section of [6].) Since the Hamiltonian formulation is the point of departure for canonical quantization, from now on we will consider only the class A models.

Let us simplify matters further by restricting ourselves to those spatially homogeneous metrics which can be diagonalized. (This can be achieved in the type VIII and IX models by exploiting the gauge freedom made available by the vector or the diffeomorphism constraint. In the remaining models there *is* a loss of generality, which, however, is mild in the sense that the restriction corresponds only to fixing certain constants of motion. For details, see [6].) Thus, we consider space-time metrics of the form:

$$^{(4)}ds^2 = -N^2(t)dt^2 + \sum_{a=1}^{3} g_{aa}(t)(\omega^a)^2 \, , \tag{2}$$

where $N(t)$ is the lapse function (see, e.g., [7]). Since the trace of the structure constants vanishes, they can be expressed entirely in terms of a symmetric, second rank matrix n^{ab}:

$$C^a{}_{bc} = \epsilon_{mbc} n^{ma}, \tag{3}$$

where ϵ_{mbc} is a 3-form on the 3-dimensional Lie algebra. The signature of n^{am} can then be used to divide the class A models into various types. In the literature, one generally uses the basis which diagonalizes n^{am} and then expresses $C^a{}_{bc}$ as [8]:

$$C^a{}_{bc} = n^{(a)} \epsilon_{abc} \qquad \text{(no sum over } a) \, . \tag{4}$$

The constants $n^{(a)}$ are then used to characterize the different class A Bianchi types. A convenient set of choices is given in the following table [8]:

	I	II	VI$_0$	VII$_0$	VIII	IX
$n^{(1)}$	0	1	1	1	1	1
$n^{(2)}$	0	0	-1	1	1	1
$n^{(3)}$	0	0	0	0	-1	1

To simplify the notation and calculations, let us use the Misner [9] parametrization of the diagonal spatial metric in (2)

$$
\mathbf{g} \equiv g_{ab} = \mathrm{diag}(g_{11}, g_{22}, g_{33}) \equiv e^{2\beta} \ .
$$
$$
\beta = \beta^0 \mathrm{diag}(1,1,1) + \beta^+ \mathrm{diag}(1,1,-2) + \beta^- \mathrm{diag}(\sqrt{3}, -\sqrt{3}, 0) \ . \tag{5}
$$

This parametrization leads to conformally inertial coordinates for the Lorentz (super) metric defined by the "kinetic term" in the scalar constraint. For Bianchi types I, II, VI$_0$ and VII$_0$, the scalar constraint (which, incidentally, can be obtained directly by applying the ADM procedure [10] to *all* class A models [8]) now takes the form

$$
C(\beta^A, p_A) := \tfrac{1}{24} N e^{-3\beta^0} \eta^{AB} p_A p_B + \tfrac{1}{2} N e^{\beta^0 + 4\beta^+} \left(n^{(1)} e^{2\sqrt{3}\beta^-} - n^{(2)} e^{-2\sqrt{3}\beta^-}\right)^2 = 0. \tag{6}
$$

Here the upper case latin indices $A, B...$ range over $0, +, -$; p_A are the momenta canonically conjugate to β^A and the matrix η^{AB} is given by $\mathrm{diag}(-1,1,1)$. To further simplify matters, one usually chooses the "Taub time gauge", i.e. one chooses the lapse function to be $N_T = 12 \exp 3\beta^0$ [11]. (This choice of gauge is also known as Misner's supertime gauge [9].) With this lapse, the scalar constraint takes the form:

$$
C_T = \tfrac{1}{2} \eta^{AB} p_A p_B + U_T = 0 \ ,
$$
$$
U_T = 6 e^{4(\beta^0 + \beta^+)} \left(n^{(1)} e^{2\sqrt{3}\beta^-} - n^{(2)} e^{-2\sqrt{3}\beta^-}\right)^2 \ . \tag{7}
$$

Consequently, the dynamics of all these Bianchi models is identical to that of a particle moving in a 3-dimensional Minkowski space under the influence of a potential U_T. Finally, because we have restricted ourselves to metrics which are diagonal, the vector constraint is identically satisfied. Thus, the configuration space is 3-dimensional and there is one nontrivial constraint (7). The system therefore has two true degrees of freedom.

3.   Quantization

To quantize these systems we must allow the two true degrees of freedom to undergo quantum fluctuations. Following the Dirac theory of constrained systems, let us consider, to begin with, wave functions $\phi(\vec{\beta})$ where $\vec{\beta} \equiv (\beta^0, \beta^+, \beta^-)$. The physical states can then be singled out by requiring that they should satisfy the quantum version of the constraint equation (7):

$$
\Box \phi - \mu^2(\vec{\beta})\phi = 0 \ , \text{ where}
$$
$$
\Box = -\left(\frac{\partial}{\partial \beta^0}\right)^2 + \left(\frac{\partial}{\partial \beta^+}\right)^2 + \left(\frac{\partial}{\partial \beta^-}\right)^2 \ , \text{ and} \tag{8}
$$
$$
\mu^2(\vec{\beta}) = 12 e^{4(\beta^0 + \beta^+)} \left(n^{(1)} e^{2\sqrt{3}\beta^-} - n^{(2)} e^{-2\sqrt{3}\beta^-}\right)^2 \ .
$$

Thus, the physical states are solutions of a "massive" Klein-Gordon equation where the mass term, however, is "position dependent": it is a potential in minisuperspace. Our first task is to endow the space of these states with the structure of a complex Hilbert space. It is here that we need to extend the Dirac theory of quantization of constrained systems.

Consider, for a moment, type I models where the potential vanishes. In this case, we are left with just the free, massless Klein-Gordon field in a 3-dimensional Minkowski space. Therefore, to endow the space of solutions with the structure of a Hilbert space, we can use the standard text-book procedure. First, decompose the fields into positive and negative frequency parts (with respect to a timelike Killing vector field) and restrict attention to the space V^+ of positive frequency fields ϕ^+. Then, introduce on V^+ the inner-product:

$$\langle \phi_1^+, \phi_2^+ \rangle := -2i\Omega(\overline{\phi_1^+}, \phi_2^+), \tag{9}$$

where "overbar" denotes complex conjugation and where Ω is the natural symplectic structure on the space of solutions to the Klein-Gordon equation:

$$\Omega(\phi_1, \phi_2) := \int_\Sigma d^2 S^A \left(\phi_2 \partial_A \phi_1 - \phi_1 \partial_A \phi_2 \right), \tag{10}$$

Σ being any (2-dimensional) Cauchy surface on (M, η^{AB}). Alternatively, we can restrict ourselves to the vector space V of *real* solutions to the Klein-Gordon equation and introduce on this space a complex structure J as follows: $J \circ \phi = i(\phi^+ - \overline{\phi^+})$. This J is a real-linear operator on V with $J^2 = -1$. It enables us to "multiply" real solutions $\phi \in V$ by complex numbers: $(a + ib) \circ \phi := a\phi + bJ \circ \phi$, which is again in V. Thus (V, J) can be regarded as a complex vector space. Furthermore, J is compatible with the symplectic structure in the sense that $\Omega(J\phi_1, \phi_2)$ is a symmetric, positive definite metric on V. Therefore, $\langle ., . \rangle := \Omega(J., .) - i\Omega(., .)$ is a Hermitian inner product on the complex vector space (V, J) and thus $\{V, \Omega, J, \langle ., . \rangle\}$ is a Kähler space [2, 3, 4]. There is a natural isomorphism between the complex vector spaces V^+ and $\{V, J\}$: $\phi^+ = \frac{1}{2}(\phi - iJ\phi)$. Furthermore, this map preserves the Hermitian inner products on the two spaces. Thus, the two descriptions are equivalent. While we introduced J in terms of positive frequency fields, one can also proceed in the opposite direction and treat J as the basic object. One would then use the above isomorphism to *define* the positive frequency fields. In fact, it turns out that the description in terms of $\{V, J\}$ can be extended more directly to the Bianchi models with non-vanishing potentials (as well as to other contexts such as quantum field theory in curved space-times). This is the strategy we will adopt.

Let us make a small digression to discuss the problem of finding the required operator J in a general context and then return to the Bianchi models. Let us suppose we are given a real vector space[1] V equipped with a symplectic structure Ω. Thus,

---

[1] Throughout this paper, our aim is to convey only the main ideas. Therefore, we will not make

$\Omega : V \otimes V \mapsto \mathbb{R}$ is a second rank, anti-symmetric, non-degenerate tensor over V. A 1-parameter family of canonical transformations on V consists of linear mappings $U(\lambda)$ from V to itself ($\lambda \in \mathbb{R}$) such that $\Omega(U(\lambda) \circ v, U(\lambda) \circ w) = \Omega(v, w)$ for all v, w in V. The generator T of this family, $T := dU(\lambda)/d\lambda$, is therefore a linear mapping from V to itself satisfying $\Omega(T \circ v, w) = -\Omega(v, T \circ w)$. It is generally referred to as an *infinitesimal canonical transformation*. These operators can be regarded as symmetries on $\{V, \Omega\}$. The mathematical question of interest to us is: can one endow V with the structure of a complex Hilbert space on which $U(\lambda)$ are unitary operators?

Since V is equipped with a symplectic structure Ω, to "Hilbertize" V, it is natural to seek a complex structure J on it which is compatible with Ω in the sense discussed above; the resulting Kähler structure can then provide the Hermitian inner-product on the complex vector space $\{V, J\}$. The issue of existence and uniqueness of such complex structures was discussed in detail in [4] and we will only report the final result here. The generating function $F_T(v)$ of the canonical transformation under consideration is simply $F_T(v) := \frac{1}{2}\Omega(v, Tv)$. If this is positive definite, then there exists a unique complex structure J which is compatible with Ω such that $U(\lambda)$ are unitary operators on the resulting Kähler space: $\langle U(\lambda) \circ v, U(\lambda) \circ w) \rangle = \langle v, w \rangle$, where, as before, the inner-product is given by

$$\langle v, w \rangle = \Omega(Jv, w) - i\Omega(v, w) . \tag{11}$$

Thus, if one has available a preferred canonical transformation on the real symplectic space $\{V, \Omega\}$ whose generating function is positive definite, one can endow it unambiguously with the structure of a complex Hilbert space.

The preferred complex structure is defined as follows. Using T, let us first introduce (only as an intermediate step) a fiducial, real inner-product $(.,.)$ on V as follows:

$$(v, w) := \tfrac{1}{2}\Omega(v, T \circ w). \tag{12}$$

This is indeed an inner product: it is symmetric because T is an infinitesimal canonical transformation and the resulting norm is positive definite because it equals the generating functional $F_T(v)$. Let $\bar{V}$ denote the Hilbert space obtained by Cauchy completing V w.r.t. $(.,.)$. It is easy to check that T^2 is a symmetric, negative operator on $\bar{V}$, whence it admits a self-adjoint extension which we also denote by T^2. Then,

$$J := -(-T^2)^{-\frac{1}{2}} \cdot T \tag{13}$$

is a well-defined operator with $J^2 = -1$. Using the expression (12) of the inner-product on $\bar{V}$, it is easy to check that J is indeed compatible with Ω. We can

---

digressions to discuss the subtle but often important issues from functional analysis. In particular, we will not specify the precise domains of various operators nor shall we discuss topologies on the infinite dimensional spaces with respect to which, for example, the symplectic structures are to be continuous.

therefore introduce a new *Hermitian* inner product $\langle .,. \rangle$ on the *complex* vector space $(\bar{V}, J)$ via (11). The Cauchy completion H of this complex pre-Hilbert space is then the required Hilbert space of physical states. (Thus, the inner-product (12) and the resulting real Hilbert space $\bar{V}$ were introduced only as mathematical tools to enable us to construct the physical Hermitian structure.) Finally, it is easy to show that J commutes with $U(\lambda)$. It then follows, from the fact that $U(\lambda)$ are canonical transformations and from the definition (11) of the Hermitian inner-product, that $U(\lambda)$ are unitary. The proof of uniqueness of J is straightforward but more involved [2, 3, 4].

In the case when V is the space of solutions to the Klein-Gordon equation in Minkowski space-time (as is the case in the Bianchi type I models) the standard complex structure (corresponding to the positive/negative frequency decomposition) can be obtained by choosing for T the operator $\mathcal{L}_t$ where t^A is any time translation Killing field in Minkowski space. More generally, one can think of J as arising from "a positive and negative frequency decomposition defined by T". More precisely, J is the unitary operator in the *polar decomposition* [12] of the operator T on the Hilbert space $\bar{V}$.

With this general machinery at hand, we can now return to the Bianchi types II, VI_0, VII_0. Denote by V the space of real solutions ϕ to the operator constraint equation (8). Elements of V represent the physical quantum states of the system. As observed earlier, V is naturally equipped with a symplectic structure Ω (see (10)) and our task is to find a compatible complex structure J. For this, we need a preferred canonical transformation on V with a positive generating function. The situation in type I models suggests that we attempt to construct this transformation using an appropriate Killing field t^A of the flat supermetric η^{AB} on the 3-dimensional mini-superspace. However, now, the constraint equation (8) involves a nontrivial potential term $\mu^2(\vec{\beta})$. Hence, for $\mathcal{L}_t$ to be a well-defined operator on V – i.e, for $\mathcal{L}_t \phi$ to be again a solution to (8) – it is essential that $\mathcal{L}_t \mu^2 = 0$. As mentioned earlier, such a Killing field is called a *conditional symmetry* [5]. Fortunately, the models under consideration do admit such a conditional symmetry[2]. An inspection of the potential term $\mu^2(\vec{\beta})$ shows that it is invariant under the action of the diffeomorphism generated by the null Killing field

$$t^A := \left(\frac{\partial}{\partial \beta^0} \right)^A - \left(\frac{\partial}{\partial \beta^+} \right)^A \tag{14}$$

of the supermetric η^{AB}. The question therefore reduces to that of positivity of the generating functional of the canonical transformation $T \circ \phi := \mathcal{L}_t \phi$. A straightforward calculation shows that the generating functional $F_T(\phi) \equiv \frac{1}{2}\Omega(\phi, T\phi)$ is simply the

---

[2]The origin of this symmetry can be traced back to the fact that each of the type I–VII_0 models admits a diagonal automorphism [8, 13] and –like all vacuum models– enjoys a scale invariance.

"conserved energy" associated with the null Killing field:

$$F_T(\phi) = \tfrac{1}{2} \int_\Sigma [\partial_A \phi \partial_B \phi - \eta_{AB}(\partial^C \phi \partial_C \phi + \mu^2 \phi^2)] t^A dS^B \, , \qquad (15)$$

where Σ is a spacelike Cauchy surface. Since the potential is nonnegative and t^A is everywhere future-directed, it follows that $F_T(\phi)$ is indeed positive. Hence, by carrying out a polar decomposition of the operator $T \equiv \mathcal{L}_t$, we obtain a Kähler structure on V which provides us with the required complex Hilbert space of physical states of the system. On this Hilbert space, the 1-parameter family of diffeomorphisms generated by the conditional symmetry t^A is unitarily implemented.

To summarize, because Bianchi types I, II, VI$_0$ and VII$_0$ admit conditional symmetries whose generating functions are positive definite, in these cases, one *can* satisfactorily complete the Dirac quantization program. In the more familiar language of positive and negative frequency decomposition, the Hilbert space of physical states consists of positive frequency solutions ϕ^+ of the quantum constraint equation (8): $\phi^+ = \tfrac{1}{2}(\phi - iJ \circ \phi)$ where ϕ is a real solution and $J = -(-T^2)^{-\frac{1}{2}} \cdot T$. The inner product (9) is constructed from the familiar probability current, i.e., the symplectic structure (10): $\langle \phi_1^+, \phi_2^+ \rangle = -2i\Omega(\overline{\phi_1^+}, \phi_2^+)$. Finally, the 1-parameter family of diffeomorphisms generated by t^A is unitarily implemented in this Hilbert space. Motions along the integral curves of t^A can be interpreted as "time evolution" in the classical theory *both* in the minisuperspace *and* in space-times: t^A is future directed in the minisuperspace and, in any space-time defined by the classical field equations, β^0 –an affine parameter of t^A– increases monotonically with physical time. In the quantum theory, this time evolution is unitary.

This mathematical structure can now be used to probe various issues in quantum cosmology. We conclude with an example.

Let ask whether there is a "preferred" state in the Hilbert space which can be taken to be the ground state. The question is well-posed because on the physical Hilbert space there is a well-defined Hamiltonian which generates time-evolution in the sense described above. On real solutions to the quantum constraint, the Hamiltonian is given by $\hat{H} \circ \phi = J\hbar\mathcal{L}_t\phi$; while in the positive frequency representation of physical states it is given by $\hat{H} \circ \phi^+ = i\hbar\mathcal{L}_t\phi^+$. We will show that $\hat{H}$ is a non-negative operator with zero only in the *continuous* part of its spectrum. This will establish that $\hat{H}$ does not admit a (normalizable) ground state. Thus, the most direct procedure to select a "preferred" state fails in all these models. Furthermore, a detailed examination of the mathematical structures involved suggests that as long as the configuration space – spanned by $\beta^\pm$ – is noncompact, there is in fact *no* preferred state in the physical Hilbert space. It would appear that to obtain such a state, which could be taken, e.g., as the wavefunction of the universe in these models, one must modify in an essential way the broad quantization program proposed by Dirac.

Finally, we outline the proof of the technical assertion made above. Note first, that the expectation value of $\hat{H}$ in any physical state is given by

$$\frac{\langle \phi, \hat{H}\phi \rangle}{\langle \phi, \phi \rangle} = \hbar \frac{\Omega(J\phi, J\mathcal{L}_t\phi)}{\Omega(J\phi, \phi)} = \hbar \frac{(\phi, \phi)}{(\phi, (-T^2)^{-\frac{1}{2}}\phi)} \geq 0, \tag{16}$$

where we have used the expression (11) of the physical, Hermitian inner product $\langle .,. \rangle$, the expression (12) of the fiducial inner product $(.,.)$ and the fact that $(-T^2)$ is a positive definite operator with respect to $(.,.)$. Since the expectation values of $\hat{H}$ are non-negative, it follows that its spectrum is also non-negative. To show that zero is in the continuous part of the spectrum, it is convenient to work with the real Hilbert space $\bar{V}$ defined by the inner product $(.,.)$. Using the fact that the potential $\mu^2(\vec{\beta})$ in (8) is smooth, non-negative and takes values that are arbitrarily close to zero, one can show that the expectation values of $(-T^2)$ on $\bar{V}$ can also approach arbitrarily close to zero. Hence zero is in the spectrum of $(-T^2)$, which implies that the spectrum of $(-T^2)^{-\frac{1}{2}}$ is unbounded above. It therefore follows from (16) that one can find physical states ϕ in which the expectation values of $\hat{H}$ are arbitrarily close to zero. Hence zero is in the spectrum of $\hat{H}$. Finally, if it were in the discrete part of the spectrum, there would exist a physical state ϕ_0 which is annihilated by $\mathcal{L}_t$. In this state, in particular, $\langle \phi_0, \hat{H}\phi_0 \rangle = 2\hbar(\phi_0, \phi_0)$ must vanish. However, it is straightforward to show that

$$2(\phi_0, \phi_0) = \int_{\Sigma} \left((\mathcal{L}_t\phi_0)^2 + \left(\frac{\partial \phi_0}{\partial \beta^-}\right)^2 + \mu^2(\vec{\beta})\phi_0^2 \right) d^2x \tag{17}$$

where Σ is any $\beta^0 = const.$ 2-plane. Hence (ϕ_0, ϕ_0) can vanish iff ϕ_0 and $\mathcal{L}_t\phi_0$ vanish on Σ, i.e., iff ϕ_0 is the zero solution to (8). Hence zero must belong only to the continuous part of the spectrum of $\hat{H}$.

4.   Discussion

In the literature on spatially homogeneous quantum cosmology, the emphasis has been on type I and type IX models. Type I is the simplest and its quantization has been well-understood for sometime now. In fact, the Hamiltonian description of this model is the same as that of the strong coupling limit of general relativity $(G_{\text{Newton}} \mapsto \infty)$, whence its quantization has been discussed also in the context of this limit. Type IX is the most complicated of the spatially compact models and exhibits, in a certain well-defined sense, chaotic behavior in the classical theory[3]. It is also the most "realistic" of the Bianchi models as far as dynamics of general relativity in strong field regions near singularities is concerned. In the early work on quantum

---

[3]The type VIII model has many of the interesting features of the type IX. However, it seems not to have drawn as much attention in quantum cosmology because it is spatially open. Many of the remarks that follow on the type IX model are applicable to the type VIII model as well.

cosmology therefore, the focus of attention was type IX models; type I was studied mainly as a preliminary step towards the quantum theory of type IX. Unfortunately, the "potential" in the scalar constraint of type IX models is so complicated that quantization attempts met with only a limited success. In particular, until recently, not a single exact solution to the quantum constraints was known, whence one could not even begin to address the issues we have discussed in this paper[4].

The reason we could make further progress in this paper is that we analysed the *intermediate models* in some detail. Although they are not as "realistic" as type IX, we have seen that these models do have an interesting structure. In particular, unlike in the type I models, the potential in the scalar constraint is non-zero, whence their dynamics is quite non-trivial already in the classical theory. For example, we still do not know the explicit form of a complete set of constants of motion in type VI_0 and type VII_0 space-times. In spite of this, we could complete the Dirac quantization program because these models admit an appropriate (i.e., future-directed) conditional symmetry. In physical terms, quantization is possible because one can define an internal time variable on these minisuperspaces in a consistent fashion.

The resulting mathematical framework is well-suited for addressing a number of conceptual issues in quantum gravity in general and quantum cosmology in particular.

First is the issue of time. Using the conditional symmetry t^A, we were able to construct a complex structure J such that the 1-parameter family of transformations on the physical quantum states ϕ induced by t^A are implemented by a 1-parameter family of unitary operators on the Hilbert space. Hence, using the affine parameter along the vector field t^A as a "time" variable, one can deparametrize the theory and express time evolution through a Schrödinger equation: it is the introduction of the complex structure – or, the use of only positive frequency fields – that provides a "square-root" of the quantum scalar constraint which can be re-interpreted as the Schrödinger equation. In the type I model, for example, the "square-root" takes the following form:

$$i\hbar \mathcal{L}_t \phi^+(\vec{\beta}) = \hbar \left((-\Delta)^{\frac{1}{2}} - i\frac{\partial}{\partial\beta^+} \right) \phi^+(\vec{\beta}) \tag{18}$$

where $\Delta = (\partial/\partial\beta^+)^2 + (\partial/\partial\beta^-)^2$ is the Laplacian in the $\beta^\pm$ plane. Thus the argument β^0 of ϕ^+ can be regarded as time and $\hat{H} \equiv \hbar((-\Delta)^{\frac{1}{2}} - i(\partial/\partial\beta^+))$ can be regarded as the Hamiltonian. (The second term in the expression for $\hat{H}$ arises simply because our evolution is along a *null* Killing field (14) rather than a timelike one.) Now, there exist in the literature several distinct approaches to the issue of time in full quantum gravity, including those that suggest that one should generalize quantum mechanics in a way in which there is in fact *no* preferred time variable. It would be

---

[4]Over the past two years, a handful of solutions have been obtained [14, 15] using new canonical variables [16] in terms of which the potential term disappears. While this development does represent progress, the solutions obtained thus far are rather special and there are too few to construct a useful Hilbert space.

fruitful indeed to apply these ideas to the minisuperspaces considered in this paper and compare the resulting quantum descriptions with the one obtained here using the conditional symmetry.

A second issue is the strategy to select inner products. Since we do not have access to any symmetry group in general relativity, it has been suggested [17, 18] that we should let the "reality conditions" determine the inner product. More precisely, the strategy is the following: Find a sufficient number of real classical observables a la Dirac –i.e. functions on the phase space which weakly commute with the constraints– and demand that the inner product between quantum states be so chosen that the corresponding quantum operators are Hermitian. In many examples, this condition suffices to pick out the inner product uniquely. In the minisuperspaces considered in this paper, on the other hand, we have selected the inner-product using a completely different strategy: we exploited the fact that these models admit a conditional symmetry of an appropriate type, thereby side-stepping the issue of isolating a *complete* set of Dirac observables. It would be interesting to identify the Dirac observables, use the "reality conditions" strategy and compare the resulting inner product with the one we have introduced. (Equivalently, the question is whether the real Dirac observables can be promoted to operators which are Hermitian with respect to the inner product we have introduced.) This program has been completed in type I and II models [19]. While the two inner products do agree in a certain sense, subtleties arise already in type II models. However, the question remains open in types VI_0 and VII_0.

A third issue is the fate of classical singularities in the quantum descriptions. Every (non-flat) classical solution belonging to the models considered here has a singularity and one sometimes appeals to "the rule of unanimity" [20] to argue that in such cases, singularities must persist also in quantum theory. On the other hand, we have found that the quantum evolution is given by a 1-parameter family of unitary operators[5]. In all cases considered in this paper, the quantum Hilbert spaces are well-defined and quantum evolution is unitarily implemented. This appears, at least at first sight, to be a violation of the rule of unanimity and it is important to understand the situation in detail. Is it the case that the classical singularities simply disappear in the quantum theory? Or, do they persist but in a "tamer" fashion? In the space-time picture provided by classical general relativity, evolution is implemented by hyperbolic equations which simply breakdown at curvature singularities. Is this loss of predictability recovered in the quantum theory? That is, in quantum theory, can we simply "evolve through the singularity"? It would be useful to apply the semi-classical methods available in the literature to these models both to gain physical insight into this issue and to probe the limitations of the semi-classical methods

---

[5]The situation is similar in full 2+1 gravity, where, *every* classical cosmological solution begins with a "little bang" where (there is no curvature singularity –hence the pre-fix "little"– but where) the spatial volume goes to zero. Inspite of this, the mathematical framework of quantum theory is complete and well-defined.

themselves. (Some results pertaining to these issues are discussed in [18, 19].)

Finally, further work is needed to achieve a more complete understanding of the quantum physics of these models. What we have constructed is the Hilbert space of physical states. Any self-adjoint operator on this space may be regarded as a quantum Dirac observable. However, unless the operator has a well-defined classical analog it is in general not possible to interpret it physically[6]. In all the models considered here, the generator of the "time translation" defined by the conditional symmetry t^A is one such observable: its classical analog is simply $p_0 - p_+$, the momentum corresponding to the conditional symmetry. However, since these models have two (configuration) degrees of freedom, the reduced phase space is four dimensional whence one expects there to exist four independent Dirac observables. The open question therefore is that of finding the three remaining observables. In the type I model, there is no potential, whence every generator of the Lorentz group defined by the flat supermetric η^{AB} is a Dirac observable. Thus, it is easy to find the complete set both classically and quantum mechanically. In the type II model, the task is already made difficult by the presence of a non-zero potential. However, now the potential is rather simple and can actually be eliminated by a suitable canonical transformation [19]. (In full general relativity, canonical transformations with the same property exist [16, 22]. However, the resulting supermetric is curved. In the type II model, the new supermetric is again flat; only the global structure of the constraint surface is different from that in the type I model.) In terms of these new canonical variables it is easy to find the full set of Dirac observables both classically and quantum mechanically. Thus, in the type I and II models, the issue of the physical interpretation is under control. In the type VI_0 and VII_0 models, on the other hand, the issue is wide open. Resolution of this issue will also shed light on the question of uniqueness of the inner product and clarify the issue of singularities discussed above.

What is the situation with respect to the more complicated Bianchi models, types VIII and IX? Now, the conditional symmetry which played a key role in quantum theory no longer exists: the origin of this symmetry can be traced to the presence of a diagonal automorphism group which happens to be zero dimensional in type VIII and IX models (see references in footnote 2). However, it is possible that these models admit some "hidden" symmetries –i.e. symmetries which are not induced by space-time diffeomorphisms. There is indeed a striking feature along these lines, first pointed out by Charlie himself [9]: in the type IX models, there exists an asymptotic constant of motion precisely at the early chaotic stage. Can one use it for quantization?

---

[6]In addition, to obtain physical interpretations in the quantum theory of constrained dynamical systems such as the ones we are considering, one has to deparametrize the theory and express phase space variables (such as the 3-metric and extrinsic curvature) in terms of the Dirac operators and the "time"variable. (see e.g. [21, 18, 19]).

Acknowledgements:

The authors would like to thank Jorma Louko for discussions and for reading the manuscript.

References

[1] Kuchař, K. 1981 *in* "Quantum Gravity II," Isham, C.J., Penrose R. and Sciama D.W. (ed.), Clarendon Press, Oxford.

[2] Ashtekar, A. and Magnon, A. 1975 *Proc. R. Soc. Lond.*, **346A**, 375.

[3] Kay, B. 1978 *Comm. Math. Phys.*, **62**, 55.

[4] Ashtekar, A. and Sen, A. 1978 Enrico Fermi Institute, University of Chicago Report.

[5] Kuchař, K. 1982 *J. Math. Phys.* **22**, 2640.

[6] Ashtekar, A. and Samuel, J. 1991 *Class. Quantum Grav.*, **8**, 2191.

[7] MacCallum, M.A.H. 1979 *in* "Physics of the Expanding Universe," Demianski, D. (ed.), Springer, Berlin.

[8] Jantzen, R.T. 1984 *in* "Cosmology of the Early Universe," Ruffini, R. and Fang, L.Z., (eds.), World Scientific, Singapore.

[9] Misner, C.W. 1972 *in* "Magic Without Magic," Klauder, J. (ed.), Freeman, San Francisco.

[10] Arnowit, R., Deser, S. and Misner, C.W. 1962 *in* "Gravitation: An Introduction to Current Research," L. Witten (ed.), Wiley, New York.

[11] Taub, A.H. 1951 *Ann. Math.*, **53**, 472.

[12] Reed, M. and Simon, B. 1975 Functional analysis, Academic Press, New York.

[13] Rosquist, K., Uggla, C.and Jantzen R.T. 1990 *Class. Quantum Grav.*, **7**, 611.

[14] Kodama, H 1988 *Prog. Theor. Phys.*, **80**, 295; 1990 *Phys. Rev.*, **D42**, 2548.

[15] Moncrief, V. and Ryan M.P. 1991 *Phys. Rev.*, **D44**, 2375.

[16] Ashtekar, A. 1987 *Phys. Rev.*, **D36**, 1587.

[17] Ashtekar, A. (Lectures prepared in collaboration with Tate, R.S.) 1991 Lectures On Non-Perturbative Canonical Gravity, World Scientific, Singapore (see chapter 10).

[18] Tate, R.S. 1992 *Ph.D. Dissertation*, Syracuse University, Syracuse University preprint SU-GP-92/8-1.

[19] Ashtekar, A., Tate, R.S. and Uggla, C. 1992 Syracuse University preprint SU-GP-92/2-6.

[20] Wheeler, J.A. 1977 *Gen. Rel. Grav.*, **8**, 713.

[21] Rovelli, C. 1990 *Phys. Rev.*, **D42**, 2638; 1991 *Phys. Rev.*, **D43**, 442.

[22] Tate, R.S. 1992 *Class. Quantum Grav.*, **9**, 101.

Quantum Cosmology

BEVERLY K. BERGER

Oakland University

1 INTRODUCTION

The presumed breakdown of the general theory of relativity (GTR) at the Planck scale without, as yet, a complete quantum theory of gravity (QG) to replace it is used to motivate consideration of a simpler problem—the quantum mechanics of cosmological models. In a pioneering paper, Misner (1969a) coined the term "quantum cosmology" (QC) for the quantization of the dynamical system whose degrees of freedom describe a spatially homogeneous universe. He also introduced the term "minisuperspace" (MSS) (Misner 1972) for the finite-dimensional configuration space of the dynamics of homogeneous cosmologies—a finite subspace of Wheeler's "superspace" (Wheeler 1968), the space of all three-geometries. The first invocation of quantized MSS models was that of DeWitt (1967) to apply Dirac (1958, 1959) quantization of gravity to a tractable system. He considered the closed Friedmann-Robertson-Walker (FRW) model (see Misner *et al* 1973). Misner (1969a, 1970, 1972) considered the application of Arnowitt, Deser, and Misner (1962) (ADM) quantization methods to the Bianchi Type cosmologies (*e.g.* Ryan and Shepley 1975, MacCallum 1975). In both treatments, it became clear that issues of time, factor ordering, and interpretation which plague the canonical quantization of gravity survive in the truncated models. It was, of course, recognized that, while perfectly valid as classical solutions to Einstein's equations, quantized cosmologies where degrees of freedom have been zeroed by hand need have no relation to a true QG theory. [A systematic attack on the validity of the MSS "approximation" has only recently begun (Kuchař and Ryan 1986, 1989).] Misner argued (1969a) that one might reasonably expect the quantum universe to be dominated by the dynamics of its spatially homogeneous mode. A similar argument was given by Hawking (1984). One hope was that the quantized Mixmaster universe (Misner 1970) could serve as an appropriate initial state for our present Universe by explaining the uniform cosmic microwave background temperature and the absence of anisotropic expansion. A modern version of "chaotic cosmology" was given by Calzetta (1989). Interest in QC waned due to intractability of all but the simplest models, questions of time and interpretation, and lack of any guiding principle to select a physically significant quantum state.

A general revival of interest in QC followed a proposal by Hartle and Hawking (1983) for a boundary condition to be satisfied by the wavefunction of the Universe. Since this boundary (actually initial) condition selects a preferred state for the quantized u-niverse, the consequences of the properties of the state can be compared to the observable Universe. For example, given a boundary proposal, the probability for anisotropic expansion (Amsterdamski 1985, Furusawa 1986a, Hawking and Lutrell 1984, Ryan 1971, Wright and Moss 1985, del Campo and Vilenkin 1989a) or inflation (Carow and Watamura 1985, Hawking 1984, Hawking and Page 1988, Moss and Wright 1984, Kazama and Nakayama 1985, Zhuk 1988, del Campo and Vilenkin 1989b) can be considered. Alternative boundary conditions were also proposed (Vilenkin 1986, Grishchuk and Sidorov 1988, Suen and Young 1989, Wada 1987, Zhuk 1988, Kiefer 1988, Conradi 1992) to address the same questions. Other topics studied include quantum effects on the singularity (Gotay and Isenberg 1980, Gotay and Demaret 1983, Louko 1987, Smith and Bergmann 1986, Lemos 1991), quantum creation of the universe itself (Graham and Szepfalusy 1990, Atkatz and Pagels 1982, Gott 1982, Rubakov 1984, Vilenkin 1982, Vilenkin 1985b), and the choice and/or origin of time (Unruh 1989, Unruh and Wald 1989, Castagnino 1989, Halliwell 1989, Zeh 1986, 1988, Mensky 1990).

Interpretation still remains an issue for QC. The difficulty, of course, occurs because the quantized universe does not fit the standard quantum mechanical framework which appears to require a classical "observer" to study the statistics of an ensemble of identically prepared systems. In the Universe, we have not only a single system, but one of which we are a part. Thus a modification of (at least the interpretation of) quantum mechanics is required to regard the Universe to be a quantum system (DeWitt 1967, Misner 1969a, Hartle 1990). A very exciting recent development has been the attempts to explain the classical world as our inevitable perceptions of the quantum Universe. (See *e.g.* Zurek 1981, 1982, Unruh and Zurek 1989, Zeh 1986a, Griffith 1984, Omnes 1988, Gell-Mann and Hartle 1990, Hu 1991.) This topic is addressed elsewhere in this volume (Hu 1992).

2 CLASSICAL THEORY

It is convenient to approach QG from the Hamiltonain formulation of GTR given by Dirac (1958) and ADM (1962). One defines (Misner et al 1973, Wald 1984) a spacelike hypersurface Σ_t (at constant coordinate time t) with intrinsic metric h_{ab} $(a, b = 1, 2, 3)$ and extrinsic curvature K_{ab}. For the case of matter described by stress-energy tensor T_{ab} the combined matter-gravitational action can be written in the 3 + 1 form (Wald 1984)

$$S = \int dt\, d^3x \left[\pi^{ab} \frac{\partial}{\partial t} h_{ab} - NC - N_a C^a + P^A \frac{\partial}{\partial t} Q_A - H(\pi, h, P, Q) \right] \qquad (1)$$

where $\pi^{ab} = \sqrt{h}\,(K^{ab} - h^{ab}K)$ is conjugate to h_{ab}. We use natural units such that $c = 16\pi G = \hbar = 1$. The last two terms in (1) schematically define the Hamiltonian formulation for the non-gravitational terms. The Lagrange multipliers N and N_a are respectively the lapse which measures the proper time separation between nearby spacelike hypersurfaces and the shift which tells how to identify points with the same spatial coordinates on nearby spacelike hypersurfaces. The constraints C and C^a are respectively the generators of time and space translations with the form

$$C = h^{-\frac{1}{2}} \left(\pi^{ab} \pi_{ab} - \tfrac{1}{2}\pi^2 \right) - h^{\frac{1}{2}}\, {}^3R + 2h^{\frac{1}{2}}\rho \qquad (2)$$

$$C^a = -2D_b \pi^{ab} + h^{\frac{1}{2}} t^b \qquad (3)$$

where $h = \det h_{ab}$, ${}^3R$ is the spatial scalar curvature, D is the covariant derivative formed from h, and ρ and t^a are respectively the 0-0 and 0-a components of T_{ab}. We note that the term containing π^{ab}'s in (2) can be written as (DeWitt 1967) $G^{abcd} \pi_{ab} \pi_{cd}$. G_{abcd} has six components at each point on Σ_t, Lorentzian signature, and is called the metric on superspace (Wheeler 1968, DeWitt 1970, Fischer and Marsden 1979), the configuration space for the Hamiltonian formulation of GTR. Each point in superspace is a 3-geometry, ${}^3\mathcal{G}$, *i.e.* the equivalence class of three dimensional spaces related by spatial diffeomorphisms. A spacetime is a trajectory in superspace. We note that (1) does not include boundary terms which do not affect the classical theory (Arnowitt et al 1962) but cannot be ignored in the quantum theory. Gibbons and Hawking (1977) propose a modified action which we shall assume has been taken into account. In the Dirac approach to the Hamiltonian formulation, Einstein's equations can be obtained by variation of the superhamiltonian $N_\mu C^\mu$ where $\mu = 0, \ldots, 3$. Since only two of the gravitational degrees of freedom (at each space point) represented by (h_{ab}, π^{ab}) are dynamical, the remaining degrees of freedom are eliminated from the solutions to the variational equations via the constraints and coordinate conditions. Alternatively, the ADM approach requires the identification of $Q^\mu[h, \pi]$ as coordinates and solving the constraints for the momenta P_μ conjugate to Q^μ. Variation of $-P_0 = H_{ADM}$ yields the equations of motion for the dynamical degrees of freedom (where P_0 is conjugate to the variable identified to be the time). (In a final step, the N^μ may be identified to relate the intrinsic coordinates Q^μ to x^μ, the original coordinate system.) At the classical level, the Dirac and ADM approaches are equivalent although implementation of the ADM reduction can be intractable.

3 CANONICAL QUANTIZATION OF GRAVITY

Quantization of constrained systems was studied by Dirac (1958, 1959) and subsequently by many others. (See Guven and Ryan 1992.) The basic idea is straightforward as can be seen from the following simple example. A system described by one degree of freedom $\{p, q\}$ with a Hamiltonian $H(p, q)$ may be written in parametrized form by defining a new degree of freedom $Q = t$ with conjugate momentum $P = -H$. With the addition of this degree of freedom, the action acquires a term $-NC$ where $C = P + H$ constrains the variations of the new degree of freedom and $N = dt/d\lambda$. Schroedinger's equation is recovered if the constraint annihilates the wavefunction— i.e. $\hat{C}\,\Psi(q, Q) = 0$ with the identification $Q = t$ and $\hat{C}$ the operator form of C (obtained from $p \to -i\partial/\partial q$, etc.). The myriad of attempts to apply this formalism to QG have encountered insurmountable difficulties (Guven and Ryan 1992, Kuchař 1990, 1992). Here we shall ignore the important issues of interpretation (Hu 1992) to consider the relevant wavefunction $\Psi(^3\mathcal{G})$ to be the amplitude for a given 3-geometry. Modulo (non-trivial) operator ordering questions (Kuchař 1987), the Dirac quantization requires the wavefunction to be annihilated by the constraints (i.e. the physical state space is defined) via $\hat{C}^\mu \Psi(^3\mathcal{G}) = 0$ where $\hat{C}^\mu$ is the operator form ($\pi^{ab} \to -i\delta/\delta h_{ab}$) of the constraints (2) and (3). The treatment of the momentum constraints is not regarded to be fundamentally more difficult than in standard gauge theories (Guven and Ryan 1992). The Hamiltonian constraint operator [which acting on the wavefunction yields the Wheeler-DeWitt equation (DeWitt 1967)] is, however, generally regarded to be problematical. For example (Kuchař 1992) C^0 is the generator of both time coordinate transformations and the dynamical evolution of the initial hypersurface, Σ_t. Its hyperbolic, rather than elliptic, form hampers the construction of a probabilistic interpretation and has thus led to proposals [actually implemented in minisuperspace (Hosoya and Morikawa 1989, Zhuk 1992, McGuigan 1988)] for a "third quantization" formalism.

As an alternatvie (Arnowitt et al 1962), the ADM reduction can be performed classically to quantize only the true dynamical degrees of freedom. This runs into the difficulty that a generic spacetime is not covered by a single coordinate patch. Even when a suitable time variable can be identified, the resulting H_{ADM} often fails to be self-adjoint so that the states of the system cannot be constructed(Misner 1972, Kuchař 1992, Guven and Ryan 1992). Attempts have been made to carry over the formalism from gauge theories—e.g. by implementing a Fadeev-Popov or BRST construction within the Feynman path integral formalism but the differences between the diffeomorphism invariance of GTR and true gauge invariance have prevented complete success of this approach (e.g. Kuchař 1992, Guven and Ryan 1992).

4 QUANTIZATION IN MINISUPERSPACE

A large number of spatially homogeneous cosmologies [e.g. Class A Bianchi Types (MacCallum 1979)] can be described completely by the Hamiltonian constraint C^0 since the momentum constraints are identically satisfied and no error is introduced by symmetrization prior to variation. It is often convenient to replace the intrinsic metric (and non-gravitational variables) by some equivalent set $\{q^A, p_A\}$, $A = 0, \ldots, N - 1$, for a system with N degrees of freedom. The Hamiltonian constraint has the generic form

$$C^0 = G^{AB}(q)\, p_A p_B + V(q) \tag{4}$$

where G^{AB} is the Lorentzian (inverse) minisuperspace metric. The potential term (in the vacuum case) is proportional to the spatial scalar curvature. Minisuperspace is the N-dimensional dynamical configuration space with axes q^A. Einstein's equations are equivalent to those obtained from the variation of $N_0 C^0$. Although (4) appears to be an energy-like expression, $N_0 C^0$ differs from a true Hamiltonian in various crucial ways. These differences are also present in the full QG and contribute major obstructions to straightforward canonical quantization. The Lorentzian signature of G_{AB} and the arbitrary sign of ${}^3R$ cause the failure of the generic C^0 to be bounded below. As has been pointed out (Hawking 1979), a Wick rotation to Euclidean time $\tau = it$ does not fix the problem. In some cases (i.e. for a bounded from below potential term), a conformal rotation may yield a bounded operator. It should be emphasized that the "time-like" component of G_{AB} is that associated with the volume expansion rate (i.e. the term proportional to the square of the Hubble parameter). Hawking (1979) proposed a path integral formulation to treat this conformal (and problematical) degree of freedom separately. A proposal for a positive definite Euclidean action (with curvature squared terms) has been given by Horowitz (1985). We remark that a procedure to obtain a well-defined (in the mathematical sense) and bounded from below Hamiltonian constraint is essential for numerical quantum cosmology or quantum gravity (e.g. Myers 1991). It is important to realize that not all the q^A's are dynamical since the freedom to change time parametrization remains (although the spatial homogeneity fixes the foliation). The ADM reduction therefore consists of identification of $t(q^A, p_A)$ as a time variable and then solving (4) for its conjugate momentum $P_t = -H_{ADM}$. In the Dirac analysis, the "gauge" (time) is fixed by choice of the lapse N. Detailed investigations relating to aspects of time reparametrization invariance and choice of time gauge (particularly in minisuperspace) have been given by Hartle and Kuchař (1984, 1986), Guven and Ryan (1992), Teitelboim (1980, 1982, 1983), Berger and Vogeli (1985), and Schleich (1989).

In minisuperspace, the Dirac quantization yields (DeWitt 1967, Misner 1970, 1972) the Wheeler-DeWitt equation

$$\hat{C}\,\Psi(q^A) = \left[: - G^{AB}(q)\frac{\delta}{\delta q^A}\frac{\delta}{\delta q^B} : + V(q) \right]\Psi(q^A) = 0 \qquad (5)$$

where : : emphasizes the presence of the factor ordering ambiguity. On occasion, arguments have been made for a particular factor ordering (*e.g.* Hawking and Page 1986). A convenient factor ordering (equivalent to a choice of minisuperspace variables) requires G_{AB} to be flat. Of course, this is not always possible but can be achieved for many of the most popular MSS models. The wave function and/or quantum propagator can also be expressed in a path integral formulation (Hartle and Hawking 1983, Narlikar and Padmanabhan 1983). Hartle and Hawking (1983) reminded us that the path integral or Wheeler-DeWitt equation requires a "boundary condition" to produce a unique wavefunction. [That the wavefunction should be unique is an issue of interpretation. If one presumes the Universe to be a quantum system in a particular state, then a "theory of initial conditions" (Hartle 1987) should single out this preferred state.] The best known boundary conditions (actually proposed for the full theory) are those due to Hartle and Hawking (1983) and Vilenkin (1986) although these are by no means the only ones. The Hartle-Hawking proposal requires the amplitude for a given 3-geometry, $\Psi(^3\mathcal{G})$, to be the sum (in the sense of the path integral) over all compact, Euclidean 4-geometries bounded by $^3\mathcal{G}$. Vilenkin suggests that one choose the solution to the Wheeler-DeWitt equation which has no incoming radiation at the boundaries of superspace. In all cases, the boundary conditions are most conveniently applied to MSS models (and, in fact, may be intractable in the full theory). (See Halliwell 1990 for an extensive bibliography.) Once the indicated wavefunction has been obtained, its properties may be examined to see if the selected state is the most physically reasonable (and relevant) among the possible states. This analysis requires an interpretation of the wavefunction. (For a review of recent work on this subject see Laflamme 1991, Hu 1992.)

The ADM reduction leads to a Schroedinger equation

$$i\frac{\delta}{\delta t}\Psi(q^I, t) = \hat{H}_{ADM}\left(-i\frac{\delta}{\delta q^I}, q^I, t\right)\Psi(q^I, t) \qquad (6)$$

$I = 1,\ldots,N-1$, rather than the Wheeler-DeWitt equation. As a natural boundary condition, one might consider an analog of Feynman's $i\varepsilon$ prescription in the path integral (Itzykson and Zuber 1978) to single out wavefunctions which represent either purely expanding or purely contracting universes and not superpositions of the two states (Berger and Vogeli 1985). In a sense [which can be made precise by introducing spinors (Teitelboim 1977, Macías *et al* 1987, D'Eath and Hughes 1988)], the ADM formulation is the "square root" of the Wheeler-DeWitt equation. The quantum theories from the two formulations are not equivalent (Schleich 1990, Guven and Ryan 1992, Kuchař 1992). For MSS models, they are asymptotically (*i.e.* semiclassically) identical and can be related in simple models by (an unnatural) choice of factor ordering. A possible way around the failure of H_{ADM} to be always self-adjoint is to apply some variant of the Fadeev-Popov (1975) procedure [or the Fadeev (1969) procedure for MSS] or the BRST approach (Guven and Ryan 1992, Kuchař 1992, Laflamme 1991) to the unreduced degrees of freedom. Several examples exist in the literature (*e.g.* Schleich 1989, Berger and Vogeli 1985). Implementation of these gauge reduction procedures need not yield the ADM formalism.

The path integral formulation has the advantage that the classical form of the action is used so that operators need never be defined. Although often evaluated for Euclidean time (Hartle and Hawking 1983), it can be argued that there is no particular advantage over the Lorentzian form since the Euclidean gravitational action is not positive definite (Brown and Martinez 1990). The path integral may be used to implement the boundary condition [by defining a suitable integration contour (Halliwell 1990b)] and to yield a semi-classical wavefunction. Alternatively, one may evaluate the path integral directly [for a suitable skeletonization of MSS (Kuchař 1983)] to obtain the wavefunction or propagator (Berger and Vogeli 1985, Schleich 1989). The path integral also lends itself to Monte Carlo simulations which could allow the treatment of more complicated (and realistic) MSS models (Berger 1988, 1989, Hartle 1985, 1986, Hawking and Wu 1985).

5 EXAMPLES
The following examples illustrate issues of time, boundary conditions, and physical properties.

5.1 The DeSitter Model
Perhaps the most widely studied quantum cosmology (*e.g.* Hartle and Hawking 1983, Vilenkin 1988 , Suen and Young 1989, Halliwell and Louko 1989) is the closed FRW universe with positive cosmological constant. The Wheeler-DeWitt equation takes the form (for scale factor a and $\lambda > 0$)

$$\left(a\frac{d}{da}a^{-1}\frac{d}{da} - a^2 + \lambda a^4\right)\Psi(a) = 0 \tag{7}$$

where a particularly convenient factor ordering has been imposed. The solution can be expressed in terms of Airy functions $\mathrm{Ai}(z)$, $\mathrm{Bi}(z)$ with $z = (4\lambda)^{-\frac{1}{3}}(1 - \lambda a^2)$. The Hartle-Hawking boundary condition selects the asymptotic behavior that dominates the Euclidean path integral as $a \to 0$ to be $\mathrm{Ai}(z)$ so that the wavefunction behaves as

$$\Psi(a) \propto \begin{cases} \exp\left(\frac{1}{2}a^2\right) & ,0 \approx a < \frac{1}{\sqrt{\lambda}} \\ \sin\left(\frac{2}{3}|z|^{\frac{3}{2}} + \frac{\pi}{4}\right) & ,a \to \infty \end{cases} \tag{8}$$

The above wavefunction decays into the classically forbidden region $a < 1/\sqrt{\lambda}$. For the Lorentzian MSS metric, outgoing waves are $\exp(-i\vec{k} \cdot \vec{x})$. As $a \to \infty$, Vilenkin's condition of no incoming radiation selects the solution $\mathrm{Ai}(z) + i\mathrm{Bi}(z)$ to give

$$\Psi(a) \propto \begin{cases} \exp\left(-\frac{1}{2}a^2\right) & ,0 \approx a < \frac{1}{\sqrt{\lambda}} \\ \exp\left[i\left(\frac{2}{3}|z|^{\frac{3}{2}} + \frac{\pi}{4}\right)\right] & ,a \to \infty \end{cases} \tag{9}$$

This solution peaks at $a = 0$ (within the classically forbidden region) and decays through the barrier. It thus represents a tunneling solution. Alternative boundary proposals choose other combinations of $\mathrm{Ai}(z)$ and $\mathrm{Bi}(z)$.

The results for this model can be extended qualitatively to those for a self-interacting scalar field in the FRW background by replacing the cosmological constant with the scalar field potential $V(\phi)$ (Vilenkin 1988). Vilenkin finds that the probability density is

$$\rho(a,\phi) \propto \begin{cases} \exp\left[2/3V(\phi)\right] & ,\text{Hartle-Hawking} \\ \exp\left[-2/3V(\phi)\right] & ,\text{Vilenkin} \end{cases} \tag{10}$$

This yields ρ large for $V(\phi)$ large for Vilenkin's boundary condition, a behavior consistant with the initiation of inflation. More detailed treatments of self-interacting scalar field models have been given by *e.g.* Carow and Watamura (1985), Hawking (1985), Hawking and Page (1988).

The ADM formulation for this problem (Berger 1986b) has the immediate difficulty that there is only one degree of freedom that is consumed by the time coordinate condition. The Schroedinger equation

$$i\frac{\partial}{\partial\Omega}\Psi(\Omega) = \pm(\lambda e^{6\Omega} - \kappa e^{4\Omega})^{\frac{1}{2}}\,\Psi(\Omega) \tag{11}$$

has a clearly non-self-adjoint operator on the RHS. Evaluation of the solutions (using a prescription for the definition of the square root in the classically forbidden region) yields exponentials which are equivalent to the asymptotic solutions to the Wheeler-DeWitt equation.

5.2 Massless Scalar Field in a Friedmann-Robertson-Walker Universe

This model is mathematically equivalent to that discussed by Hosoya and Morikawa (1989), Zhuk (1992), and Berger (1985). It is particularly simple for the variable choice of $\Omega = \ln a$ for a the FRW scale factor, and ϕ the (spatially homogeneous) scalar field amplitude with p_Ω and p the respective conjugate momenta. Classically, p is a constant of the motion. The Wheeler-DeWitt equation is

$$\left[\frac{\partial^2}{\partial \Omega^2} - \frac{\partial^2}{\partial \phi^2} + \zeta \, e^{\xi \Omega}\right] \Psi(\phi, \Omega) = 0. \tag{12}$$

Various possibilities for these constants include p as the magnitude of the momentum in a higher dimensional model [e.g. $p^2 = p_+^2 + p_-^2$ with $\zeta = 0$ is Bianchi I (Misner 1972)], $\zeta = -\kappa$, $\xi = 4$ for a spatial scalar curvature term, and $\zeta = \lambda$, $\xi = 6$ for a cosmological constant term. Factor ordering issues have been avoided by choosing variables with a flat MSS metric. For $\zeta < 0$ there is a maximum of expansion at $\Omega = \frac{1}{\xi} \ln \frac{p^2}{|\zeta|}$. The general solution is a superposition of mode functions $\mathcal{F}_p(\phi, \Omega)$ where

$$\mathcal{F}_p(\phi, \Omega) = \begin{cases} \exp(ip_\Omega \Omega \pm ip\phi), & \zeta = 0 \\ Z_{i\frac{2p}{\xi}}\left[\frac{2\sqrt{\zeta}}{\xi} \exp(\frac{\xi}{2}\Omega)\right] \exp(\pm ip\phi), & \zeta \neq 0 \end{cases} \tag{13}$$

for $Z_\nu(x)$ any solution to Bessel's equation for order ν. For $\zeta > 0$, the mode functions will be oscillatory as $\xi \Omega \rightarrow \infty$ but will be growing and decaying exponentials for $\zeta < 0$. The former case is analogous to the DeSitter model. The comparison for this more general model has been given by Zhuk (1992). He finds (for the case $p = 0$, $\zeta > 0$)

$$\Psi(\Omega, \phi) = \begin{cases} I_0(\frac{2\sqrt{\zeta}}{\xi} \exp \frac{\xi}{2}\Omega) & , \text{Hartle-Hawking} \\ H_0^{(2)}(\frac{2\sqrt{\zeta}}{\xi} \exp \frac{\xi}{2}\Omega) & , \text{Vilenkin} . \end{cases} \tag{14}$$

For the case $\zeta < 0$, however, it is not clear how to apply the Hartle-Hawking or Vilenkin proposals since a natural condition is imposed by the classically forbidden region beyond the maximum of expansion. If a curvature term is added to the case $\zeta > 0$, the classically forbidden region exists close to the singularity as it does in the

DeSitter model. One might therefore argue that the cosmological models be restricted to have a barrier between the large Universe and the Big Bang.

The ADM quantization for these models can be performed with the natural intrinsic gauge choice $\Omega = t$ to yield the Schroedinger equation

$$i\frac{\partial}{\partial\Omega}\Psi(\Omega,\phi) = \pm\left(\hat{p}^2 + \zeta\exp\xi\Omega\right)^{\frac{1}{2}}\Psi(\Omega,\phi) \quad. \tag{15}$$

(Of course, there are problems with an intrinsic time, particularly for Ω-time if $\zeta < 0$. For more fundamental questions see Teitelboim 1980, 1982, 1983, Guven and Ryan 1992, and Kuchař 1992.) In this simple model, sense can be made of the operator on the RHS of (15) by considering its spectral representation. Another possibility is Misner's proposal (1972) to consider only the square of (15). The options indicated by $\pm$ can be interpreted as the time direction for the cosmological evolution. For $\zeta > 0$, H_{ADM} is self-adjoint. The general solution to (15) is again a superposition of modes f_p where

$$f_p(\Omega,\phi) = \exp\left[\mp i\int^{\Omega}\left(p^2 + \zeta\,e^{\xi x}\right)^{\frac{1}{2}}dx + ip\phi\right] \quad. \tag{16}$$

Although (16) appears to differ significantly from (13), the asymptotic behavior of the former is identical to some combinations of that in the latter (as $\Omega \to \infty$). If $\zeta < 0$, a prescription to define the square root must be given.

To illustrate the role of boundary conditions, consider the simplest model of this class— $\zeta = 0$ in (12). The solution to the Wheeler-DeWitt equation can be constructed from the basis functions $\exp[ip(\Omega \mp \phi)]$ for all p. There is no particular reason to choose among them. Based on the results (13) and (14), the Hartle-Hawking proposal would yield a real wavefunction and the Vilenkin boundary condition one proportional to $\exp[-i\,|p|\,(\Omega \mp \phi)]$ as $\Omega \to \infty$ The operator on the right hand side of (15) for this case is defined by its spectral representation as $|p|$. The solution in the p-representation is proportional to $\exp(\mp i\,|p|\,\Omega)$. But the Fourier transform to return to the ϕ-representation gives a basis of solutions proportional to $\exp(\mp i\,|p|\,\Omega)\cos|p|\,\phi$. However, we could also use the ϕ-time gauge. This gives the result that the basis of solutions is proportional to $\cos|p|\,\Omega\exp(\mp i\,|p|\,\phi)$. It appears that the ADM formulation tends to imply its own natural boundary conditions. Solutions which are the same in both intrinsic time gauges are analogous to the Hartle-Hawking solutions.

5.3 Mixmaster Universe

One of the original cosmologies studied by Misner (1969a) is the vacuum Bianchi IX or Mixmaster universe. It is described by the Hamiltonian constraint

$$2C^0 = -p_\Omega^2 + p_+^2 + p_-^2 + e^{4\Omega}V(\beta_+, \beta_-) \tag{17}$$

where $\beta_\pm$ (and conjugate momenta $p_\pm$) describe the universe anisotropy and

$$V(\beta_+, \beta_-) = e^{-8\beta_+} + 2e^{4\beta_+}[\cosh(4\sqrt{3}\beta_-) - 1] - 4e^{-2\beta_+}\cosh(2\sqrt{3}\beta_-). \tag{18}$$

The complicated structure of the anisotropy potential led to original attempts (Misner 1969a, 1970, 1972, Ryan, 1971, 1972) to study only the qualitative behavior of the solution. Hawking and Lutrell (1984), Amsterdamski (1985), and Wright and Moss (1985) considered semi-classical analysis using the path integral and the Hartle-Hawking boundary condition. Numerical solutions were considered by Furusawa (1986), Berger (1989), and Graham and Sepfaluszy (1990). Recently, an exact solution to the Wheeler-DeWitt equation (for a particular factor ordering) has been found independently by Kodama (1988), Moncrief and Ryan (1992), and Graham (1991). (It is not yet clear whether these are in fact the same solution.) The wavefunction appears to conform to the shape of the minisuperspace potential. Its projection onto the $\beta_\pm$-plane fills the triangular well (Misner 1969c, Ryan and Shepley 1975) near the singularity and is a gaussian peaked at isotropy near the maximum of expansion where the potential approximates a harmonic oscillator. This may be interpreted to mean that there is significant amplitude for anisotropy near the singularity with the late time universe almost certainly isotropic. This wavefunction, while not a complete solution to the Wheeler-DeWitt equation, can certainly serve as a test for the numerical methods necessary to probe more realistic models.

6 CONCLUSIONS

It has only been possible in this limited space to give a brief summary of the motivation behind the long history of QC and some early and recent results. Quantized homogeneous cosmologies are studied as laboratories for formalisms to be used to quantize the gravitational field and to explore possible quantum effects in the Early Universe. In the former context, they have been used to illustrate the Dirac (DeWitt 1967) and ADM quantization procedures (Misner 1969a) and to emphasize the differences between general relativity and gauge theories (Guven and Ryan 1992, Kuchar 1992). In the latter context, a prescription has been sought to single out a prefered state of the Universe (Hartle and Hawking 1983, Vilenkin 1986) and to find a way to interpret our classical world as a "measurement" of the quantum Universe (Laflamme 1991, Hu 1992). Although there is no reason to assume that quantized cosmologies approximate quantized gravity (Kuchař and Ryan 1986, 1989), QC remains an essential tool to understand how to approach the real problem.

ACKNOWLEDGEMENT
This work was supported in part by National Science Foundation Grant PHY9107162 to Oakland University. The author wishes to thank the Institute of Geophysics and Planetary Physics at Lawrence Livermore National Laboratory for hospitality.

REFERENCES
Amsterdamski P 1985 *Phys Rev* **D31** 3073

Arnowitt R, Deser S, and Misner C W 1962 in **Gravitation: An Introduction to Current Research** edited by L Witten (NewYork: Wiley)

Atkatz D and Pagels H 1982 *Phys Rev* **D25** 2065

Berger B K 1985 *Phys Rev* **D32** 2485

Berger B K 1986b unpublished

Berger B K 1988 *J Gen Rel Grav* **20** 755

Berger B K 1989 *Phys Rev* **D39** 2426

Berger B K and Vogeli C N 1985 *Phys Rev* **D32** 2477

Brown J D 1990 *Phys Rev* **D41** 1125

Brown J D and Martinez E A 1990 *Phys Rev* **D42** 1931

Calzetta E 1989 *Class Quant Grav* **6** L227

del Campo S and Vilenkin A 1989a *Phys Lett* **B224** 45

del Campo S and Vilenkin A 1989b *Phys Rev* **D40** 688

Carow U and Watamura S 1985 *Phys Rev* **D32** 1290

Castagnino M 1989 *Phys Rev* **D39** 2216

Castagnino M A and Mazzitelli F D 1989 *Int J Theor Phys* **28** 1043

Chmielowski P and Page D N 1988 *Phys Rev* **D38** 2392

Conradi H 1992 *Phys Rev* **D46** 612

D'Eath P D and Hughes D 1988 *Phys Lett* **B214** 498

DeWitt B S 1967 *Phys Rev* **160** 1113

DeWitt B S 1970 in **Relativity** edited by M Carmeli *et al* (New York: Plenum)

DeWitt B S 1981 in **Quantum Gravity 2: A Second Oxford Symposium** edited by C J Isham, R Penrose, and D W Sciama (Oxford: Clarendon)

Dirac P A M 1958 *Proc Roy Soc (London)* **A246** 217

Dirac P A M 1959 *Phys Rev* **114** 924

Faddeev L 1969 *Teor Mat Fiz* **1** 3

Faddeev L D and Popov V N 1975 *Sov Phys -Usp* **16** 777

Fakir R 1990 *Phys Rev* **D41** 3012

Fischer A E and Marsden J E 1979 in **General Relativity, an Einstein Centenary Survey** edited by S W Hawking and W Israel (Cambridge:

Cambridge University)

Furusawa T 1986 *Prog Theor Phys* **75** 59; **76** 67

Gell-Mann M and Hartle J B 1990 in **Complexity, Entropy, and the Physics of Information** edited by W H Zurek (Reading: Addison-Wesley)

Gibbons G W and Hawking S W 1977 *Phys Rev* **D15** 2752

Gonzalez-Diaz P F 1985 *Phys Lett* **159B** 19

Gotay M J and Demaret J 1983 *Phys Rev* **D28** 2402

Gotay M J and Isenberg J A 1980 *Phys Rev* **D22** 235

Gott J R 1982 *Nature* **295** 304

Graham R 1991 *Phys Rev Lett* **67** 1381

Graham R and Szepfalusy P 1990 *Phys Rev* **D42** 2483

Griffith R 1984 *J Stat Phys* **36** 219

Grishchuk L P and Sidorov Yu V 1988 *Zh Eksp Teor Fiz* **94** 29 (*Sov Phys JETP* **67** 1533)

Guven J and Ryan M P Jr 1992 *Phys Rev* **D45** 3559

Halliwell J J 1987 *Phys Rev* **D36** 3626

Halliwell J J 1988 *Phys Rev* **D38** 2468

Halliwell J J 1989 in **Proceedings of the Osgood Hill Meeting on Conceptual Problems in Quantum Gravity** edited by A Ashtekar and J Stachel (Boston: Birkhauser)

Halliwell J J 1990 *Int J Mod Phys A* **5** 2473

Halliwell J J 1990b in **Proceedings of the Jerusalem Winter School on Quantum Cosmology and Baby Universes** edited by T Piran (Singapore: World Scientific)

Halliwell J J and Hartle J B 1990 *Phys Rev* **D41** 1815

Halliwell J J and Louko J 1989 *Phys Rev* **D39** 2206; **D40** 1868

Hartle J B 1985 *J Math Phys* **26** 804

Hartle J B 1986 *J Math Phys* **27** 287

Hartle J B 1987 in **Proceedings of the Thirteenth Texas Symposium on Relativistic Astrophysics** edited by M P Ulmer (Singapore: World Scientific)

Hartle J B 1990 in **Proceedings of the 12th International Conference on General Relativity and Gravitation** edited by N Ashby, D F Bartlett, W Wyss (Cambridge: Cambridge University)

Hartle J B and Hawking S W 1983 *Phys Rev* **D28** 2960

Hartle J B and Kuchař K V 1984 *J Math Phys* **25** 57

Hartle J B and Kuchař K V 1986 *Phys Rev* **D34** 2323

Hawking S W 1979 in **General Relativity, an Einstein Centenary Survey** edited by S W Hawking and W Israel (Cambridge: Cambridge University)

Hawking S W 1984 *Nucl Phys* **B239** 257

Hawking S W and Luttrell J C 1984 *Phys Lett* **143B** 83

Hawking S W and Page D N 1986 *Nucl Phys* **B264** 185

Hawking S W and Page D N 1988 *Nucl Phys* **B298** 789

Hawking S W and Wu Z C 1985 *Phys Lett* **B151** 15

Horowitz G 1985 *Phys Rev* **D31** 1169

Hosoya A and Morikawa M 1989 *Phys Rev* **D39** 1123

Hu B L 1991 in **Thermal Fields and Their Applications** edited by H Ezawa *et al* (Amsterdam: North-Holland)

Hu B L 1992 this volume

Itzykson C and Zuber J B 1978 **Quantum Field Theory** (New York: McGraw-Hill)

Kazama Y and Nakayama R 1985 *Phys Rev* **D32** 2500

Kiefer C 1988 *Phys Rev* **D38** 1761

Kodama H 1988 *Prog Theor Phys* **80** 1024

Kuchař K 1983 *J Math Phys* **24** 2122

Kuchař K 1987 *Phys Rev* **D34** 3044

Kuchař K 1990 in **Einstein Studies Vol 2** edited by A Ashtekar and J Stachel (Boston: Birkhauser)

Kuchař K 1992 in **General Relativity and Relativistic Astrophysics** edited by G Kunstatter, D Vincent, and J Williams (Singapore: World Scientific)

Kuchař K and Ryan M P 1986 in **Gravitational Collapse and Relativity** edited by H Sato and T Nakamura (Singapore: World Scientific)

Kuchař K V and Ryan M P Jr 1989 *Phys Rev* **D40** 3982

Laflamme R 1991 *Lectures presented at XXII GIFT International Seminar on Theoretical Physics*

Lemos N A 1991 *Class Quant Grav* **8** 1303

Louko J 1987 *Class Quant Grav* **4** 581

Louko J 1987 *Phys Rev* **D35** 3760

MacCallum M 1975 in **Quantum Gravity** edited by C J Isham *et al* (Oxford: Clarendon)

MacCallum M 1979 in **General Relativity, an Einstein Centenary Survey** edited by S W Hawking and W Israel (Cambridge: Cambridge University)

Macías A, Obregón O, and Ryan M P Jr 1987 *Class Quant Grav* **4** 1477

McGuigan M 1988 *Phys Rev* **D38** 3031

Misner C W 1969a *Phys Rev* **186** 1319

Misner C W 1969b *Phys Rev* **186** 1328

Misner C W 1969c *Phys Rev Lett* **22** 1071

Misner C W 1970 in **Relativity** edited by M Carmeli *et al* (New York: Plenum)

Misner C W 1972 in **Magic without Magic** edited by J Klauder (San Francisco: Freeman)

Misner C W, Thorne K S , and Wheeler J A 1973 **Gravitation** (San Francisco:

Freeman)

Moss I G and Wright W A 1984 *Phys Rev* **D29** 1067

Myers E 1991 *The Unbounded Action and the 'Density of States' in Nonperturbative Quantum Gravity* preprint

Narlikar J V and Padmanabhan T 1983 *Phys Rep* **100** 151

Omnes R 1988 *J Stat Phys* **53** 893, 933, 957

Poletti S 1989 *Class Quant Grav* **6** 1943

Rubakov V A 1984 *Phys Lett* **B148** 280

Ryan M P Jr 1971 *Ann Phys (N Y)* **65** 506

Ryan M P Jr 1972 **Hamiltonian Cosmology** (New York: Springer-Verlag)

Ryan M P and Moncrief V 1991 *Phys Rev* **D44** 2375

Ryan M P Jr and L C Shepley L C 1975 **Homogeneous Relativistic Cosmologies** (Princeton: Princeton University)

Schleich K 1989 *Phys Rev* **D39** 2192

Schleich K 1990 *Class Quant Grav* **7** 1529

Smith G J and Bergmann P G 1986 *Phys Rev* **D33** 3570

Suen W M and Young K 1989 *Phys Rev* **D39** 2201

Teitelboim C 1977 *Phys Rev Lett* **38** 1106

Teitelboim C 1980 *Phys Lett* **96B** 77

Teitelboim C 1982 *Phys Rev* **D25** 3159

Teitelboim C 1983 *Phys Rev* **D28** 297

Unruh W 1989 *Int J Theor Phys* **28** 1181

Unruh W G and Wald R M 1989 *Phys Rev* **D40** 2598

Vilenkin A 1982 *Phys Lett* **B117** 25

Vilenkin A 1985a *Phys Rev* **D32** 2511

Vilenkin A 1985b *Nucl Phys* **B252** 141

Vilenkin A 1986 *Phys Rev* **D33** 3560

Vilenkin A 1988 *Phys Rev* **D37** 888

Vilenkin A 1989 *Phys Rev* **D39** 1116

Wada S 1987 *Phys Rev Lett* **59** 2375

Wald R M 1984 **General Relativity** (Chicago: University of Chicago)

Wheeler J A 1968 in **Battelle Rencontres** edited by C M DeWitt and J A Wheeler (New York: Benjamin)

Wright W A and Moss I G 1985 *Phys Lett* **154B** 115

Zeh H D 1986 *Phys Lett* **A116** 9

Zeh H D 1988 *Phys Lett* **A126** 311

Zhuk A 1988 *Class Quant Grav* **5** 1357

Zhuk A 1992 *Integrable Multidimensional Quantum Cosmology* preprint

Zurek W H 1981 *Phys Rev* **D24** 1516

Zurek W H 1982 *Phys Rev* **D26** 1862

A Pictorial History of some Gravitational Instantons*

Dieter Brill and Kay-Thomas Pirk[†]

Abstract

Four-dimensional Euclidean spaces that solve Einstein's equations are interpreted as WKB approximations to wavefunctionals of quantum geometry. These spaces are represented graphically by suppressing inessential dimensions and drawing the resulting figures in perspective representation of three-dimensional space, some of them stereoscopically. The figures are also related to the physical interpretation of the corresponding quantum processes.

1. Introduction

Understanding General Relativity means to a large extent coming to terms with its most important ingredient, geometry. Among his many contributions, Charlie has given us new variations of this theme [1], fascinating because geometry is so familiar on two-dimensional surfaces, but so remote from intuition on higher-dimensional spacetimes. The richness he uncovered is shown nowhere better than in the 137 figures of his masterful text [2].

Today quantum gravity [3] leads to new geometrical features. One of these is a new role for Riemannian (rather than Lorentzian) solutions of the Einstein field equations: such "instantons" can describe in WKB approximation the tunneling transitions that are classically forbidden, for example because they correspond to a change in the space's topology. In order to gain a pictorial understanding of these spaces we can try to represent the geometry as a whole with less important dimensions suppressed; an alternative is to follow the ADM method and show a history of the tunneling by slices of codimension one.

---

*Research supported in part by the National Science Foundation.

[†]Department of Physics, University of Maryland, College Park, MD 20742, USA and Max Planck Institut für Astrophysik, 8046 Garching, FRG. The second author was supported by a Feodor-Lynen fellowship of the Alexander von Humboldt-Foundation.

We can readily go from equation to picture thanks to computer plotting routines, from the simpler ones as incorporated in spreadsheet programs [4] to the more powerful versions of *Mathematica*.[1] We therefore decided it would be fun to see how well the computer can draw pictures associated with tunneling and instantons. In Section 2 we recall the idea behind these by an example of a two-dimensional potential. In the following sections we present and interpret several general relativity instantons, both sliced and unsliced.

2.   Tunneling in Several Dimensions

The one-dimensional potential (Fig. 1a)

$$V(x) = x^2 - x^3 \qquad (1)$$

is typical of the class to which tunneling arguments are often applied. The exponential

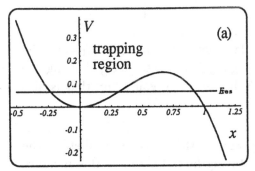
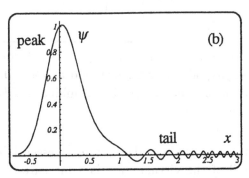

Figure 1. (a) Plot of the potential $V(x)$ with a resonance at $E_{res} = 0.06605$. (Here we have put $\hbar^2/m = .01$.) (b) Virtual state wavefunction $\psi(x)$ in this potential.

"barrier penetration" coefficient, which governs the probability of escape for a particle that is initially trapped near $x = 0$, is related to the "resonance" (virtual state) solution of the time-independent Schrödinger equation in this potential (Fig. 1b). The maximum of this wavefunction occurs in the trapping region near the relative minimum of the potential, and the oscillating "tail" in the exterior region is very small compared to this maximum. This property characterizes the virtual state. If the potential barrier is sufficiently high, virtual states occur at rather well-defined energies.

A priori the virtual state has nothing directly to do with the tunneling problem: it is stationary, real-valued (at $t = 0$) and yields no net flux into or out of the potential's

---

[1] *Mathematica* is a trademark of Wolfram Research Inc., 100 Trade Center Drive, Champaign, Il 61820-7237, USA. We also used the program *showstereo.m* available by e-mail from math-source@wri.com. We thank Peter Hübner (MPI-Astro) and Dr.-Ing. Werner Rupp (DASA) for help with computer resources

trapping region. (These same properties also make this wavefunction easy to draw!) To obtain a wavefunction with nonzero flux we need to use neighboring energy states. The properties of their wavefunctions change rapidly as the energy is varied away from resonance: to first order in the variation the exterior phase changes, and to second order the interior amplitude decreases. Fig. 2 shows two nearby energy states whose exterior phase differs by $\pi/2$. From these we can build an outgoing wave. Because

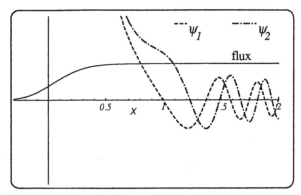

Figure 2. Wavefunctions ψ_1 (dashed) and ψ_2 (dash-dotted) at energies $E_1 = 0.0663$ slightly above resonance, and at $E_2 = 0.0658$ slightly below resonance, showing the rapid phase change. The combination $\psi_1 - i\psi_2$ has a net outgoing flux that is also shown (solid).

the amplitude of these waves is essentially the same as that of the virtual state, the tail of the virtual state is a measure of the outgoing flux. Of course this measure is rather crude, giving us only the main exponential factor in the decay probability. Just how long it takes for the wave to leak out depends on how rapidly the properties of the wavefunction change with energy; thus this "prefactor" to the exponential is not determined by the virtual state alone, and is rather more difficult to compute.

In one dimension there is only one way for the particle to leak out of the potential in Fig. 1a (namely, toward positive x). To find in more detail "how" the particle gets out of the trapping region, and the analogous issues in tunneling of fields, we must consider at least two-dimensional potentials. Usually one treats the rotationally symmetric case: angular momentum is then conserved, and the problem reduces to a one-dimensional effective potential motion for each angular momentum. Because of the symmetry no direction of tunneling away from the trapping region is preferred over any other. But when the potential is not symmetric, some tunneling "paths" can be considerably more probable than others.[2]

---

[2] An optical analogon is the observation of "frustrated total internal reflection"[5]. The potential barrier can be provided by the air space between the hypotenuse faces of two 90° prisms. More light (and at larger wavelengths) gets through where this barrier is narrowest.

A simple two-dimensional potential, in which the classical and the quantum mechanical problem can be solved because the corresponding equations are separable, is given by

$$V(x,y) = x^2 + y^2 - x^3 - 0.8y^3 = V_1(x) + V_2(y). \qquad (2)$$

(Here the factor 0.8 was chosen only for convenience of plotting.) Fig. 3 shows a plot of this potential. It has a trapping region that is separated from the "exterior"

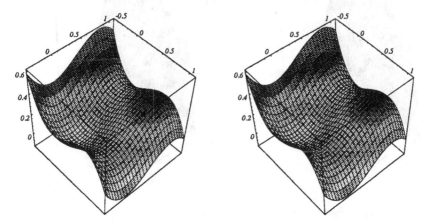

Figure 3. 3D plot of the potential $V(x,y)$. This, and some of the subsequent figures, are stereoscopic pairs similar to those found in the text *Methods of Theoretical Physics* by Morse and Feshbach. We know no better instructions how to view these figures than those found in the preface of that text.

region ($x \gg 0, y \gg 0$) by a barrier with two saddle points of different heights. The virtual state in this potential is simply the product,

$$\psi(x,y) = \psi_1(x)\psi_2(y)$$

where ψ_i is the virtual state in potential V_i. This two-dimensional wave function is plotted in Fig. 4a. In Fig. 4b the square of the wave function is indicated by increasing gray levels. We note that there is a path, namely along the x-axis, on which the wavefunction is generally (and particularly at the end of the exponential decrease) larger than along other paths leading from the trapping region to the exterior. This is the "most probable escape path" [6] that may be said to describe "how" the particle most likely tunnels through the barrier, in this sense: if the system were prepared in the virtual state and a position measurement were made in the barrier region, the particle would most likely be found near the most probable escape path. (Of course, after the measurement the particle would no longer be in the virtual state.)

The tunneling behavior of a virtual state is well approximated by a WKB wavefunction, and the analogous problem in *classical* mechanics [7]. For example, the most

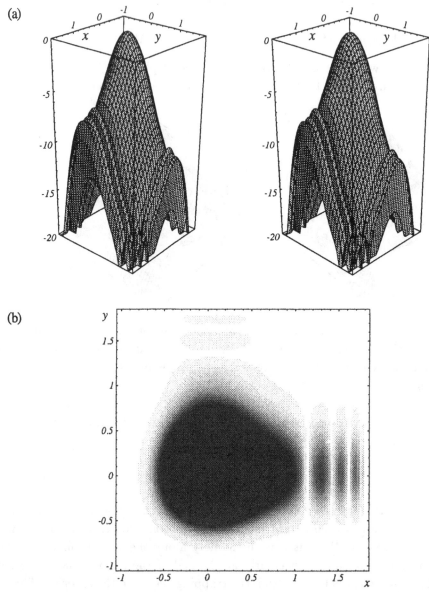

Figure 4. (a) 3D log square plot of the virtual state wavefunction for the potential of Fig. 3. As in Fig. 2, the interference maxima and minima have little to do with the physical outgoing wave, in the case that this wavefunction is used to represent the decay of a virtual state. (b) Gray-level plot of the square of the wavefunction, showing the "most probable escape path" as a locus of relatively high density crossing the barrier region along the x-axis.

probable escape path is given by the "bounce" solution of particle motion in *imaginary time* — or equivalently, real time motion in the upside-down potential. The reader is encouraged to consider Fig. 3a turned upside down (for the less agile reader we have provided Fig. 5) and to imagine how a particle released at the central maximum would move in the resulting potential. Fig. 6 shows some of the orbits in such

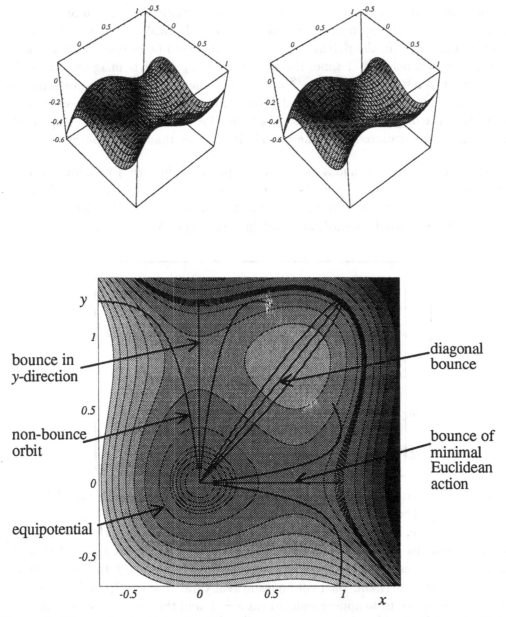

Figure 6. The equipotentials of $-V(x, y)$ and some particle orbits in this potential.

a potential. The bounce solutions are those that have a strict turning point, where the particle velocity vanishes; thus they are the only ones that reach the exterior zero equipotential, where energy conservation allows the particle that started at rest in the center to stop tunneling and again "become a classical particle."

Along the bounce path we can solve a one-dimensional time-independent Schrödinger equation to find the fall-off of the wavefunction, and hence obtain the tail amplitude and the exponential factor in the decay rate. To essentially the same approximation we can use the WKB estimate of this factor via the classical "Euclidean" action S_E of the bounce path (in the potential $-V$). Since this is to be computed for a motion in imaginary ("Euclidean") time, the action $S = iS_E$ itself is imaginary, and the lowest order WKB wavefunction, $e^{iS} = e^{-S_E}$ becomes a decreasing exponential. If we compute the action for the complete bounce, from the origin to the escape point and back to the origin, we obtain twice the action for the most probable escape path, which when exponentiated gives the probability (rather than the amplitude).

In fact, we can get a continuous picture of the particle's history if we allow the time parameter to change between real and imaginary at turning points. Fig. 7 shows such a picture of the nonrelativistic penetration history of the barrier of Eq. (1) (but with time plotted upward, as usual in relativity). We note that the infinite

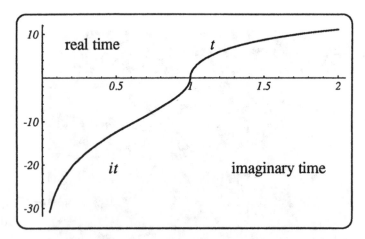

Figure 7. The semiclassical history of the decay from the virtual state of the 1D potential $V(x)$. The particle starts at the classical ground state at $x = 0$, but moves in *imaginary* time, it, starting at $it = -\infty$, $x = \cosh^{-2}(it/\sqrt{2m})$. (This motion does not correspond to any passage of real, physical time, but signals the probabilistic nature of the process.) At the bounce (or "nucleation") time $it = 0$ the motion stops momentarily at $x = 1$, and then continues in real time, $x = \cos^{-2}(t/\sqrt{2m})$. In real time the particle seems to appear suddenly at $x = 1$, and then continues on a classical escape orbit.

imaginary time needed for the tunneling has nothing to do with the real time needed for the physical process (which is vanishingly short in this approximation). The WKB wavefunction is given by the action evaluated along this entire history. The Euclidean part of the motion contributes an imaginary part to the action, and the real-time motion contributes a real part. Thus the WKB wavefunction exhibits exponential decrease in the Euclidean region, and oscillation in the real time region. (However, this wavefunction does not have continuous derivatives at the turning points; this problem is usually solved by finding "transition formulas", but we neglect it in this lowest approximation.)

In general, as in our example, there may be more than one bounce and probable escape path. Because the probability is an exponential, the one with the least Euclidean action is generally overwhelmingly more probable (however, also see [8]). Of course, in order to get a non-zero decay probability, this least action must be finite (this is part of the definition of a bounce or instanton solution). Of the three bounces shown in Fig. 6, the diagonal one has the largest action (most improbable). In fact, it is really a combination of the other two bounces, rather than a different way for the particle to escape. This can be seen from the behavior of nearby particle orbits: they converge toward this bounce, showing that there is a second zero mode in the second variation of the action (the first zero mode is given by an infinitesimal time translation). Because the orbits intersect, the second variations vanish at the turning points, so that there will be two variations that lower the action. Only bounces with one negative mode should be counted as independent decay channels.

Another kind of Euclidean solution ("instanton") is useful for potentials in which the escape region is replaced by a second trapping region whose minimum is degenerate with the first (Fig. 8a). If one starts out with a wavepacket concentrated in one of the regions, it will at a later time be concentrated in the other region and vice versa, so that in general it has a fluctuating behavior. The two lowest energy eigenstates that most importantly contribute to this behavior are approximated by a sum resp. a difference between two WKB wavefunctions (of the type e^{S_E} and e^{-S_E}). These wavefunctions can again be evaluated by finding the action of a Euclidean solution of the classical equations of motion. But this instanton does not exhibit a bounce. Instead the particle takes an infinite imaginary time to move away from the center of the first region, and an infinite time to reach the center of the second region (Fig. 8b). A corresponding two-dimensional potential and its lowest energy wavefunction is shown in Fig. 9a. Again the wavefuntion is relatively large on the instanton path, making it the "most probable connecting path" (Fig. 9b).

Thus, in a multidimensional setting we may again say that the instanton tells us "how" the particle gets from one region to the other during a fluctuation. The instanton action will again give us the main exponential factor in the probability that a particle initially present in one region will be found in the other. In particular, the

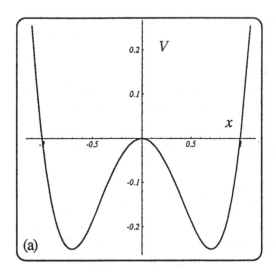
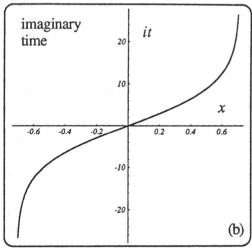

Figure 8. (a) The potential $V(x) = -x^2 + x^4$ has two degenerate minima, separated by a tunneling region. (b) The semiclassical history of the tunneling in this potential, $x = (1/\sqrt{2})\tanh(it/\sqrt{m})$. The particle starts from one of the minima at $it = -\infty$, and reaches the other minimum at $it = \infty$, with the entire motion taking place in imaginary time.

existence of an instanton with finite action indicates that the fluctuation in question does take place. For the details of the fluctuation, such as its frequency, one would again need information about more than one energy eigenstate.[3]

One assumes that an analogous semiclassical approximation is valid in field theory. That is, to see whether a classically forbidden transition is possible in the quantum theory one tries to find a finite action solution of the Euclidean field equations that connects the relevant classical initial and final states of the transition. If the connection is by a bounce solution, in which the initial state is reached only asymptotically, but the final state occurs at the turn-around "point" (surface of imaginary-time symmetry, or nucleation surface [9]), the transition is interpreted as a decay. If the connection is by an instanton (without a bounce), in which both the initial and final states are reached only asymptotically, the transition is a fluctuation. In general relativity, Euclidean solutions of the Einstein equations are Riemannian (rather than Lorentzian) manifolds.

---

[3]In this connection, fluctuation means a transition between two or several states that are "classically allowed" and have the same energy. We do not mean the kind of virtual fluctuation that produces, for example, a virtual pair in the vacuum state. (One could however say that the barrier makes the virtual fluctuation real by preventing an immediate return to the initial state.)

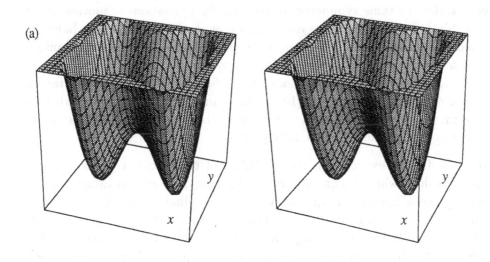

(a)

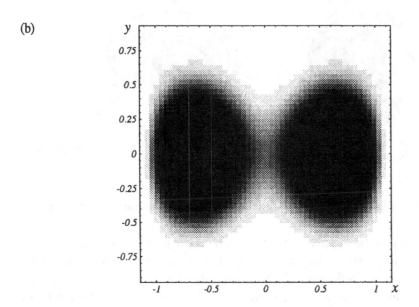

(b)

Figure 9. (a) A two-dimensional potential, $V(x, y) = -x^2 + x^4 + y^2$ with two degenerate minima. (b) The ground state wavefunction is relatively large along the most probable connecting path.

3.   Gravitational Bounces

If the classical initial state for tunneling has symmetry, the WKB tunneling state may or may not exhibit the same symmetry. If the Euclidean equations of motion admit a tunneling solution with the same symmetry, we expect this solution to have the lowest action. (If the symmetry is broken by the tunneling, then there is no unique instanton of smallest action, and one should sum over all of them.) For example, a constant electric field is invariant under boosts in the field direction, and the instanton describing pair production by this field [10] has the same invariance. The electric field is also invariant under translation, but the instanton is not. However, the tunneling events related by translation are all equally probable.

In accordance with this expectation the "bubbles" of true vacuum expanding into false vacuum, both of which vacua are Lorentz invariant, do exhibit invariance under the homogeneous Lorentz group, and the corresponding Euclidean solution is invariant under the rotation group $O(4)$. A simple example of a similar gravitational situation is provided by the "tunneling from nothing" into a deSitter universe [11], shown in Fig. 10 (with two dimensions suppressed). As in Fig. 7, in the lower part

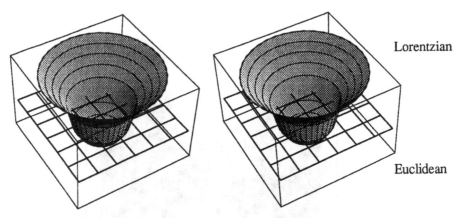

Figure 10. The semiclassical history of deSitter space, in the same spirit as Fig. 7.

time is imaginary, and in the upper part it is real. Each part solves the equation $G_{\mu\nu} = \Lambda g_{\mu\nu}$ on its appropriate, Riemannian resp. Lorentzian, manifold. The nucleation surface is the equator; reflection of either part about this surface would give a complete manifold of each type, either a Riemannian bounce (a complete 4-sphere) or a complete Lorentzian deSitter universe (that bounces at some *minimum* radius).

To obtain a history of the deSitter evolution we normally slice the upper part of Fig. 10 by horizontal, spacelike surfaces. For a history of the tunneling it is natural to slice the lower part similarly by horizontal planes. This history indeed shows nothing as long as the plane is below the "south pole" of the hemisphere; when the plane touches the pole, the universe originates as a point, then expands into increasing 3-spheres (represented in the figure by circles) until it reaches the maximum radius of the virtual evolution, which is also the minimum radius of the deSitter space, and the real evolution starts.

We have tacitly assumed that Fig. 10 is to be rotated about two other axes to get the 4-manifolds (4-sphere, deSitter space) that we really intended. But we can just as well imagine the product of Fig. 10 with S^2, to generate the Nariai metric [12] by tunneling from nothing. In this case the nucleation surface, represented by the same horizontal plane in Fig. 10 as before, has the topology $S^2 \times S^1$ of a wormhole universe. The Euclidean part of the action turns out to be larger (corresponding to smaller tunneling probability) than for tunneling into the more symmetric deSitter space [9] (cf. the y−axis vs. the x−axis bounce of Fig. 6).

Further examples of $O(4)$ symmetric bounces are associated with vacuum decay. Since the ordinary 4-dimensional Minkowski space vacuum is stable [13], we have to consider compactified higher-dimensional cases. To be realistic the spacetime should be at least 5-dimensional, but in our figures we will have to suppress at least two of these dimension: each point will represent a 2-sphere (with metric $d\Omega^2$ if it is a unit 2-sphere).[4]

Figure 11a shows the "5-dimensional Schwarzschild instanton,"

$$ds^2 = (1 - (R/r)^2)d\phi^2 + (1 - (R/r)^2)^{-1}dr^2 + r^2 d\Theta^2 + r^2 \sin^2 \Theta d\Omega^2, \quad r \geq R \quad (3)$$

used by Witten [14] to discuss the decay of the Kaluza-Klein vacuum, with ϕ the compactification direction. (Since the picture is only qualitatively correct, it could also represent the 4-dimensional Schwarzschild instanton, which is usually taken to describe the thermal properties of a Schwarzschild black hole, but which in this connection would describe the decay of the 4-dimensional vacuum, compactified in one direction [15].) In the lower part of this figure we have plotted r and Θ in a vertical plane, and ϕ in the orthogonal horizontal direction. Since the latter is to be identified with period 2π, the front and back boundaries of the Figure are to be identified. That this is possible without singularity is shown in Fig. 11b, where this identification is performed on a horizontal slice of Fig. 11a, resulting in a "test tube" that is smoothly closed on one end, which represents the metrically correct embedding of the r, ϕ section of this metric.[5] The upper part of the Figure shows the analytic continuation to a Lorentzian manifold.

---

[4]Actually, points related by symmetry about the vertical plane perpendicular to the paper correspond to the same 2-sphere.

[5]Since the test tube is topologically $\mathbf{R}^2$, the whole metric of Eq. (3) is $\mathbf{R}^2 \times S^3$, but one should remember that the $\mathbf{R}^2$ does not have the standard infinity, but instead becomes cylindrical — like the test tube.

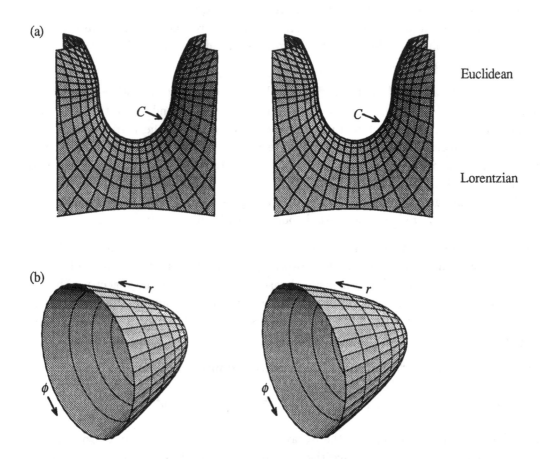

Figure 11. (a) The semiclassical history of the decay of a compactified "false" 5D vacuum geometry is represented by the 3D region bounded above by the surface shown. Each point of this region represents a two-sphere, obtained by rotating in two additional dimensions about the center of the figure. The front and back parts of the surface are to identified in such a way that points along the middle curve C are regular (non-conical) origins at $r = R$ in the r, ϕ plane. (b) The actual geometry obtained from a slice of Fig. 11a by a plane perpendicular to the paper, with the identification carried out.

The surface $\Theta = \pi/2$ in the Witten instanton (3) (or any rotation of it) is a nucleation surface; it represents the final state of the tunneling. Since the instanton is asymptotically flat, the initial state, which it reaches asymptotically, is the vacuum. Once we have chosen a nucleation surface, the initial vacuum is most appropriately represented by the hyperplane at large distance from the origin that does not intersect the nucleation surface. One can then fill in other sections between these that give a reasonable tunneling history [16]. At first the topology of these sections is $R^3 \times S^1$, like that of a spacelike surface of the compactified vacuum. When the minimum r on such a section reaches R, the section represents the singular instance at which the topology changes. Later sections have the topology $\mathbf{R}^2 \times S^2$, like that of the final state. In the Figure this looks like two disconnected spaces moving apart, but they are of course connected through the two dimensions that were suppressed[4], so that in fact a spherical hole has appeared in the final state, and expands to infinity in the subsequent Lorentzian development.

As a final example we show the creation of two magnetically charged Wheeler wormhole mouths by the magnetic field in an initially Melvin universe [17]. This remarkable topology-changing bounce has two axes of symmetry: one is the line joining the wormholes, and the other corresponds to invariance under boosts of the created pair. We suppress the former to reduce the Euclidean space dimensions the three, and show only the latter as a symmetry of the three-dimensional space in which the figure is drawn, namely the rotational symmetry of Fig. 12a about the horizontal axis, part of which is labeled "horizon." (Except for this symmetry the Figure is qualitative only.[6]) For ease of visualization we show only the lower, Euclidean half of the history — the space within the rectangular box and outside the semispherical cavities. As in the other figures, the upper part should consist of the analytic continuation of the lower half to a Lorentzian spacetime. Points on the inner surface of the two cavities are to be identified by reflection about the mid-plane. The top horizontal plane is the nucleation surface. It contains the axis of symmetry, where the rotational Killing vector vanishes. The part between the cavities is the wormhole's (Euclidean) Killing horizon. Its Lorentzian extension will as usual consist of two lightlike horizon surfaces, on which the boost Killing vector that is the continuation of the rotational Killing vector is lightlike. (The part of the axis outside the cavities, not drawn in the Figure, corresponds to the "Rindler horizon" that separates the two created wormhole mouths.)

On each horizontal section of Fig. 12a imagine a vector field like that representing a laminar fluid flow from left to right, avoiding the cavities. This can be taken as a representation of the magnetic field.[7] In the bottom surface this field is approximately constant and represents the field lines of the (asymptotic) Melvin universe's central region [19]; away from this central region the Melvin magnetic flux actually falls off with distance. At the nucleation surface some of this flux crosses the wormhole horizon and so has become trapped in the Wheeler wormhole. To show this in more detail we have drawn in Fig. 12b only the top, nucleation surface, but with the two

---

[6]For another way of representing this geometry — in which two dimensions are suppressed — see Banks et al. [18].

[7]In a spacetime the electromagnetic field tensor should in general be represented by a honeycomb structure [2], not by lines. But once one has chosen a "spacelike" surface, the usual electric and magnetic field lines of course make sense.

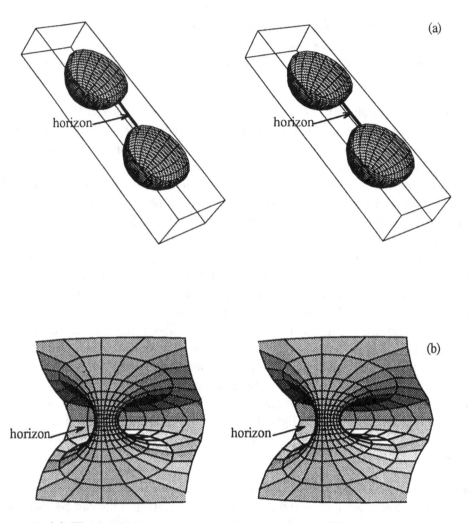

Figure 12. (a) The imaginary-time part of the history of Wheeler wormhole production. Beyond the top (nucleation) surface the real-time history could be represented
in 2+1 Minkowski space, with the rotational symmetry replaced by a boost symmetry. (b) The wormhole at the nucleation surface is the geometry of the top surface of
part (a), with the identification of the two circles carried out. (These circles are the
vertical, circular grid line that is partically hidden behind the wormhole's throat.)
The grid lines can be taken to be equipotentials and field lines of the wormhole's
field; to them should be added those of the Melvin universe background to get the
net equipotentials and fields.

circular curves in it identified as required. The wormhole's Killing horizon now occurs, as expected, at the narrowest part (throat) of the wormhole.

It is now not difficult to imagine other horizontal slices of Fig. 12a to obtain the rest of the Euclidean history of the wormhole formation. The singular slice is the one that first touches the cavities, so that only a pair of points are to be identified, rather than two circular curves. It is the first slice in which some flux has been trapped, and this trapped flux is conserved in the subsequent development. This flux is maximal in the sense that, after the wormhole mouths have moved far enough apart in their Lorentzian development to be compared to single black holes, they correspond to extremal Reissner-Nordström black holes. It is interesting that the spacelike distance through the wormhole is finite on the nucleation surface; it increases as the wormhole mouths move apart after nucleation, and asymptotically approaches the infinite value that one expects from the "horn" structure of the isolated Reissner-Nordström black hole [20]. It seems that other interesting features of the Lorentzian analytic continuation have been explored only to a limited extent, and they are too far removed from the instanton to be treated here [18].

4.   Fluctuation Instantons

The solution of the Euclidean Einstein-Maxwell equations,

$$ds^2 = V^{-2}dt^2 + V^2(dx^2 + dy^2 + dz^2) \qquad V = \sum m_i/r_i, \qquad *F = d(V^{-1}) \wedge dt \quad (4)$$

can be interpreted as an instanton [21]. Here r_i, $i = 1...n$ are the distances in flat Euclidean 3-space from n different origins to the field point. This solution is similar to the multi-extremal Reissner-Nordström solution, except that it has no asymptotically flat region; instead it becomes cylindrical in the limit $r \to \infty$, just as it does in each "horn", $r_i \to 0$. If we place all the origins in the xy plane and suppress the $z-$ and $t-$directions, the geometry of the remaining dimensions can be shown embedded in flat Euclidean space as in Fig. 13. (The Gaussian curvature of this 2-surface is negative, reaching zero asymptotically; it appears remarkably difficult to find an accurate and unique embedding of such a surface, so Fig. 13 is correct only in its main features.) The way *Mathematica* sliced this figure suggests that this is an instanton that interpolates between a single universe and two or several daughter (or baby) universes. Indeed, the geometry and fields of Eq. (4) approach the Euclidean version of a Bertotti-Robinson universe [22] in all asymptotic regions. The fact that there is no bounce or nucleation surface, and that all connections to Lorentzian solutions are made in the asymptotic regions, indicates that this is a fluctuation-type instanton.

Many of the details of this instanton have been given elsewhere [21], where it is argued that it should also describe the quantum fluctuations of the region near the horizon of an extremal Reissner-Nordström black hole. Here we want to show the result of

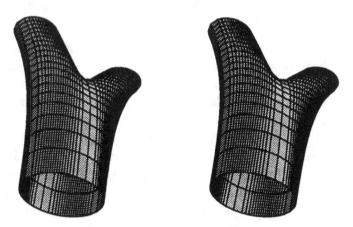

Figure 13. Imaginary time history of a fluctuation by which a maximally charged black hole throat splits into two.

slicing this instanton for $n = 2$ and $m_1 = m_2$ by 3-surfaces. A simple choice consists of the surfaces on which V is constant. These slices are shown in Fig. 14. They are metrically accurate (except for the suppression of the axis direction). For small V, and hence large r, these surfaces of constant V have the topology of a single sphere. For large V, one or the other of the r_i has to be small, so one obtains two separate spheres. The critical, singular surface occurs when $V = m/d$, where d is the distance in Euclidean 3-space between the two origins. Depending on the interpretation, Fig. 14 then gives the imaginary-time history of the break-up of a Bertotti-Robinson universe, or of an extremal Reissner-Nordström black hole's horizon.

5. Conclusions

When instanton solutions were first investigated in general relativity, they were regarded primarily as a mathematical device, not to be interpreted in the same way as Lorentzian solutions — it was a bold step even to draw a Riemannian and a Lorentzian manifold connected in the same diagram. Since then it has become clear that many of the traditional methods work in both arenas. In particular, we have seen above that slicing by 3-dimensional surfaces can help the physical interpretation, just as it did for classical relativity when ADM pioneered this method.

Figure 14. (next page) Slices of constant V of the instanton of Fig. 13, with only one direction suppressed, (a) "before" the topology change (b) singular slice at the instant of topology change (c) "after" the topology change.

(a)

(b)

(c)

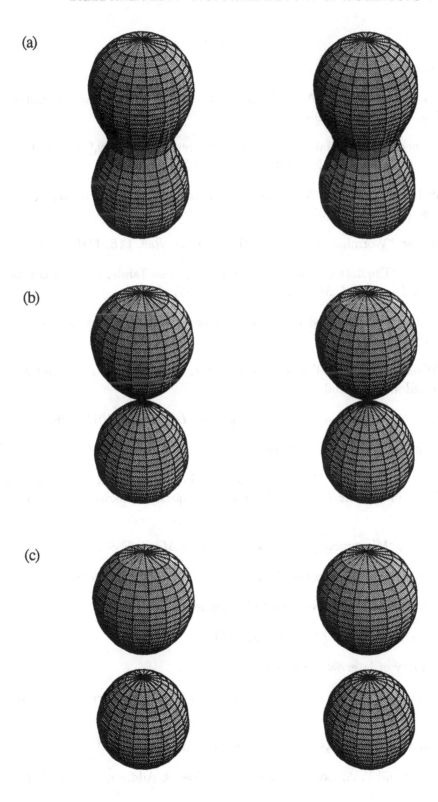

References

[1]   A few examples are:

C. W. Misner and J. A. Wheeler, "Classical physics as geometry", *Ann. Phys. (NY)* **2**, 525 (1957)

C. W. Misner, "Differential geometry and differential topology," in *Relativity, Groups, and Topology* Gordon and Breach, New York 1964

D. Finkelstein and C. W. Misner, "Some new conservation laws", *Ann. Phys. (NY)* **6** 230 (1959)

C. W. Misner, "Wormhole initial conditions", *Phys. Rev.* **118**, 1110 (1960)

C. W. Misner, "The flatter regions of Newman, Unti and Tamburino's generalized Schwarzschild space," *J. Math. Phys.* **4**, 924 (1963)

C. W. Misner, "The method of images in geometrostatics," *Ann. Phys. (NY)* **24**, 102 (1963)

C. W. Misner, "Taub-NUT space as a counterexample to almost anything" *Lect. Applied Math.* **8**, 160 (1967)

C. W. Misner, "Mixmaster universe" *Phys. Rev. Letters* **22**, 1071 (1969)

C. W. Misner, "Absolute zero of time" *Phys. Rev.* **186**, 1328 (1969)

C. W. Misner and J. A. Wheeler, "Conservation laws and the boundary of a boundary" in *Gravitation: Problems and Prospects*, Naukova Dumka, Kiev (1972)

C. W. Misner, "Minisuperspace" in *Magic without Magic*, Freeman 1972

[2]   C. W. Misner, K. S. Thorne and J. A. Wheeler, *Gravitation*, Freeman 1973

[3]   Feynman quantization was first applied to general relativity in

C. W. Misner, *Rev. Mod. Phys.* **29**, 497 (1957)

For a summary of later work see, for example,

S. W. Hawking in *General Relativity, an Einstein Centenary Survey*, Cambridge 1979

[4]   Using spreadsheets for physics was pioneered by Misner, see e.g.

C. W. Misner and P. A. Cooney, *Spreadsheet Physics*, Addison-Wesley, 1991

[5] see for example Heavens and Ditchburn, *Insight into Optics*, Wiley 1991

[6] Banks, Bender and Wu, *Phys. Rev.* D8, 3346 and 3366 (1973)

[7] S. Coleman, *Phys. Rev.* D15, 2929 (1977)

[8] S. Carlip, *Phys. Rev.* D46, 4387 (1992)

[9] G. W. Gibbons and J. B. Hawking, *Phys. Rev.* D42, 2458 (1990)

[10] A. Vilenkin, *Phys. Rev.* D27, 2848 (1983)

C. Stephens, *Ann. Phys. (USA)* 193, 255 (1989) and *Phys. Lett.* A142, 68,1989

[11] A. Vilenkin, *Phys. Lett.* 117B, 25 (1982)

[12] H. Nariai, *Sci. Rep. Tohoku Univ., Series 1* 34, 160 (1950); *ibid.,* 35, 62 (1951)

[13] D. Brill and S. Deser, *Ann. Phys. (USA)* 50, 548 (1968)

P. Schoen and S. T. Yau, *Phys. Rev. Lett.* 42 547 (1979)

E. Witten, *Comm. Math. Phys.* 80, 381 (1981)

[14] E. Witten, *Nucl. Phys.* B195, 481 (1982)

[15] A. S. Lapides, *Phys. Rev.* D22, 1837 (1980)

[16] D. Brill, *Found. Phys.* 16, 637 (1986)

[17] D. Garfinkle and A. Strominger, *Phys. Lett.* B256, 146 (1991)

[18] Banks, O'Loughlin, and Strominger, *Black Hole Remnants and the Information Puzzle*, Preprint RU-92-40 (hep-th/9211030)

[19] M. A. Melvin, *Phys. Lett.* 8, 65 (1964)

[20] B. Carter, "Black hole equilibrium states" in *Black Holes*, Gordon and Breach (1973)

[21] D. Brill, *Phys. Rev.* D46, 1560 (1992)

[22] T. Levi-Civita, R. C. Acad. Lincei (5) 26, 519 (1917)

B. Bertotti, *Phys. Rev.* 116, 1331 (1959)

I. Robinson, Bull. Akad. Polon. 7, 351 (1959)

No Time Machines from Lightlike Sources in 2+1 Gravity

S. Deser *   Alan R. Steif *†

Abstract

We extend the argument that spacetimes generated by two timelike particles in $D=3$ gravity (or equivalently by parallel-moving cosmic strings in $D=4$) permit closed timelike curves (CTC) only at the price of Misner identifications that correspond to unphysical boundary conditions at spatial infinity and to a tachyonic center of mass. Here we analyze geometries one or both of whose sources are lightlike. We make manifest both the presence of CTC at spatial infinity if they are present at all, and the tachyonic character of the system: As the total energy surpasses its tachyonic bound, CTC first begin to form at spatial infinity, then spread to the interior as the energy increases further. We then show that, in contrast, CTC are entirely forbidden in topologically massive gravity for geometries generated by lightlike sources.

Among the many fundamental contributions by Charlie Misner to general relativity is his study of pathologies of Einstein geometries, particularly NUT spaces, which in his words are "counterexamples to almost everything"; in particular they can possess closed timelike curves (CTC). As with other farsighted results of his which were only appreciated later, this 25-year old one finds a resonance in very recent studies of conditions under which CTC can appear in apparently physical settings, but in fact require unphysical boundary conditions engendered by identifications very similar to those he discovered. In this paper, dedicated to him on his 60th birthday, we review and extend some of this current work. We hope it brings back pleasant memories.

1.   Introduction

Originally constructed by Gödel [1], but foreshadowed much earlier [2], spacetimes possessing CTC in general relativity came as a surprise to relativists. The shock was

---

*The Martin Fisher School of Physics, Brandeis University, Waltham, MA 02254, USA. This work was supported by the National Science Foundation under grant #PHY88–04561.

†Present address: DAMTP, Cambridge University, Silver St., Cambridge, CB3 9EW, United Kingdom.

softened by the fact that these solutions required unphysical stress tensor sources, and in this sense should not have been so unexpected: it is after all a tautology that *any* spacetime is the solution of the Einstein equations with *some* stress tensor, as often emphasized by Synge. Indeed, Einstein himself, while elaborating general relativity, apparently worried about geometries with loss of causality and CTC, and hoped that they would be excluded by physically acceptable sources.[1] Almost two decades after Gödel's work, Misner in his pioneering studies [3] of NUT space showed how CTC could be generated as global effects, by taking local expressions for a metric and making appropriate identifications among the points.

Very recently, the subject of CTC was revived in two quite different contexts. The first, which we shall not discuss, involves tunneling through wormholes in $D=4$ gravity. The second, which will be our subject here, concerns solutions to $D=3$ Einstein theory with point sources or equivalently, $D=4$ gravity with infinite parallel cosmic strings (since the latter system is cylindrically symmetric). We will therefore operate entirely in the reduced dimensionality. The dramatic simplification at $D=3$ is that the Einstein and Riemann tensors are equivalent, so spacetime is locally flat wherever sources are absent; consequently there is neither gravitational radiation nor any Newtonian force between particles. For these reasons, the sign of the Einstein constant is not physically determined, unlike in $D=4$. We will adopt the usual sign here, but mention the opposite sign, "ghost" Einstein theory, at the end. Local flatness in $D=3$ means that all properties are encoded in the global structure, *i.e.*, in the way the locally flat patches are sewn together. Indeed, we will use this geometric approach to analyze the CTC problem. However, for orientation we will begin with a brief discussion in terms of the analytic form of the metric.

Consider the general solution outside a localized physical source as given by the "Kerr" metric [4]

$$ds^2 = -d(t + J\theta)^2 + dr^2 + \alpha^2 r^2 d\theta^2 . \qquad (1.1)$$

Our units are $\kappa^2 = c = 1$, and $\alpha \equiv 1 - M/2\pi$. The constants of motion are the energy M and angular momentum J (space translations not being well defined [5]). This interval is manifestly locally flat in terms of the redefined coordinates $\Theta = \alpha\theta$, $T = t + J\theta$, but the global content lies in the different range, $0 \leq \Theta \leq 2\pi\alpha$, of Θ corresponding to the usual conical identification, and in the time-helical structure resulting from the fact that the two times T and $T+2\pi J$ are to be identified whenever a closed spatial circuit is completed. The interval (1.1) can clearly support CTC; for example, the interval traced by a circle at constant r and t,

$$\Delta s^2 = (2\pi\alpha)^2(r^2 - J^2/\alpha^2) < 0 , \qquad (1.2)$$

is timelike for $r < |J|/\alpha$. However, the relevant physical question is whether the constituent particles are ever confined within this radius; otherwise the CTC criterion (1.2) ceases to apply. To be sure, if we simply insert the metric (1.1) into the

---

[1] We thank John Stachel for telling us this.

Einstein equations, it is valid down to $r = 0$; the "source" is a spinning particle with $T_0^0 \sim m\delta^2(r)$, $T_0^i \sim J\,\epsilon^{ij}\partial_j\,\delta^2(r)$. But we do not accept classical spinning particles as physical, precisely because of their singular stress tensors, any more than we do Gödel's sources. Instead, one must check whether a system of moving spinless particles with *orbital* angular momentum can support CTC. This was the question that was initially raised in [4] and answered in the negative, on the simple physical grounds that the point particles, being essentially free, will — both initially and finally — be dispersed so that the constant J would have been exceeded at $t = \pm\infty$ by the radius at which the exterior metric (1.1) is valid. Thus, CTC, if present at all, would have to appear and then disappear spontaneously in time in an otherwise normal Cauchy evolution, and it seemed unlikely that this violation of Cauchy causality would occur in a finite time region for an otherwise non-pathological system.

It was therefore quite surprising when, a year ago, Gott [6] gave an explicit construction of a geometry generated by an apparently acceptable source consisting of two massive particles passing by each other at subluminal velocities, in which CTC appear only during a limited time interval [7]. However, it was then shown both that, in these spaces CTC will also be present at spatial infinity, which constitutes an unphysical boundary condition, and that the spacetimes have an imaginary (tachyonic) total mass [8]. This is in contrast to the globally flat space of special relativity, where a collection of subluminal particles cannot of course be tachyonic. Indeed, the fact that in $D=3$ everything lies in the global properties raises a cautionary, and as we shall see, decisive, note. Let us illustrate this with one simple object lesson for the case of static sources. It is clear from (1.1) that a single particle cannot have a mass greater than 2π (in our units); indeed, $m = 2\pi$ corresponds to a cylindrical rather than conical 2-space. One might suppose that, since there are no interactions in $2 + 1$ dimensions, two stationary particles should give rise to a perfectly well-defined metric as long as each one separately satisfies the above inequality. This in fact is not so; the sum of the two masses must also not exceed 2π. If it does, the total mass must then jump to the value 4π and at least one further particle is required to be present, the total system now having an S_2 — rather than an open — topology: G_0^0 is essentially the Euler density of the 2-space [4]. This example reflects the presence of effective global constraints in $2 + 1$ dimensions, even though the theory is locally trivial, so that a source distribution consisting of several individually acceptable particles is *not* thereby guaranteed to be itself physically acceptable. The moral applies to the Gott pair and, as we shall see, also to its lightlike extension. We emphasize that the pathology here is not merely that there is a total spacelike momentum, but more importantly, that the latter implies a "boost-identified" exterior geometry, namely one in which CTC will always be present at spatial infinity. But if one allows pathology in the boundary conditions of any system, then it is no surprise that it will be present in the interior as well! Indeed, this is just the sort of behavior that the Misner identifications [3] gave, and can be seen in the metric form of the interval as well. For, to say that the effective source is a tachyon, really means that the exterior geometry is that generated by an effective pointlike stress-tensor which replaces the $T^{00} \sim \delta(x)\delta(y)$ of a particle with a $T^{yy} \sim \delta(x)\delta(t)$, etc. Consequently, in (the cartesian coordinate form of) the Kerr line element (1.1), the (x,y) space is replaced by (x,t) with a jump in t replaced by one in y [8]. The resulting metric shows that CTC do not really appear and disappear spontaneously in some finite region where the particles pass each other, but rather that they are always present at spatial infinity; thence they close in (very rapidly!) on the finite interaction region.

Attempts to remedy these difficulties by adding more particles [9] to the system fail; it has been shown that the total momentum of any system containing the Gott time machine is necessarily tachyonic [10]; thus, two particles constituting the Gott system cannot arise from the decay of a pair of static particles (of allowed mass less than 2π) since the latter's momentum is timelike [11]. If the mass of the initial static particles exceeds 2π, the universe closes; but as was shown in [12], a closed universe will end in a big crunch just before the CTC appear. Since there is no spatial infinity, the pathology there has been transmuted into a singularity!

In this paper, we extend the Gott construction to systems involving lightlike particles ("photons"). Here too, CTC will appear just as the the system becomes tachyonic. In Section 2, we review the geometries due to a two-photon source and to the "mixed" system consisting of one photon and one massive particle [13]. These systems will be our testing ground for the existence of CTC. In Section 3, we calculate the mass for these two-particle systems, thereby obtaining the condition for their total momentum to be non-tachyonic. In Section 4, the condition for CTC to arise is derived and is shown to coincide exactly with the condition that the system be tachyonic. Furthermore, it will be manifest that (since they first occur there) CTC exist at spatial infinity if they are present at all. In Section 5, the analysis is extended to a more general model, topologically massive gravity. Its two-photon solution is constructed and is shown to exclude CTC for all positive values of the photons' energies. This is true for the ghost Einstein theory as well.

2. Spacetimes Generated by Lightlike Sources

In this section, we review two systems, involving lightlike sources, from which we will attempt to build a time machine. The first consists of two non-colliding photons, the second of one photon and one massive particle. Each can be obtained by pasting together the appropriate one-particle solutions, which we first describe.

In $D=3$, a vacuum spacetime is specified by the way in which locally flat patches are sewn together. Different patches are identified using Poincaré transformations, since these define the symmetries of flat space. This method of constructing solutions, in which the particle parameters (mass, velocity, and location) determine the transformation generators, was presented[2] in [4]. This procedure is, of course, completely equivalent to the standard analytic approach of obtaining the metric from the field equations.

The simplest example is the conical spacetime describing a particle of mass m at rest at the origin of the $x-y$ plane. This solution is obtained by excising a wedge of angle m with vertex at the origin and identifying the two edges according to $x' = \Omega_m x$, where Ω_m is a rotation by m

$$\Omega_m = \begin{pmatrix} 1 & 0 & 0 \\ 0 & \cos m & \sin m \\ 0 & -\sin m & \cos m \end{pmatrix} , \tag{2.1}$$

whose rows and columns are labelled by (t, x, y). This description is completely equivalent to the metric form (1.1) with $J=0$.

The geometric description of the one-photon solution can be found from the analytic solution, or by an infinite boost of the conical static metric [13]. Consider a single photon moving along the x-axis with energy E and energy-momentum tensor $T_{\mu\nu} = E\delta(u)\delta(y)l_\mu l_\nu$, $l_\mu = \partial_\mu u$ where $u = t-x$, $v = t+x$ are the usual lightcone coordinates. The Einstein equations $G_{\mu\nu} = T_{\mu\nu}$ can be solved with a plane-wave ansatz

$$ds^2 = ds_0^2 + F(u, y)du^2 , \tag{2.2}$$

where $ds_0^2 = -dudv + dy^2$ is the flat metric. This ansatz simplifies the Einstein tensor to $G_{\mu\nu} = -\frac{1}{2}\frac{\partial^2 F}{\partial y^2} l_\mu l_\nu$, and reduces the Einstein equations to the ordinary differential equation $\frac{\partial^2 F}{\partial y^2} = -2E\delta(u)\delta(y)$. Solving for F yields the general one-photon solution:

$$ds^2 = ds_0^2 - 2Ey\theta(y)\delta(u)du^2 \tag{2.3}$$

up to a homogeneous solution, of the form $F = B(u)y + C(u)$, that can be absorbed by a coordinate transformation.

If we now apply the coordinate transformation $v \to v - 2Ey\theta(y)\theta(u)$, the metric becomes

$$ds^2 = \theta(u)\{-dud(v - 2Ey\theta(y)) + dy^2\} + \theta(-u)\{-dudv + dy^2\} . \tag{2.4}$$

In this form, the geometric description of the one-photon solution becomes clear. It corresponds to making a cut along the $u = 0$, $y > 0$ halfplane extending from the photon's worldline to infinity and then identifying v on the $u = 0^-$ side with $v - 2Ey$ on the $u = 0^+$ side. It is easily checked that the points being identified are in fact related by the Lorentz transformation

$$N_E = \begin{pmatrix} 1 + \frac{1}{2}E^2 & -\frac{1}{2}E^2 & E \\ \frac{1}{2}E^2 & 1 - \frac{1}{2}E^2 & E \\ E & -E & 1 \end{pmatrix} . \tag{2.5}$$

---

[2]Such procedures are described more formally in [14].

This matrix corresponds to the $\beta \to 1$, $m \to 0$ fixed energy $E = \frac{m}{\sqrt{1-\beta^2}}$, limit of the boost-conjugated rotation matrix $\Lambda_\beta \Omega_m \Lambda_\beta^{-1}$. Here Λ_β is a Lorentz boost in the x-direction,

$$\Lambda_\beta = \frac{1}{\sqrt{1 - \beta^2}} \begin{pmatrix} 1 & \beta & 0 \\ \beta & 1 & 0 \\ 0 & 0 & 1 \end{pmatrix}. \tag{2.6}$$

[This geometric construction of the one-photon solution is analogous to that of the Aichelburg–Sexl one-photon geometry [15] in $D=4$, our null boost being the analog of their null shift.] The above formulation is not unique however; an equivalent one, more analogous to the conical solution, is obtained by boosting the cone along its bisector rather than perpendicular to it [16]. The physics is of course independent of such choices.

The solution for two non-colliding photons can now be constructed by pasting together the individual one-photon solutions. [It is of course not possible to construct the two-photon solution in this way in $D=4$, since spacetime is not flat between sources.] We consider two non-parallel[3] photons in their center-of-momentum frame, where the photons are taken to be moving with energy E respectively in the positive x-direction along $y = a$, $(a > 0)$ and in the negative x-direction along $y = -a$. The spacetime associated with the first photon is obtained by making a cut along the $u = t - x = 0$, $y > a$ halfplane and then identifying the point (x, x, y) on the $u = 0^-$ side with the point $(x - E(y - a), x - E(y - a), y)$ on the $u = 0^+$ side. For the second photon, one makes a cut along $v = t + x = 0$, $y < -a$ and identifies $(-x, x, y)$ on $v = 0^-$ with $(-x + E(y + a), (x - E(y + a), y)$ on $v = 0^+$. The complete two-photon geometry then consists of these two one-photon solutions simply pasted together along the $y = 0$ plane. In contrast to the massive Gott pair, no relative boost between the two particles' halfspaces is necessary, since the photons' motion is already encoded in the identification made on their respective halfplanes.

The second system, consisting of a photon and a massive particle, can also be obtained by pasting together the respective one-particle solutions. We use a frame in which the massive particle is at rest at the origin in the $x - y$ plane, and the photon is moving in the positive x-direction along $y = a > 0$. For the static massive particle we excise from the $y < a/2$ halfspace a wedge of angle m whose vertex is at the origin and which is oriented in the negative y-direction, then identify the edges as usual. For the photon, we simply translate the one-photon solution given above from $y = 0$ to $y = a$. If $m < \pi$, then the orientation of the wedge ensures that there is no intersection with the photon's halfplane, so that the two-particle solution is obtained by pasting together these two halfspaces along $y = a/2$, again with no relative boost. If $m > \pi$, the solution is no longer obtainable by simple gluing, as the tails of the two sources would now overlap.

---

[3]The solution for parallel photons can also be constructed, but it does not admit CTC for the same reason as that given below for the one-photon solution.

3.   Total Energy and Tachyon Conditions

In this section, we calculate the total mass of the two previously described systems. In the following section, we will see that CTC arise precisely when the total mass becomes imaginary, *i.e.*, the system becomes tachyonic and non-physical Misner identifications emerge. The mass can be found by composing the one-particle identifications and writing the result for the complete system in the form of the general spacetime identification

$$x' = a + \mathcal{L}(x - a) + b \, . \tag{3.1}$$

The spatial vector $a = (0, \mathbf{a})$ describes the location of the center-of-mass; the direction and magnitude of the timelike vector b define, respectively, the time axis and the time-shift along it. For a system of total mass M and velocity β in the x direction, $\mathcal{L}$ will be a Lorentz transformation of the form $\mathcal{L} = \Lambda_\beta \Omega_M \Lambda_\beta^{-1}$, implying in particular that

$$\cos M = \tfrac{1}{2}(\operatorname{Tr} \mathcal{L} - 1) \, . \tag{3.2}$$

The system is non-tachyonic provided M is real, implying that the right-hand side lies in the range $[-1, 1]$. We now proceed to calculate $\mathcal{L}$, and from it M, in terms of the constituent parameters of the systems constructed in the previous section.

For the two-photon system, the one-particle identifications are given by

$$
\begin{aligned}
x_1' &= a + N_E(x_1 - a) \\
x_2' &= -a + \Omega_\pi N_E \Omega_{-\pi}(x_2 + a) \, .
\end{aligned}
\tag{3.3}
$$

The conjugation of N_E by a π-rotation in the second equation reflects the fact that the second photon is moving in the negative x-direction. Composing the two identifications in (3.3), we find $\mathcal{L} = N_E \Omega_\pi N_E \Omega_{-\pi}$, and $\operatorname{Tr} \mathcal{L} = 3 - 4E^2 + E^4$. Comparing with (3.2), we obtain the condition

$$E > E_{\max} = 2 \tag{3.4}$$

for the system to be tachyonic. This condition can also be formally obtained as the limit of the original Gott condition [6] for two masses m moving subluminally: there, the tachyonic threshold is given by

$$\frac{\sin \tfrac{1}{2}m}{\sqrt{1 - \beta^2}} > 1 \, . \tag{3.5}$$

Clearly, the limit $m \to 0$, $\beta \to 1$ in this equation (with the energy fixed) yields (3.4).

For the mixed system, the one-particle identifications are given by

$$
\begin{aligned}
x_1' &= a + N_E(x_1 - a) \, , \\
x_2' &= \Omega_m x_2 \, .
\end{aligned}
\tag{3.6}
$$

Composing these yields $\mathcal{L} = N_E \Omega_m$; comparison of its trace with (3.2) implies

$$\cos M = \cos m - (\sin m)E + \tfrac{1}{4}(1 - \cos m)E^2 . \tag{3.7}$$

The criterion for tachyonic M can be expressed as a condition on E for fixed m,

$$E > E_{\max} = 2 \, \frac{\sin m + \sqrt{2(1 - \cos m)}}{1 - \cos m} . \tag{3.8}$$

4.   Closed Timelike Curves

We now find the conditions for CTC to be present in the two-photon and mixed systems. We first show that the one-photon spacetime (like the conical spacetime for a particle of non-zero mass) does not admit CTC. This may not be obvious, since the identification involves a timeshift, which as in the case of the Kerr solution (1.1), could potentially lead to CTC. Recall that the one-photon solution is characterized by the shift $(x, x, y) \rightarrow (x', x', y') = (x - Ey, x - Ey, y)$ upon crossing the $u = t - x = 0$, $y > 0$ null halfplane from $u = 0^-$ to $u = 0^+$. A CTC, γ, would have to cross this halfplane to take advantage of the timeshift (there being no CTC within flat spacetime patches). Irrespective of where γ enters the halfplane, the shifted point from which γ emerges is separated from the entry point by a lightlike interval. Therefore only particles travelling faster than the speed of light can complete the loop in time, showing that the one-photon solution does not admit CTC.

Now consider the two-photon solution described by the identifications (3.3). Since, as shown above, the one-photon solution does not permit CTC, we can restrict ourselves to curves γ that enclose both photons' worldlines and therefore intersect both of their halfplanes. In order that it not become spacelike, γ must be directed opposite to the photon whose plane it is about to cross; this sense will automatically yield a gain in time upon crossing the halfplanes. Label the point at which γ intersects the $u = 0$, $y > a$ halfplane by $x_1^\mu = (x_1, x_1, y_1)$ on the $u = 0^-$ side and hence by $x_1^{\mu'} = (x_1 - E(y_1 - a), x_1 - E(y_1 - a), y_1)$ on the $u = 0^+$ side, and the point at which γ intersects the other halfplane by $x_2^\mu = (-x_2, x_2, y_2)$ on the $v = 0^-$ side and hence by $x_2^{\mu'} = (-x_2 + E(y_2 + a), x_2 - E(y_2 + a), y_2)$ on the $v = 0^+$ side. For γ to be timelike, the total traversed distance,

$$d = |\mathbf{x_2} - \mathbf{x_1'}| + |\mathbf{x_1} - \mathbf{x_2'}| \tag{4.1}$$

$$= \sqrt{(x_1 - x_2 - E(y_1 - a))^2 + (y_1 - y_2)^2} + \sqrt{(x_1 - x_2 + E(y_2 + a))^2 + (y_1 - y_2)^2},$$

must be less than the total elapsed time,

$$T = (t_2 - t_1') + (t_1 - t_2') = E(y_1 - y_2 - 2a) . \tag{4.2}$$

For a given T, we can find the minimum value of d as a function of its arguments. The extremization occurs at $x_1 - x_2 = T/2$ and $y_1 + y_2 = 0$, with the result that $d_{min} = 4y_1$, with $T = 2E(y_1 - a)$. Therefore CTC will be present if

$$y_1/a > E/(E - 2) . \qquad (4.3)$$

Recalling that $y_1 > a > 0$, we see that the lowest allowed value is $E = 2$, precisely the tachyon threshold E_{max} of (3.4); there, y_1 (and therefore also $-y_2$) becomes infinite, i.e., CTC first arise at spatial infinity. As the energy increases, the CTC spread into the interior as well, but they are always present at spatial infinity, if present at all. [The requirement that γ be everywhere future-directed imposes no relevant conditions: The individual time segments $(t_2 - t_1')$ and $(t_1 - t_2')$ must each be positive, implying the inequalities $E(y_1 - a) > (x_1 + x_2) > E(y_2 + a)$. At the minimum, they read $|x_1 + x_2| < E(y_1 - a)$, and are easily satisfied, since $(x_1 + x_2)$ is otherwise unconstrained.]

Let us now find the condition for which the "mixed" system admits CTC. Since neither individual one-particle solution alone admits CTC, a potential CTC, γ, must again enclose both particles, thereby intersecting both the photon's halfplane and the static particle's wedge. Let the point of intersection with the halfplane be labelled by $x_1^\mu = (x_1, x_1, y_1)$ on the $u = 0^-$ side and by $x_1^{\mu'} = (x_1 - E(y_1 - a), x_1 - E(y_1 - a), y_1)$ on the $u = 0^+$ side, and that with the wedge by $x_2^\mu = (t_2, y_2\tan \frac{m}{2}, y_2)$ on one edge and by the rotated values $x_2^{\mu'} = (t_2, -y_2\tan \frac{m}{2}, y_2)$ on the other edge. The curve γ is timelike provided the distance traversed,

$$d = \sqrt{(x_1 - E(y_1-a) - y_2 \tan \tfrac{m}{2})^2 + (y_1 - y_2)^2} + \sqrt{(x_1 + y_2 \tan \tfrac{m}{2})^2 + (y_1 - y_2)^2} , \quad (4.4)$$

is less than the elapsed time $T = E(y_1 - a)$. Here, we minimize d with respect to x_1 and y_2 for fixed T (or y_1); the extremum occurs at $x_1 = T/2$, $y_2 = y_1\cos \frac{m}{2} - \frac{1}{2}E(y_1 - a)\sin \frac{m}{2}$ and is given by $d_{min} = 2a\sin \frac{m}{2} + (y_1 - a)(E\cos \frac{m}{2} + 2\sin \frac{m}{2})$. Therefore, existence of CTC requires

$$y_1/a > E/\left(E - 2 \frac{\sin \frac{m}{2}}{1 - \cos \frac{m}{2}} \right) . \qquad (4.5)$$

Since $y_1 > 0$, E must equal or exceed the threshold tachyon value E_{max} of (3.8), as is easily seen using half-angle formulas. Again, y_1 is infinite at $E = E_{max}$, and as the energy increases, CTC begin to move into the finite region. [Here the requirements that the travel segments be future-directed reduce to $E(y_1 - a) > (x_1 - t_2) > 0$, which can always be fulfilled by adjusting t_2.]

5.   Topologically Massive Gravity

Topologically massive gravity (TMG) [17] is of interest because, in contrast to pure $D=3$ gravity, it is a dynamical theory. In [13] exact solutions for lightlike sources, including (for certain orientations) two-photon solutions, were found. Here we show that they do not admit CTC for any values of the photons' energies. [We cannot construct the analog of the "mixed" system since the exact solution for a massive source is not known in TMG.] The field equations for TMG are the ghost (*i.e.*, with the opposite sign of κ^2) Einstein equations, to which is added the conformally invariant, conserved, symmetric Cotton tensor $C^{\mu\nu} \equiv \epsilon^{\mu\alpha\beta}D_\alpha(R_\beta{}^\nu - \frac{1}{4}\delta_\beta^\nu R)$:

$$E_{\mu\nu} \equiv G_{\mu\nu} + \tfrac{1}{\mu}C_{\mu\nu} = -\kappa^2 T_{\mu\nu} \,. \tag{5.1}$$

Here μ is a parameter (whose sign is arbitrary) with dimensions of mass or inverse length. These equations can be solved exactly for a photon source [13], using the plane-wave ansatz (2.2). For a photon with energy E moving along the positive $x-$axis, the resulting spacetime metric is given by

$$ds^2 = -dudv + dy^2 + 2\kappa^2 E f(y)\delta(u)du^2, \quad f \equiv (y + \tfrac{1}{\mu}(e^{-\mu y} - 1))\theta(y) \,. \tag{5.2}$$

Observe that as $\mu \to \infty$, $f(y) \to y\theta(y)$ and (5.2) reduces to the ghost gravitational one-photon solution (*i.e.*, (2.3) with the opposite sign of κ^2). Like its Einstein counterpart, the metric (5.2) can also be obtained by a cut-and-paste procedure, albeit not by using Poincaré transformations, since the spacetime is not flat along the $u = 0$, $y > 0$ null halfplane. After applying the coordinate transformation $v \to v + 2\kappa^2 E f(y)\theta(u)$ to remove the $\delta(u)$ factor in (5.2), it takes the form

$$\begin{aligned} ds^2 &= -dudv + dy^2 - 2\kappa^2 E\theta(u)f'(y)dudy \\ &= \theta(u)\{-dud(v + 2\kappa^2 E f(y)) + dy^2\} + \theta(-u)\{-dudv + dy^2\} \,. \end{aligned} \tag{5.3}$$

Clearly, this corresponds to identifying v on $u = 0^-$ with $v + 2\kappa^2 E f(y)$ on $u = 0^+$.

We can construct the two-photon spacetime, in the convenient frame where one photon moves in the positive x-direction along $y = a > 0$ and the other in the negative x-direction along $y = -a$, by pasting together the one-photon solutions along the $y = 0$ hyperplane. This pasting is possible since, as in Einstein gravity, each of the one-particle solutions is both flat and has zero extrinsic curvature on this hyperplane. The resulting spacetime consists in making cuts along the $u = 0$, $y > a$ and $v = 0$, $y < -a$ halfplanes and then identifying the point (x_1, x_1, y_1) on the $u = 0^-$ side with $(x_1 + \kappa^2 E f(y_1 - a), x_1 + \kappa^2 E f(y_1 - a), y_1)$ on the $u = 0^+$ side, and the point $(-x_2, x_2, y_2)$ on the $v = 0^-$ side with $(-x_2 - \kappa^2 E f(-y_2 - a), x_2 + \kappa^2 E f(-y_2 - a), y_2)$ on the $v = 0^+$ side. [We note that if the directions of both photons were reversed (corresponding to the parity operation $x \to -x$), then this simple pasting prescription is not possible, since the individual one-photon solutions are no longer flat on the $y = 0$ hyperplane. This reflects the parity violation implicit in the dependence of the field equations (5.1) on $\epsilon_{\mu\nu\rho}$.] The absence of CTC for all positive values of E can be seen as follows. Since $f(y) \geq 0$, the time shift upon crossing a halfplane has opposite sign relative to that of the pure gravity case, (2.3). Hence, to gain a timeshift one would have to cross the (null) halfplane from the $u > 0$ side, rather than from the $u < 0$ side, which is impossible for any timelike (or lightlike) curve. These conclusions obviously also apply to ghost Einstein gravity[4] as well, since the latter is just the $\mu \rightarrow \infty$ limit of TMG.

6.   Conclusion

We have examined, in the case where one or both of their sources are lightlike, the physical difficulties associated with geometries that permit CTC in $2 + 1$ Einstein gravity. We first obtained the total energy in terms of the constituent parameters, using the geometric approach in which flat patches are identified through null boosts, and found the conditions for the systems to be tachyonic. We then showed that, as the energy of a system first surpasses its tachyonic bound, CTC initially emerge at spatial infinity, then spread into the interior, but always remain present at infinity. This is, of course, the manifestation of the unphysical spatial boundary conditions that are the price paid for CTC, and are analogous to the Misner identifications that give rise to CTC in NUT space. We also demonstrated, using the known two-photon solution in TMG, that this dynamical model, and therefore also its limit, ghost Einstein theory, never admits CTC.

All sources of the $2 + 1$ Einstein equations considered to date thus share the property that if they are physical—nontachyonic—they do not engender acausal geometries. These results add evidence in favor of both Einstein's original hope and its recent avatar [18], that this is a universal property of general relativity.

7.   Acknowledgements

We thank G. 't Hooft for useful conversations.

References

[1] K. Gödel, *Rev. Mod. Phys.* **21** (1949) 447.

[2] K. Lanczos, *Zeits. f. Phys.* **21** (1924) 73; W.J. van Stockum, *Proc. R. Soc. Edin.* **57** (1937) 135; W. B. Bonnor, *J. Phys. A* **13** (1980) 2121.

[3] C.W. Misner, in *Lectures in Applied Mathematics*, vol. 8, J. Ehlers, ed., American Mathematical Society, Providence, R.I., 1967; C.W. Misner and A. Taub, *Soviet Physics JETP* **28** (1969) 122.

[4] S. Deser, R. Jackiw, G. 't Hooft, *Ann. Phys.* **152** (1984) 220.

---

[4]As we mentioned at the outset, there is no *a priori* physical requirement that κ^2 be positive within the $D=3$ picture, except if one wishes to regard it as a reduction from $D=4$ in order to make contact with cosmic strings.

[5] M. Henneaux, *Phys. Rev. D* **29** (1984) 2766; S. Deser, *Class. Quantum Grav.* **2** (1985) 489.

[6] J. Gott, *Phys. Rev. Lett.* **66** (1991) 1126.

[7] A. Ori, *Phys. Rev. D* **44** (1991) R2214; C. Cutler, *Phys. Rev. D* **45** (1992) 487.

[8] S. Deser, R. Jackiw, and G. 't Hooft, *Phys. Rev. Lett.* **68** (1992) 267.

[9] D. Kabat, *Phys. Rev. D* **46** (1992) 2720.

[10] S. Carroll, E. Farhi, and A. Guth, MIT preprint CTP-2117 (1992).

[11] S. Carroll, E. Farhi, and A. Guth, *Phys. Rev. Lett.* **68** (1992) 263, (E)3368.

[12] G. 't Hooft, *Class. Quantum Grav.* **9** (1992) 1335.

[13] S. Deser and A. Steif, *Gravity Theories with Lightlike Sources in D=3*, *Class. Quantum Grav.*, (in press).

[14] D.C. Duncan and E.C. Ihrig, *Gen. Rel. Grav.* **23** (1991) 381.

[15] P. C. Aichelburg and R. U. Sexl, *J. Gen. Rel. Grav.* **2** (1971) 303.

[16] G. 't Hooft, private communication.

[17] S. Deser, R. Jackiw, S. Templeton, *Ann. Phys.* **140** (1982) 372.

[18] S.W. Hawking, *Phys. Rev. D* **46** (1992) 603.

Inhomogeneity and Anisotropy Generation in FRW Cosmologies

G. F. R. Ellis *   *D. R. Matravers* [†]

Abstract

Misner [1] suggested kinetic theory ideas could explain the smoothing of the universe at early times. Recently [2], [3] the inverse has been investigated: the generation of Bianchi I anisotropy in cosmologies with an exact FRW geometry, again for kinetic theory reasons. At the time it was stated that more general anisotropy and inhomogeneity could be generated by the same methods. In this paper the more general case is addressed to show how the model can be produced and to identify some open questions connected with it.

1.   Introduction

One of the major contributions Charles Misner has made to cosmology was the first serious investigation of the chaotic cosmology idea [1], namely that the universe started off in a very chaotic state and then developed (by physical processes) towards a smooth state. This idea has since been taken up with a vengeance by the Inflationary Universe school of thought, utilising Guth's insight [4] that the vast expansion associated with the false vacuum (a scalar field) could provide the required smoothing mechanism.

The resulting theory is very interesting but perhaps over-stated [5]. In particular it does not overcome the Stewart remark [6] that no matter what smoothing mechanisms one might find there are always initial data that will lead to universes more lumpy than the observed universe (simply run the equations backwards from any considered present state to find the needed initial conditions). Furthermore entropy arguments have been adduced by Penrose [7] to suggest that the universe must have started off in a smooth state rather than the very chaotic state suggested by many inflationists. The Hartle-Hawking quantum cosmology programme [8] leads to a similar conclusion

---

*Department of Applied Mathematics, University of Cape Town, Cape Town 7700, South Africa.

[†] School of Mathematical Studies, Portsmouth Polytechnic, Portsmouth PO1 2EG England. We thank the FRD (South Africa) for support of this work.

(the universe is envisaged as starting off in a lowest energy state rather than one that is a highly chaotic).

It is interesting therefore to return to the alternative idea that the universe could have started off smooth and then developed inhomogeneity and anisotropy. This paper remarks on some intriguing exact solutions that make good this possibility. How realistic they are in physical terms is not clear, but in looking at the issue of the origin of the universe one should consider the phase space of all possibilities rather than some chosen restricted subset of the allowed initial conditions. This paper is a contribution in terms of exploring that phase space.

We assume that the evolution of the universe is described at early times by a solution to the Einstein-Boltzmann system of equations, i.e., the matter is described by a kinetic theory model and the relation between the matter and gravitation (the geometry) is given by Einstein's equations[1]. The key point then is that models of the evolution of the universe can be constructed which evolve from homogeneity and isotropy, i.e. having FRW geometry, to an inhomogeneous and anisotropic geometry. They do so by having a distribution of particles which

(1) initially is inhomogeneous and anisotropic but is compatible with the FRW geometry, that is, it does not communicate the anisotropy to the space-time geometry, this being permitted because

(2) it is initially effectively collision-free, as might occur, for instance, in regimes where the temperature is so high that the potentially interacting particles experience asymptotic freedom [2]. Thus we seek solutions of the Liouville (collision-free) equation for the particles, at these epochs.

The final element is that

(3a) as the universe expands and the temperature drops, collisions begin to have an appreciable effect on the distribution function, changing its harmonic structure through the Boltzmann collision term,

or alternatively

(3b) initially we have effectively a zero rest-mass distribution, and as the universe expands and the temperature drops below some threshold temperature the distribution is no longer effectively of zero-rest mass particles.

In either case, at that stage inhomogeneity in the particle distribution is communicated to the space-time geometry. Hence a transition is made by that geometry from

---

[1] Another of Misner's contributions was some of the most sustained early kinetic theory calculations of complex effects in cosmology [1].

[2] The alternative of detailed balancing would also do were it not expected to isotropise the distribution function (but note that it has not been rigorously proved that this will in fact happen).

the isotropic and spatially homogeneous FRW form to a form that can contain seeds for galactic formation.

Discussions of the basic idea have been presented before in [2] and [3]. In the first of these papers a Bianchi I geometry is used to demonstrate the generation of anisotropy. Inhomogeneity generation is mentioned in both, but is not dealt with in detail. Here we point out that recently published exact solutions can be used in that discussion, and raise new possibilities for application of the ideas.

2.   The matter description

The formalism to be used is established in [2], [3] and the references they cite, so only the essentials will be given here.

At the base of our analysis is a "3+1" decomposition relative to a preferred 4-velocity. The matter in the Universe[3] can define such a preferred velocity in various ways. We assume that a particular choice has been made, and denote the resulting vector field by u^a. It can be used to produce a $(3 + 1)$ split in the spacetime and to write the four momentum p^a of a particle as,

$$p^a = Eu^a + \lambda e^a \tag{1}$$

where (u^a, e^a) form an orthonormal dyad, i.e., $e^a e_a = 1$ and $e^a u_a = 0$. The scalars λ and E are the magnitude of the momentum and the energy of the particle relative to an observer with velocity u^a. If the particle has mass m then

$$p^a p_a = -m^2 = E^2 - \lambda^2 \tag{2}$$

The distribution function [9] $f(x^i, p^a)$ for a gas of identical particles (attention will be confined to identical particles for simplicity) can be written; $f(x^i, p^a) = g(x^i, m, E, e^a)$ using the decomposition (1). The function g is expanded in covariant harmonics,

$$g = F + F_a e^a + F_{ab} e^a e^b + \ldots\ldots \tag{3}$$

where the harmonics $F_{ab\ldots r}(x^i, m, E)$ are symmetric and trace free tensors orthogonal to u^a.

In a $k = 0$ FRW spacetime with length scale factor $S(t)$ the Liouville equation for the harmonics takes the form;

$$SE\left(\frac{\partial F_{ab\ldots r}}{\partial t}\right) - \dot{S}(E^2 - m^2)\left(\frac{\partial F_{ab\ldots r}}{\partial E}\right) = 0 \tag{4}$$

and it is easy to show that a solution is given by

$$G_{ab\ldots r}(X) = F_{ab\ldots r}(x^i, m, E) \tag{5}$$

---

[3]A capital U will be used for the observed Universe.

where
$$X = (E^2 - m^2)^{1/2} S(t) \tag{6}$$

In [10] a solution to the Einstein-Liouville system of equations is constructed using the harmonics $Fab....r$ subject to the conditions,

if $m \neq 0$ (case (a))
$$G_a = 0 \tag{7}$$
$$G_{ab} = 0 \tag{8}$$

or

if $m = 0$ (case (b))
$$G_a = 0 \tag{9}$$
$$\int X^3 G_{ab}(X) dX = 0 \tag{10}$$

The condition (10) permits non-zero solutions for G_{ab}. The zeroth harmonic and all other harmonics higher than the second are essentially arbitrary; subject to suitable convergence of the expansion and $f(x^i, p^a)$ being positive.

In [2] the spacetime is assumed to start off with FRW geometry and a distribution function which has Bianchi I symmetry; it is therefore spatially homogeneous but anisotropic. As the universe cools, in case (a) the particles whose distribution initially satisfied the conditions (7,8) start to collide, and so anisotropies in the higher moments can cascade down to the lower moments; consequently the distribution which initially satisfied conditions (7, 8), will no longer do so, in general. Alternatively, in case (b), particles whose distribution initially satisfied the conditions (9,10) pick up mass; consequently the required conditions change to (7, 8), and the latter will not be satisfied in general. In either case, failure to satisfy condition (8) causes an anisotropic pressure which pushes the geometry away from the FRW form, the anisotropy generating shear. This then feeds backs through the Boltzmann equations, tending to increase the anisotropy in the distribution function; but on the other hand the expansion of the universe tends to decrease it. Depending on the details of the situation, the result may or may not be to isotropise the higher moments.

In the next section the assumptions of the distribution function are relaxed and general anisotropy and inhomogeneity are introduced.

3.   An anisotropic and inhomogeneous case

In [11] Maharaj and Maartens have obtained solutions to the Einstein-Liouville equations which have the properties we require. The spacetime has FRW symmetry but the distribution functions are in general inhomogeneous and anisotropic. The solutions, given in isotropic coordinates, depend on constants of the motion Y_l constructed from the Killing vectors, and a constant C, generated by conformal Killing vectors if $m = 0$. They give both a general solution to the Liouville equation, and a more specific one that is easier to analyse, because it is spherically symmetric about the origin of coordinates.

The analysis carried out in the previous case goes through in the case of the specific solution, except for the part which we will now discuss. In place of the solution (5) to the Liouville equation we use the Maharaj and Maartens solution which has the form,

$$F_{ab....r}(x^i, m, E) = H_{ab....r}(t, r, m, X) \tag{11}$$

where X is defined as above, $r^2 = x^2 + y^2 + z^2$ and the r and t enter through the constant of motion $C = C(X, r, t)$. The conditions that give rise to (7,8,9,10) become,

$$\int_0^\infty X^3 H_a(t, r, X, m) dX = 0 \tag{12}$$

and

$$\int_0^{infty} X^4 [x^2 + m^2 S^2(t)]^{-1/2} H_{ab}(t, r, X, m) dX = 0. \tag{13}$$

These conditions are weaker than those in the previous discussion. Because H is explicitly dependent on t, condition (13) does not require that H_{ab} be zero everywhere when $m \neq 0$. Maharaj and Maartens use $H_{ab....r}$ to construct exact solutions to the Einstein-Boltzmann equations[4] in which the distribution functions are inhomogeneous because of the dependence on r and anisotropic because of the dependence on the directions e^a. However the conditions on $H_{ab....r}$ are different depending on whether m is zero or not and thus in general the same function will not satisfy condition (13) in both cases. Hence the above argument that the universe will be driven away from an FRW geometry holds, in both cases (that is, this can happen either through collisions becoming effective, or through mass becoming non-negligible).

The above argument holds for the explicit form of the distribution function of Maharajah and Maartens, spherically symmetric about the origin of coordinates. However the further point is that as long as the solution is effectively collision-free, as they assume, one can essentially *linearly add such solutions together to get new, more complex solutions of the Einstein-Liouville equations* in the FRW regime.

The first point is that the Liouville equation is linear in the distribution function, as are equations (12, 13); consequently we can add distribution functions linearly (with constant coefficients), still obtaining solutions of the Liouville equation (indeed this is what underlies the harmonic decomposition (3)). There is a resultant non-linearity in the field equations when the resulting energy densities determine the time-evolution of the solution (through equation (22) of [11]); however that is the limit of the non-linear effects, and will cause no serious problem in practice.

---

[4]Their analysis really needs extension to make explicit how their function $h(C)$ is expanded in the form (11); we assume no problems arise in carrying this out.

The second point is that while (14) is spherically symmetric about the origin of co-ordinates $r = 0$, there is of course nothing special about that point; so we can add together as many solutions of the Liouville equation (14) as we like, with each one based on a different origin. This will give much more complex resulting behaviour in the epoch when the inhomogeneities have cascaded down to the lower moments; for whereas the solutions discussed in the previous paragraph would generate one spherically symmetric inhomogeneity, the procedure produced here will generate a distribution of spherically symmetric inhomogeneities, arbitrarily placed and in prin-ciple capable of giving any desired mass distribution.

Given that for the continuation the initial conditions are that ω, σ_{ab}, u^a and the conformal curvature are zero, equation (4.16) of Ellis [12] shows that the introduc-tion of an anisotropic pressure π_{ab} will generate shear. This also follows from the phenomenological equations of kinetic theory which relate shear to the anisotropic pressure. Reference to equation (4.12) of Ellis et al [13] shows that once one has shear the decoupled equations which hold in the FRW, $k = 0$ regime become coupled and the moments get mixed. Hence inhomogeneity and isotropy in the higher moments of the distribution can be passed down to the zeroth, first and second moment both through the collision term and through the Liouville operator; it can then affect the geometry through Einstein's equations. Even the density and pressure will in general become inhomogeneous. Clearly the geometry cannot remain FRW.

Even more general solutions will be obtained on using the generic solution of Maartens and Maharaj, rather than the restricted one employed here. The key point is that this will allow tilted matter flows (cf. [10]) that include the possibility of rotation, which cannot occur even with the very general solution just discussed (if all the component solutions have no rotation, as is the case [11], then the combined one will still have no rotation).

In [2] the continuation of the solution after the onset of the anisotropic pressure was achieved by assuming Bianchi I symmetry in the collisions and matching a Bianchi I model to an FRW model across a spacelike boundary at the onset time t_0 using the Darmois conditions. The present situation is more complicated because the con-tinuation is to be inhomogeneous and anisotropic and we do not have much chance of constructing an exact solution once the collisions have introduced the anisotropy. However we can at least give an argument that shows the solution will continue regularly into this new regime.

4. Continuation of the model

Here we deal with continuation of the model after the onset of the anisotropic pressure. In general the fluid flow will change smoothly to become imperfect, inhomogeneous, anisotropic, and accelerating. These new properties derive from the higher moments of f through coupling in the Boltzmann equation. The implicit equation of state will be time-dependent, determined through the Boltzmann equation.

It is useful to phrase the problem in mathematical terms. The equations that must be solved to produce a continuation are,

the Einstein equations,

$$R_{ab} - \frac{1}{2} g_{ab} R = \kappa T_{ab} \qquad (14)$$

the stress tensor equation

$$T^{ab} = \int p^a p^b f \pi \qquad (15)$$

and the Boltzmann equation

$$p^a \left(\frac{\partial f}{\partial x^a} - \Gamma^c_{ab} p^b \frac{\partial f}{\partial p^c} \right) = C \qquad (16)$$

where C is a collision function the exact form of which depends on the process by which the transition occurs and the initial conditions on the geometry given above. The initial conditions are set by the FRW regime and this ensures that they will be consistent.

The existence of a solution that continues after the onset of the anisotropic pressure, if a suitable collision functional is used, is assured by the work of Bichteler, Bancel and Choquet-Bruhat ([14] and references cited there). The required theorem is,

There exists (subject to mild conditions) a domain Ω in R^4, a metric g^{ab} and a function f on $\Omega \times R^3$ such that,

1. g^{ab} and f are solutions to the Einstein-Boltzmann equations

2. g^{ab} and f take the given Cauchy data.

This solution is unique in Ω and depends continuously on the data. (Ω may be chosen to be globally hyperbolic for the metric g^{ab} and to admit a Cauchy surface).

The mild conditions refer to the Sobolev class in which the solutions fall. The model they use for the collision functional is for binary, elastic collisions and probably can be used here. Thus the inhomogeneous initial conditions produced by the mechanisms discussed above will then continue to produce an inhomogeneous space-time with the kinds of seeds that could in principle lead to a plausible distribution of galaxies.

5.   Some Questions

The primary question is how one could arrive at initial conditions of the type described here with a spatially homogeneous and isotropic space-time but a distribution function with less symmetry. We do not have a mechanism that can do so; this would have to be given as initial conditions. Thus the motivation for studying these examples is not that we can claim they are particularly physically significant, but that we desire to study the full phase space of possibilities.

Given this position, a number of questions arise. We list them together with tentative responses.

1. The continuation appears to allow great deal of freedom depending only on the choice of the higher moments of f, the collision process chosen and how it is modelled. Is it really this open or are there other constraints? Our present position is that it is indeed this open.

2. If the model is to be useful it must evolve in the future to near FRW at a large enough scale. Can this be achieved? Our position is yes it can, provided the expansion of the universe dominates over the shear generated. There will be a large family of models for which this is true.

3. Once the shear is created it cannot go away but it must decay to a small value. Unless the f isotropises it will continue to drive the shear up once it is invoked by the collision process. Does f isotropise? The response is the same as in the previous case.

4. What form must f have if the model is to generate the present matter distribution - e.g. can we choose f to give voids and galaxy clusters? This is a rather difficult inverse problem; however it might be possible to successfully answer it in the case of a single void.

5. Can we obtain perturbed almost-FRW solutions that demonstrate this kind of evolution in detail? This should be possible, and may be illuminating.

6. What effect would this have on the Cosmic Background calculations? They are probably not affected if all this happens before last scattering which it should do.

Indeed the thrust of the paper is almost precisely the opposite of that in Misner's pioneering paper [1]. However the spirit is the same as in the paper by Misner on the Taub-NUT universes [15]: The models here are put forward in the hope they will help us against being too dogmatic about what is and what is not possible in General Relativity Theory.

References

[1] Misner C W, (1968). The Isotropy of the Universe. *Astrophys Journ* **151**, 431-457.

[2] Matravers D R and Ellis G F R, (1989) Evolution of anisotropies in Friedmann cosmologies, *Class. Quantum Grav.* **6**, 369 - 381

[3] Ellis G F R and Matravers, D R, (1990) A note on the evolution of anisotropy in a Robertson-Walker cosmology, *Class. Quantum Grav.* **7** 1869 - 1873.

[4] Guth A H, (1981). *Phys Rev* **D23**, 347.

[5] Ellis, G F R (1991): "Standard and Inflationary Cosmologies". In *Gravitation, Proceedings of Banff Summer Research Institute on Gravitation* [August 1990], Ed R Mann and P Wesson, (World Scientific), 3-53.

[6] Stewart J M (1969). Non-equlibrium processes in the early universe. *Mon Not Roy Ast Soc* **145**, 347-356.

[7] Penrose R, (1989). *The Emperor's New Mind* (Oxford University Press), Chapter 7.

[8] Hartle J B and Hawking S W (1983). *Phys Rev* **D 28**, 2960.

[9] Ehlers J (1971) General Relativity and Kinetic Theory. In *General Relativity and Cosmology XLVII Enrico Fermi Summer School Proc* ed R W Sachs (New York: Academic) pp 1 - 70.

[10] Ellis G F R, Matravers and Treciokas R (1983) An Exact Anisotropic Solution of the Einstein-Liouville Equations, *Gen. Rel. Grav.*, **15**, 10 , 931 - 944.

[11] Maharaj S D and Maartens R, (1987) Exact Inhomogeneous Einstein-Liouville Solutions in Robertson-Walker Space-Times. *Gen. Rel. Grav.* **19**, 5, 499 - 509.

[12] Ellis G F R, (1971) Relativistic Cosmology. *General Relativity and Cosmology XLVII Enrico Fermi Summer School Proc* ed R W Sachs (New York: Academic) pp 104 - 82

[13] Ellis G F R, Treciokas R and Matravers D R (1983) Anisotropic solutions to the Einstein-Boltzmann equations: 1: General Formalism *Ann Phys* **150** 455 - 486.

[14] Bancel D and Choquet-Bruhat Y, (1973) Existence, Uniqueness, and Local Stability for the Einstein-Maxwell-Boltzmann System *Commun Math Phys* **33** 83 - 96

[15] Misner C W (1967): Taub-NUT Universe as a counter-example to almost anything. In *Relativity Theory and Astrophysics I: Relativity and Cosmology*, ed. J. Ehlers. Lectures in Applied Maths, Volume 8 (American Mathematical Society), 160-169.

Misner, Kinks, and Black Holes

*David Finkelstein**

I first met Charles Misner at the genial general relativity meetings that James Anderson and Ralph Schiller organized at Stevens Institute of Technology in the mid-'50s. His birthday reminds me of our collaboration in topological physics in 1959, when we found the topological spin-statistics connection and gravitational kinks. Misner's contributions to the discovery of the relativistic theory of the black hole are not adequately appreciated. Let me say a little about these matters here.

They all hang on the thread of anomalous spin. The first anomalous spin was the spin 1/2 of the electron. It was anomalous in that the very possibility of spin 1/2 was initially overlooked by quantum theorists. Then experiment and Uhlenbeck & Goudsmit forced it to our attention. Wigner explained this spin by examining how the electron wavefunction behaved under the Wigner waltz W: a path in the rotation group describing a continuous rotation of the physical system through 2π. In three dimensions W cannot shrink continuously to a point (W is nontrivial), but W^2, a 4π rotation, can (W^2 is trivial). It followed that W can change quantum phase by π. This happens for spin 1/2.

In 1954, fresh out of graduate school, I set out to find a unified theory of the known particles and forces, as I imagined one was supposed to do. As an undergraduate my ambition was to carry out Von Neuman's quantum set theory program. What I describe next began a decade detour from that program, to which I later returned.

Misner and Wheeler had proposed gravity as a unified field, in the theory they called geometrodynamics. Before that Bopp and Haag had shown that some rather natural rotator-type fields combined integer and fractional spin in their particle spectra. This led me to wonder if we were overlooking another anomalous spin 1/2 of the non-linear gravitational field. The existence of spin 1/2 in nature is a major clue in the hunt for unity. For example it might eliminate Einstein's various unified fields. I hoped to find

---
*Physics Department, Georgia Institute of Technology, Atlanta, Georgia 30332-0430

quantum gravity devoid of anomalous spin, eliminate it as a candidate for a unified theory, and work with a clearer conscience on discrete quantum structures.

To be sure, in weak-field approximations gravity is a spin-2 field. The gravitational potentials, however, obey a nonlinear signature constraint: they must describe one time and three space dimensions. Like the rotation group and the Bopp-Haag rotators, the 10-dimensional manifold M^{10} of metric tensors of signature 1−3 has a nontrivial topology. The linear approximation must break down seriously for large amplitudes.

Such nontrivial topology is the first requirement for anomalous spin. The configuration space of any 3-dimensional system with anomalous spin 1/2 must contain a nontrivial loop whose square is trivial, to match W. This is the spin-topology connection between the Wigner waltz W and the Poincaré (first homotopy) group π of the system.

As a bridging exercise between non-relativistic mechanics and general relativity, I had tested all special-relativistic point rotators for anomalous spin, by infinitesimal rather than global methods. Three kinds tested positive, incidentally. For example, a point particle carrying a 2-form $F_{\mu\nu}$ with $F_{\mu\nu} F^{\mu\nu}=0$ supports anomalous spin 1/2; but not with $F_{\mu\nu} F^{\mu\nu} = 1$, nor a null 1-form.[1]

Call the infinite-dimensional function space of all asymptotically flat Minkowskian manifolds M^{∞}. This is an infinite dimensional function space of gravitational metric-tensor fields. The first question is: What nontrivial loops exist in M^{∞}? If none, then gravity had no anomalous spin and could be scratched.

To answer this one had to first divide M^{∞} into its connected topological components, or "pieces" for short, and then search each piece for loops that lead to anomalous spin. These tasks were too much for my tools. I brought them to Misner at Princeton in 1958 because he was approachable, sharp, and his work in geometrodynamics showed that he knew much more topology than I did.

In fact his topological publications had all used homology, and these are questions in homotopy, a whole different kind of topology. He changed tools with no perceptible pause. He was behind his desk in his office when I put my problem. As I listened he explained and sketched. Within minutes he converted my question about loops in M^{∞} to more tractable questions about bags in the ten-dimensional manifold M^{10} of symmetric Minkowskian metric forms. Then he showed how this M^{10} could be retracted into an M^3, the projective sphere, without losing or creating any bags. Finally he reached into the bookcase behind him and pulled out Steenrod's *Topology of Fiber Bundles*. In effect we had the following dialogue:

Q: "How many components (topologically self-connected, mutually disconnected parts) does M^∞ have?"

Misner: "One for every integer $K=0, \pm1, \pm2, ...$"

Q: "How many loops does each piece of M^∞ have?"

Misner: "One nontrivial loop, whose square is trivial."

Misner had reduced the first question to the third homotopy group of the projective sphere; and the second question, to the fourth homotopy group. The answers were exactly the ones needed for a topologically conserved gravitational structure to exist and to exhibit anomalous spin 1/2. Misner and I continued discussions by mail while I was on sabbatical from Stevens at CERN in 1959. This led to our one paper together, which included the idea of kinks or topological solitons, both gravitational and general, the sine-Gordon equation, and the topological spin-statistics connection.[2]

Let me describe what came out of Misner's topological deductions. His first answer forced us to recognize the existence of kinks in the gravitational field, topological defects like the twist in a Mobiüs strip, which can move about in spacetime but cannot be removed without cutting. K is the number of kinks. K was a new kind of topological particle number, and one based on the physically real field of gravity, not a hypothetical possibility.

It further meant that gravity (in a spacetime of trivial topology) had just one basic form of kink. Steenrod represented it by a map of a three-dimensional sphere (representing space) into itself (representing M^3 and the light cones). A loop in this space is a map of the unit four-dimensional sphere into the unit three-dimensional one, the Freudenthal suspension of the Hopf map of the three-dimensional sphere into the two. In its simplest form, the kink depends only on three of the four spacetime coordinates and is spherically symmetric in them. If one draws it as a field of twisting light cones, it exhibits a "unidirectional membrane", as I named this stationary horizon. In static coordinates it sprouted a singularity where $g_{00} = 0$, as Birkhoff's theorem says it must, and its singularity had the same kind of pole as the Schwarzschild singularity. This suggested that the Schwarzschild singularity is actually such a membrane, and that such a membrane will form when matter condenses enough.[3] So I could explain the physical meaning of the Schwarzschild singularity as a direct product of the search for anomalous spin. Because of its singularity at $r = 0$, the full Kruskal manifold is no kink, but with appropriate sources the Schwarzschild exterior solution can be fitted to a non-singular kinked interior.

Misner's second answer permitted gravity kinks to have anomalous spin 1/2. But for something like quantum gravity to work as a general theory, its kinks would have to obey the experimentally observed connection between spin and statistics. Since spin is associated with the Wigner homotopy W, statistics should be associated with a homotopy too, an operator X that continuously exchanges two kinks without overlap. One may then "prove" the topological spin-statistics connection

$$W \sim X$$

("W is homotopic to X") with a belt demonstration.

In two dimensions there can be nontrivial loops that first become trivial when traversed four (or n) times. Therefore, according to the spin-topology connection, there is anomalous spin 1/4 (and $1/n$) in 2 dimensions. But no isolated physical system can exhibit spin 1/4, for much the same reason that a physical rigid rotator made of integer-spin atoms cannot exhibit spin 1/2. When the crystal or molecule dissociates into a 3-dimensional gas, it will have its ordinary spin values, not anomalous ones. Therefore, by conservation of angular momentum, it must have them while it is solid. Forces do not change topology, right? Therefore spin 1/4 is impossible.

That is how I overlooked the anyon and its anomalous spin. I did not notice that a crystal structure—say a defect—an have anomalous spin 1/4 if there is a compensating spin −1/4 in the crystal. Neither of these spins 1/4 is an isolated physical system, and therefore neither is subject to the deduction of the previous paragraph. The whole crystal, which may be isolated, has ordinary spin. The anyon can still have spin 1/4, or $1/n$ for that matter, relative to the crystal. Those who took spin 1/4 seriously [4] were right.

The moral of this story dates from Planck, who first saw quantum degrees of freedom freeze and melt: *Forces change effective topology*. This moral is important to remember for spacetime.

Nowadays Thouless has invoked anomalous quantum numbers of topological origin for the fractional Hall effect and Wilczek has brought them into hot superconductivity. It is now clear that some physical variables are topological. There are so many topological variables present in spacetime that perhaps we need no others.

So the relativistic black hole first arose as a particle model, with the astrophysical applications hardly mentioned. Later Hawking discovered that black holes were also black bodies, radiating thermally with increasing temperature as they lost energy and shrank. Hawking's calculation is based on a classical theory of gravity and has the same provocative status as the classical proof that a hydrogen atom will radiate increasingly until it collapses into a point singularity. It challenges us to make a quantum theory and look for stable modes. Only a quantum theory of gravity can tell whether after an incandescent youth the atom or the star will enter a stable non-radiative old age.

Elementary particles and dark matter may yet turn out to be spacetime defects, or even micro-black-holes.

Kinks, the sine-Gordon soliton, the topological spin-statistics connection, and the physical meaning of the Schwarzschild singularity all followed swiftly from the exchange between physics and topology in Misner's office in Princeton, one of the more intense five minutes of my life in research. Our discovery of the gravitational kink and its anomalous spin was a high point on the trail to the black hole and beyond.

REFERENCES

1. D. Finkelstein, *Phys. Rev.* **100**, 1924 (1955).

2. D. Finkelstein and C. W. Misner, *Ann. Phys.* **6**, 230-243 (1959).

3. M. Kruskal, *Phys. Rev.* **119**, 1743 (1960). D. Finkelstein, *Phys. Rev.* **110**, 965 (1958).

4. G. A. Goldin, R. Menikoff, and D. H. Sharp, *J. Math. Phys.* **22**, 1664 (1981); F. Wilczek, *Phys. Rev. Lett.* **48**, 1144 (1982).

THE QUANTUM MECHANICS OF CLOSED SYSTEMS

J.B. Hartle

Department of Physics, University of California Santa Barbara, CA 93106-9530

ABSTRACT

A pedagogical introduction is given to the quantum mechanics of closed systems, most generally the universe as a whole. Quantum mechanics aims at predicting the probabilities of alternative coarse-grained time histories of a closed system. Not every set of alternative coarse-grained histories that can be described may be consistently assigned probabilities because of quantum mechanical interference between individual histories of the set. In "Copenhagen" quantum mechanics, probabilities can be assigned to histories of a subsystem that have been "measured". In the quantum mechanics of closed systems, containing both observer and observed, probabilities are assigned to those sets of alternative histories for which there is negligible interference between individual histories as a consequence of the system's initial condition and dynamics. Such sets of histories are said to decohere. We define decoherence for closed systems in the simplified case when quantum gravity can be neglected and the initial state is pure. Typical mechanisms of decoherence that are widespread in our universe are illustrated.

Copenhagen quantum mechanics is an approximation to the more general quantum framework of closed subsystems. It is appropriate when there is an approximately isolated subsystem that is a participant in a measurement situation in which (among other things) the decoherence of alternative registrations of the apparatus can be idealized as exact.

Since the quantum mechanics of closed systems does not posit the existence of the quasiclassical domain of everyday experience, the domain of the approximate aplicability of classical physics must be explained. We describe how a quasiclassical domain described by averages of densities of approximately conserved quantities could be an emergent feature of an initial condition of the universe that implies the approximate classical behavior of spacetime on accessible scales.

0 PREFACE

Charlie Misner was one of the pioneers of quantum cosmology — the effort to understand the universe as a whole as a quantum mechanical system. The minisuperspace models that he introduced and analysed with such characteristic elegance and clarity remain standard arenas in which to test ideas of the subject (Misner, 1969, 1972). As he was well aware, in the applications of quantum mechanics to cosmology one must confront the characteristic features of quantum theory in a

104

striking and unavoidable manner. A central problem is simply obtaining a coherent formulation of quantum mechanics for closed systems such as the universe. This essay on the occasion of his 60th birthday is a pedagogical summary of the efforts of Murray Gell-Mann and myself to provide such a formulation (Gell-Mann and Hartle, 1990ab).

1 INTRODUCTION

It is an inescapable inference from the physics of the last sixty years that we live in a quantum mechanical universe — a world in which the basic laws of physics conform to that framework for prediction we call quantum mechanics. We perhaps have little evidence of peculiarly quantum mechanical phenomena on large and even familiar scales, but there is no evidence that the phenomena that we do see cannot be described in quantum mechanical terms and explained by quantum mechanical laws. If this inference is correct, then there must be a description of the universe as a whole and everything in it in quantum mechanical terms. The nature of this description and its observable consequences are the subject of quantum cosmology.

Our observations of the present universe on the largest scales are crude and a classical description of them is entirely adequate. Providing a quantum mechanical description of these observations alone might be an interesting intellectual challenge, but it would be unlikely to yield testable predictions differing from those of classical physics. Today, however, we have a more ambitious aim. We aim, in quantum cosmology, to provide a theory of the initial condition of the universe which will predict testable correlations among observations today. There are no realistic predictions of any kind that do not depend on this initial condition, if only very weakly. Predictions of certain observations may be testably sensitive to its details. These include the large scale homogeneity and isotropy of the universe, its approximate spatial flatness, the spectrum of density fluctuations that produced the galaxies, the homogeneity of the thermodynamic arrow of time, and the existence of classical spacetime. Recently, there has been speculation that even the coupling constants of the effective interactions of the elementary particles at accessible energy scales may be probabilistically distributed with a distribution which may depend, in part, on the initial condition of the universe (Hawking, 1983, Coleman, 1988, Giddings and Strominger, 1988). It is for such reasons that the search for a theory of the initial condition of the universe is just as necessary and just as fundamental as the search for a theory of the dynamics of the elementary particles. They may even be the same searches.

The physics of the very early universe is likely to be quantum mechanical in an essential way. The singularity theorems of classical general relativity suggest that an early era preceded ours in which even the geometry of spacetime exhibited significant quantum fluctuations. It is for a theory of the initial condition that describes this era, and all later ones, that we need to spell out how to apply quantum mechanics to cosmology. Recent years have seen much promising progress in the search for a theory of the quantum initial condition. However, it is not my purpose to review these developments here.* Rather, I shall argue that this somewhat obscure branch of astrophysics may have implications for the formulation and interpretation

---

* For a recent review of quantum cosmology see Halliwell (1991).

of quantum mechanics on day-to-day scales. My thesis will be that by looking at the universe as a whole one is led to an understanding of quantum mechanics which clarifies many of the long standing interpretative difficulties of the subject.

The Copenhagen frameworks for quantum mechanics, as they were formulated in the '30s and '40s and as they exist in most textbooks today, are inadequate for quantum cosmology. Characteristically these formulations assumed, as *external* to the framework of wave function and Schrödinger equation, the quasiclassical domain we see all about us. Bohr (1958) spoke of phenomena which could be alternatively described in classical language. In their classic text, Landau and Lifschitz (1958) formulated quantum mechanics in terms of a separate classical physics. Heisenberg and others stressed the central role of an external, essentially classical, observer.[*] Characteristically, these formulations assumed a possible division of the world into "observer" and "observed", assumed that "measurements" are the primary focus of scientific statements and, in effect, posited the existence of an external "quasiclassical domain". However, in a theory of the whole thing there can be no fundamental division into observer and observed. Measurements and observers cannot be fundamental notions in a theory that seeks to describe the early universe when neither existed. In a basic formulation of quantum mechanics there is no reason in general for there to be any variables that exhibit classical behavior in all circumstances. Copenhagen quantum mechanics thus needs to be generalized to provide a quantum framework for cosmology.

In a generalization of quantum mechanics which does not *posit* the existence of a quasiclassical domain, the domain of applicability of classical physics must be *explained*. For a quantum mechanical system to exhibit classical behavior there must be some restriction on its state and some coarseness in how it is described. This is clearly illustrated in the quantum mechanics of a single particle. Ehrenfest's theorem shows that generally

$$M\frac{d^2\langle x\rangle}{dt^2} = \left\langle -\frac{\partial V}{\partial x}\right\rangle. \tag{1}$$

However, only for special states, typically narrow wave packets, will this become an equation of motion for $\langle x\rangle$ of the form

$$M\frac{d^2\langle x\rangle}{dt^2} = -\frac{\partial V(\langle x\rangle)}{\partial x}. \tag{2}$$

For such special states, successive observations of position in time will exhibit the classical correlations predicted by the equation of motion (2) *provided* that these observations are coarse enough so that the properties of the state which allow (2) to replace the general relation (1) are not affected by these observations. An exact determination of position, for example, would yield a completely delocalized wave packet an instant later and (2) would no longer be a good approximation to (1). Thus, even for large systems, and in particular for the universe as a whole, we can expect classical behavior only for certain initial states and then only when a sufficiently coarse grained description is used.

---

[*] For a clear statement of this point of view, see London and Bauer (1939).

If classical behavior is *in general* a consequence only of a certain class of states in quantum mechanics, then, as a particular case, we can expect to have classical spacetime only for certain states in quantum gravity. The classical spacetime geometry we see all about us in the late universe is not property of every state in a theory where geometry fluctuates quantum mechanically. Rather, it is traceable fundamentally to restrictions on the initial condition. Such restrictions are likely to be generous in that, as in the single particle case, many different states will exhibit classical features. The existence of classical spacetime and the applicability of classical physics are thus not likely to be very restrictive conditions on constructing a theory of the initial condition.

It was Everett who, in 1957, first suggested how to generalize the Copenhagen frameworks so as to apply quantum mechanics to cosmology.* Everett's idea was to take quantum mechanics seriously and apply it to the universe as a whole. He showed how an observer could be considered part of this system and how its activities — measuring, recording, calculating probabilities, etc. — could be described within quantum mechanics. Yet the Everett analysis was not complete. It did not adequately describe within quantum mechanics the origin of the "quasiclassical domain" of familiar experience nor, in an observer independent way, the meaning of the "branching" that replaced the notion of measurement. It did not distinguish from among the vast number of choices of quantum mechanical observables that are in principle available to an observer, the particular choices that, in fact, describe the quasiclassical domain.

In this essay, I will describe joint work with Murray Gell-Mann (Gell-Mann and Hartle, 1990ab) which aims at a coherent formulation of quantum mechanics for the universe as a whole that is a framework to explain rather than posit the quasiclassical domain of everyday experience. It is an attempt at an extension, clarification, and completion of the Everett interpretation. It builds on many aspects of the, so called post-Everett development, especially the work of Zeh (1971), Zurek (1981, 1982), and Joos and Zeh (1985). At important points it coincides with the, independent, earlier work of Bob Griffiths (1984) and Roland Omnès (*e.g.*, as reviewed in Omnès, 1992).

Our work is not complete, but I hope to sketch how it might become so. It is by now a very long story but I will try to describe the important parts in simplified terms.

2 PROBABILITIES IN GENERAL AND PROBABILITIES IN QUANTUM MECHANICS

Even apart from quantum mechanics, there is no certainty in this world and therefore physics deals in probabilities. It deals most generally with the probabilities for alternative time histories of the universe. From these, conditional probabilities can be constructed that are appropriate when some features about our specific history are known and further ones are to be predicted.

To understand what probabilities mean for a single closed system, it is best to understand how they are used. We deal, first of all, with probabilities for *single*

---

* The original reference is Everett (1957). For a useful collection of reprints see DeWitt and Graham (1973).

events of the *single* system. When these probabilities become sufficiently close to zero or one there is a definite prediction on which we may act. How sufficiently close to 0 or 1 the probabilities must be depends on the circumstances in which they are applied. There is no certainty that the sun will come up tomorrow at the time printed in our daily newspapers. The sun may be destroyed by a neutron star now racing across the galaxy at near light speed. The earth's rotation rate could undergo a quantum fluctuation. An error could have been made in the computer that extrapolates the motion of the earth. The printer could have made a mistake in setting the type. Our eyes may deceive us in reading the time. Yet, we watch the sunrise at the appointed time because we compute, however imperfectly, that the probability of these things happening is sufficiently low.

In quantum cosmology we must search for alternatives that have probabilities near zero or one to have definite predictions with which to test theory. Various strategies can be employed to identify such situations. Acquiring information and considering the conditional probabilities based on it is one such strategy. Current theories of the initial condition of the universe predict almost no probabilities near zero or one without further conditions. The "no boundary" wave function of the universe, for example, does not predict the present position of the sun on the sky. However, it will predict that the conditional probability for the sun to be at the position predicted by classical celestial mechanics given a few previous positions is a number very near unity.

Another strategy to isolate probabilities near 0 or 1 is to consider ensembles of repeated observations of identical subsystems in the closed system. There are no genuinely infinite ensembles in the world so we are necessarily concerned with the probabilities for deviations of the behavior of a finite ensemble from the expected behavior of an infinite one. These are probabilities for a single feature (the deviation) of a single system (the whole ensemble).

The existence of large ensembles of repeated observations in identical circumstances and their ubiquity in laboratory science should not, therefore, obscure the fact that in the last analysis physics must predict probabilities for the single system that is the ensemble as a whole. Whether it is the probability of a successful marriage, the probability of the present galaxy-galaxy correlation function, or the probability of the fluctuations in an ensemble of repeated observations, we must deal with the probabilities of single events in single systems. In geology, astronomy, history, and cosmology, most predictions of interest have this character. The goal of physical theory is, therefore, most generally to predict the probabilities of histories of single events of a single system.

Probabilities need be assigned to histories by physical theory only up to the accuracy they are used. Two theories that predict probabilities for the sun not rising tomorrow at its classically calculated time that are both well beneath the standard on which we act are equivalent for all practical purposes as far as this prediction is concerned. It is often convenient, therefore, to deal with approximate probabilities which satisfy the rules of probability theory up to the standard they are used.

The characteristic feature of a quantum mechanical theory is that not every set of alternative histories that may be described can be assigned probabilities. Nowhere is this more clearly illustrated than in the two-slit experiment illustrated in Figure

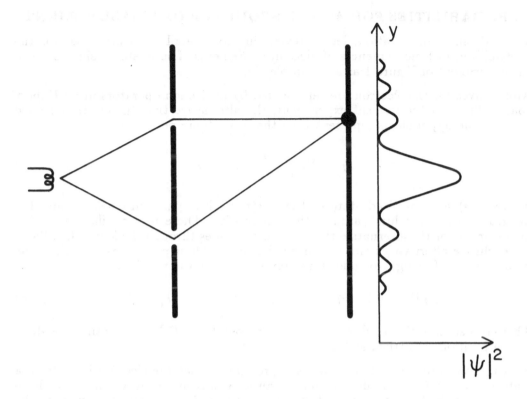

Fig. 1: *The two-slit experiment. An electron gun* 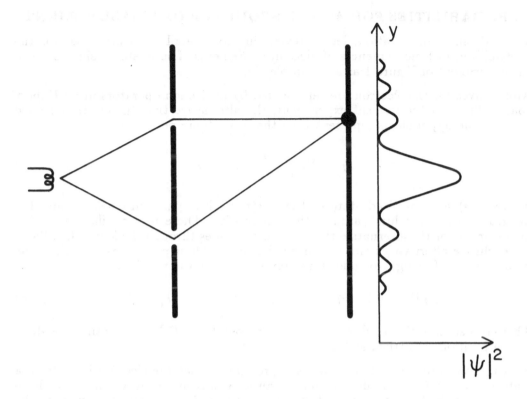 *emits an electron traveling towards a screen with two slits, its progress in space recapitulating its evolution in time. When precise detections are made of an ensemble of such electrons at the screen it is not possible, because of interference, to assign a probability to the alternatives of whether an individual electron went through the upper slit or the lower slit. However, if the electron interacts with apparatus that measures which slit it passed through, then these alternatives decohere and probabilities can be assigned.*

1. In the usual "Copenhagen" discussion if we have not measured which of the two slits the electron passed through on its way to being detected at the screen, then we are not permitted to assign probabilities to these alternative histories. It would be inconsistent to do so since the correct probability sum rule would not be satisfied. Because of interference, the probability to arrive at y is not the sum of the probabilities to arrive at y going through the upper or lower slit:

$$p(y) \neq p_U(y) + p_L(y) \tag{3}$$

because

$$|\psi_L(y) + \psi_U(y)|^2 \neq |\psi_L(y)|^2 + |\psi_U(y)|^2 \quad . \tag{4}$$

If we *have* measured which slit the electron went through, then the interference is destroyed, the sum rule obeyed, and we *can* meaningfully assign probabilities to these alternative histories.

A rule is thus needed in quantum theory to determine which sets of alternative histories may be assigned probabilities and which may not. In Copenhagen quantum mechanics, the rule is that probabilities are assigned to histories of alternatives of a subsystem that are *measured* and not in general otherwise.

3 PROBABILITIES FOR A TIME SEQUENCE OF MEASUREMENTS

To establish some notation, let us review in more detail the usual rules for the probabilities of time sequences of ideal measurements of subsystem using the two-slit experiment of Figure 1 as an example.

Alternatives for the electron are represented by projection operators in its Hilbert space. Thus, in the two slit experiment, the alternative that the electron passed through the upper slit is represented by the projection operator

$$P_U = \Sigma_s \int_U d^3x \, |\vec{x}, s\rangle\langle\vec{x}, s| \tag{5}$$

where $|\vec{x}, s\rangle$ is a localized state of the electron with spin component s, and the integral is over a volume around the upper slit. There is a similar projection operator P_L for the alternative that the electron goes through the lower slit. These are exclusive alternatives and they are exhaustive. These properties, as well as the requirements of being projections, are represented by the relations

$$P_L P_U = 0 \, , \quad P_U + P_L = 1, \quad P_L^2 = P_L, \quad P_U^2 = P_U \, . \tag{6}$$

There is a similarly defined set of projection operators $\{P_y\}$ representing the alternative positions of arrival at the screen.

We can now state the rule for the joint probability that the electron initially in a state $|\psi(t_0)\rangle$ at $t = t_0$ is determined by an ideal measurement at time t_1 to have passed through the upper slit and measured at time t_2 to arrive at point y on the screen. If one likes, one can imagine the case in which the electron is in a narrow wave packet in the horizontal direction with a velocity defined as sharply as possible consistent with the uncertainty principle. The joint probability is negligible unless t_1 and t_2 correspond to the times of flight to the slits and to the screen respectively.

The first step in calculating the joint probability is to evolve the state of the electron to the time t_1 of the first measurement

$$|\psi(t_1)\rangle = e^{-iH(t_1-t_0)/\hbar}|\psi(t_0)\rangle \, . \tag{7}$$

The probability that the outcome of the measurement at time t_1 is that the electron passed through the upper slit is:

$$(\text{Probability of } U) = \left\| P_U |\psi(t_1)\rangle \right\|^2 \tag{8}$$

where $\| \cdot \|$ denotes the norm of a vector in the electron's Hilbert space. If the outcome was the upper slit, and the measurement was an "ideal" one, that disturbed the electron as little as possible in making its determination, then after the measurement the state vector is reduced to

$$\frac{P_U |\psi(t_1)\rangle}{\|P_U |\psi(t_1)\rangle\|} \, . \tag{9}$$

This is evolved to the time of the next measurement

$$|\psi(t_2)\rangle = e^{-iH(t_2-t_1)/\hbar} \frac{P_U|\psi(t_1)\rangle}{\|P_U|\psi(t_1)\rangle\|} \ . \tag{10}$$

The probability of being detected at point y on the screen at time t_2 *given* that the electron passed through the upper slit is

$$(\text{Probability of } y \ given \ U) = \left\| P_y|\psi(t_2)\rangle \right\|^2 \ . \tag{11}$$

The *joint* probability that the electron is measured to have gone through the upper slit *and* is detected at y is the product of the conditional probability (11) with the probability (8) that the electron passed through U. The latter factor cancels the denominator in (10) so that combining all of the above equations in this section, we have

$$(\text{Probability of } y \ and \ U) = \left\| P_y e^{-iH(t_2-t_1)/\hbar} P_U e^{-iH(t_1-t_0)/\hbar} |\psi(t_0)\rangle \right\|^2 \ . \tag{12}$$

With Heisenberg picture projections this takes the even simpler form

$$(\text{Probability of } y \ and \ U) = \left\| P_y(t_2) P_U(t_1) \ |\psi(t_0)\rangle \right\|^2 \ . \tag{13}$$

where, for example,

$$P_U(t) = e^{iHt/\hbar} P_U e^{-iHt/\hbar} \ . \tag{14}$$

The formula (13) is a compact and unified expression of the two laws of evolution that characterize the quantum mechanics of measured subsystems — unitary evolution in between measurements and reduction of the wave packet at a measurement.[*] The important thing to remember about the expression (13) is that everything in it — projections, state vectors, Hamiltonian — refer to the Hilbert space of a subsystem, in this example the Hilbert space of the electron that is measured.

In "Copenhagen" quantum mechanics, it is measurement that determines which histories of a subsystem can be assigned probabilities and formulae like (13) that determine what these probabilities are. We cannot have such rules in the quantum mechanics of closed systems. There is no fundamental division of a closed system into measured subsystem and measuring apparatus. There is no fundamental reason for the closed system to contain classically behaving measuring apparatus in all circumstances. In particular, in the early universe none of these concepts seem relevant. We need a more observer-independent, measurement-independent, quasiclassical domain-independent rule for which histories of a closed system can be assigned probabilities and what these probabilities are. The next section describes this rule.

---

[*] As has been noted by many authors, *e.g.*, Groenewold (1952) and Wigner (1963) among the earliest.

4 POST-EVERETT QUANTUM MECHANICS

To describe the rules of post-Everett quantum mechanics, I shall make a simplifying assumption. I shall neglect gross quantum fluctuations in the geometry of space-time, and assume a fixed background spacetime geometry which supplies a definite meaning to the notion of time. This is an excellent approximation on accessible scales for times later than 10^{-43} sec after the big bang. The familiar apparatus of Hilbert space, states, Hamiltonian, and other operators may then be applied to process of prediction. Indeed, in this context the quantum mechanics of cosmology is in no way distinguished from the quantum mechanics of a large isolated box, perhaps expanding, but containing both the observed and its observers (if any).

A set of alternative histories for a closed system is specified by giving exhaustive sets of exclusive alternatives at a sequence of times. Consider a model closed system initially in a pure state that can be described as an observer and two-slit experiment, with appropriate apparatus for producing the electrons, detecting which slit they passed through, and measuring their position of arrival on the screen (Figure 2). Some alternatives for the whole system are:

1. Whether or not the observer decided to measure which slit the electron went through.

2. Whether the electron went through the upper or lower slit.

3. The alternative positions, $y_1, \cdots, y_N$, that the electron could have arrived at the screen.

This set of alternatives at a sequence of times defines a set of histories whose characteristic branching structure is shown in Figure 3. An individual history in the set is specified by some particular sequence of alternatives, *e.g.*, measured, upper, y_9.

Many other sets of alternative histories are possible for the closed system. For example, we could have included alternatives describing the readouts of the apparatus that detects the position that the electron arrived on the screen. If the initial condition corresponded to a good experiment there should be a high correlation between these alternatives and the position that the electron arrives at the screen. In a more refined model we could discuss alternatives corresponding to thoughts in the observer's brain, or to the individual positions of the atoms in the apparatus, or to the possibilities that these atoms reassemble in some completely different configuration. There are a vast number of possibilities.

Characteristically the alternatives that are of use to us as observers are very coarse grained, distinguishing only very few of the degrees of freedom of a large closed system. This is especially true if we recall that our box with observer and two-slit experiment is only an idealized model. The most general closed system is the universe itself, and, as I hope to show, the only realistic closed systems are of cosmological dimensions. Certainly, we utilize only very, very coarse-grained descriptions of the universe as a whole.

I would now like to state the rules that determine which coarse-grained sets of histories may be assigned probabilities and what those probabilities are. The essence of the rules I shall describe can be found in the work of Bob Griffiths (Griffiths, 1984).

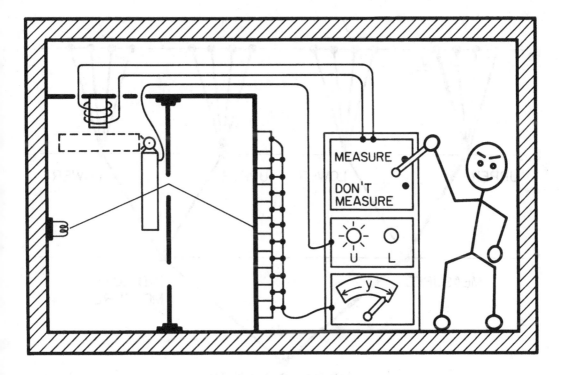

Fig. 2: *A model closed quantum system containing an observer together with the necessary apparatus for carrying out a two-slit experiment. Alternatives for the system include whether the observer measured which slit the electron passed through or did not, whether the electron passed through the upper or lower slit, the alternative positions of arrival of the electron at the screen, the alternative arrival positions registered by the apparatus, the registration of these in the brain of the observer, etc., etc., etc. Each exhaustive set of exclusive alternatives is represented by an exhaustive set of orthogonal projection operators on the Hilbert space of the closed system. Time sequences of such sets of alternatives describe sets of alternative coarse-grained histories of the closed system. Quantum theory assigns probabilities to the individual alternative histories in such a set when there is negligible quantum mechanical interference between them, that is, when the set of histories decoheres.*

A more refined model might consider a quantity of matter in a closed box. One could then consider alternatives such as whether the box contains a two-slit experiment or does not as well as alternative positions of atoms.

The general framework was extended by Roland Omnès (as reviewed in Omnès, 1992) and was independently, but later, arrived at by Murray Gell-Mann and myself (Gell-Mann and Hartle, 1990a). The idea is simple: The failure of probability sum rules due to quantum interference is the obstacle to assigning probabilities.

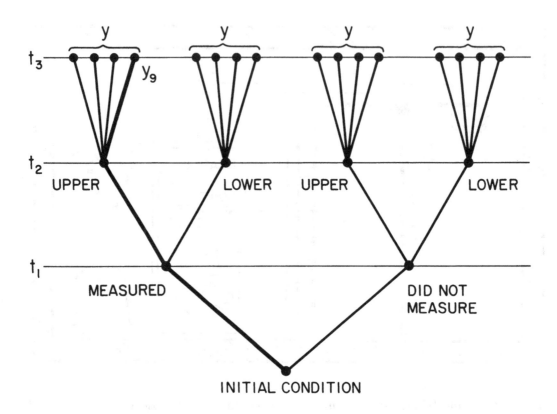

Fig. 3: *Branching structure of a set of alternative histories. This figure illustrates the set of alternative histories defined by the alternatives of whether the observer decided to measure or did not decide to measure which slit the electron went through at time t_1, whether the electron went through the upper slit or through the lower slit at time t_2, and the alternative positions of arrival at the screen at time t_3. A single branch corresponding to the alternatives that the measurement was carried out, the electron went through the upper slit, and arrived at point y_9 on the screen is illustrated by the heavy line.*

The illustrated set of histories does not decohere because there is significant quantum mechanical interference between the branch where no measurement was carried out and the electron went through the upper slit and the similar branch where it went through the lower slit. A related set of histories that does decohere can be obtained by replacing the alternatives at time t_2 by the following set of three alternatives: (a record of the decision shows a measurement was initiated and the electron went through the upper slit); (a record of the decision shows a measurement was initiated and the electron went through the lower slit); (a record of the decision shows that the measurement was not initiated). The vanishing of the interference between the alternative values of the record and the alternative configurations of apparatus ensures the decoherence of this set of alternative histories.

Probabilities can be assigned to just those sets of alternative histories of a closed system for which there is negligible interference between the individual histories in the set as a consequence of the *particular* initial state the closed system has, and for which, therefore, all probability sum rules *are* satisfied. Let us now give this idea a precise expression.

Sets of alternatives at one moment of time are represented by sets of orthogonal projection operators. Employing the Heisenberg picture these can be denoted $\{P_{\alpha_k}^k(t_k)\}$. The superscript k denotes the set of alternatives being considered at time t_k (for example, the set of alternative position intervals $\{y_1, \cdots, y_N\}$ at which the electron might arrive at the screen at time t_3), α_k denotes the particular alternative in the set (for example y_9) and t_k is the time. The set of P's satisfy

$$\sum_{\alpha_k} P_{\alpha_k}^k(t_k) = 1 \ , \quad P_{\alpha_k}^k(t_k) P_{\alpha_k'}^k(t_k) = \delta_{\alpha_k \alpha_k'} P_{\alpha_k}^k(t_k) \tag{15}$$

showing that they represent an exhaustive set of exclusive alternatives.

Sets of alternative histories are defined by giving sequences of sets of alternatives at definite moments of time, e.g., $\{P_{\alpha_1}^1(t_1)\}$, $\{P_{\alpha_2}^2(t_2)\}, \cdots, \{P_{\alpha_n}^n(t_n)\}$. Different choices for $\{P_{\alpha_1}^1(t_1)\}$, $\{P_{\alpha_2}^2(t_2)\}$, etc. describe different sets of alternative histories of the closed system. An individual history in a given set corresponds to a particular sequence $(\alpha_1, \cdots, \alpha_n) \equiv \alpha$ and, for each history, there is a corresponding chain of projection operators

$$C_\alpha \equiv P_{\alpha_n}^n(t_n) \cdots P_{\alpha_1}^1(t_1) \ . \tag{16}$$

For example, in the two-slit experiment in a box illustrated in Figure 2, the history in which the observer decided at time t_1 to measure which slit the electron goes through, in which the electron goes through the upper slit at time t_2, and arrives at the screen in position interval y_9 at time t_3, would be represented by the chain

$$P_{y_9}^3(t_3) P_U^2(t_2) P_{\text{meas}}^1(t_1) \tag{17}$$

in an obvious notation. The only difference between this situation and that of the "Copenhagen" quantum mechanics of measured subsystems is the following: The sets of operators $\{P_{\alpha_k}^k(t_k)\}$ defining alternatives for the closed system act on the Hilbert space of the closed system that includes the variables describing any apparatus, observers, and anything else. The operators defining alternatives in Copenhagen quantum mechanics act only on the Hilbert space of the measured subsystem.

When the initial state is pure, it can be resolved into *branches* corresponding to the individual members of any set of alternative histories. The generalization to an impure initial density matrix is not difficult (Gell-Mann and Hartle, 1990a), but for simplicity we shall assume a pure initial state throughout this article. Denote the initial state by $|\Psi\rangle$ in the Heisenberg picture. Then

$$|\Psi\rangle = \sum_\alpha C_\alpha |\Psi\rangle = \sum_{\alpha_1, \cdots, \alpha_n} P_{\alpha_n}^n(t_n) \cdots P_{\alpha_1}^1(t_1) |\Psi\rangle \ . \tag{18}$$

This identity follows by applying the first of (15) to all the sums over α_k in turn. The vector

$$C_\alpha |\Psi\rangle \tag{19}$$

is the branch corresponding to the individual history α and (18) is the resolution of the initial state into branches.

When the branches corresponding to a set of alternative histories are sufficiently orthogonal the set of histories is said to *decohere*. More precisely a set of histories decoheres when

$$\langle\Psi|C_{\alpha'}^\dagger C_\alpha|\Psi\rangle \approx 0 , \quad \text{for any} \quad \alpha'_k \neq \alpha_k . \tag{20}$$

We shall return to the standard with which decoherence should be enforced, but first let us examine its meaning and consequences.

Decoherence means the absence of quantum mechanical interference between the individual histories of a coarse-grained set.* Probabilities can be assigned to the individual histories in a decoherent set of alternative histories because decoherence implies the probability sum rules necessary for a consistent assignment. The probability of an individual history α is

$$p(\alpha) = \|C_\alpha|\Psi\rangle\|^2 . \tag{21}$$

To see how decoherence implies the probability sum rules, let us consider an example in which there are just three sets of alternatives at times t_1, t_2, and t_3. A typical

---

* The term "decoherence" is used in several different ways in the literature. For those familiar with other work, therefore, a comment is in order to specify how we are employing the term in this simplified presentation. We have followed our previous work (Gell-Mann and Hartle, 1990a and 1990b) in using the term "decoherence" to refer to a property of a set of alternative time *histories* of a closed system. A decoherent set of histories is one for which the quantum mechanical interference between individual histories is small enough to guarantee an appropriate set of probability sum rules. Different notions of decoherence can be defined by utilizing different measures of interference. The weakest notion is just the consistency of the probability sum rules that is called "consistency" by Griffiths (1984) and Omnès (1992) and that term is used by some to refer to all measures of interference. Vanishing of the real part of (20) is a sufficient condition for the consistency of the probability sum rules called the "weak decoherence condition". We are using the stronger condition (20) because it characterizes widespread and typical mechanisms of decoherence. Eq (20) has been called the "medium decoherence condition". "Decoherence" in the context of this paper thus means the medium decoherence of sets of histories.

In the literature the term "decoherence" has also been used to refer to the decay in time of the off-diagonal elements of a reduced density matrix defined by tracing the full density matrix over a given set of variables (Zurek, 1991). The two notions of "decoherence of reduced density matrices" and "decoherence of histories" are not generally equivalent but also not unconnected in the sense that in particular models certain physical processes can ensure both. [See, e.g. the remarks in Section II.6.4 of (Hartle, 1991a)].

sum rule might be

$$\sum_{\alpha_2} p(\alpha_3, \alpha_2, \alpha_1) = p(\alpha_3, \alpha_1) \ . \tag{22}$$

We show (20) and (21) imply (22). To do that write out the left hand side of (22) using (21) and suppress the time labels for compactness.

$$\sum_{\alpha_2} p(\alpha_3, \alpha_2, \alpha_1) = \sum_{\alpha_2} \left\langle \Psi | P^1_{\alpha_1} P^2_{\alpha_2} P^3_{\alpha_3} P^3_{\alpha_3} P^2_{\alpha_2} P^1_{\alpha_1} | \Psi \right\rangle \ . \tag{23}$$

Decoherence means that the sum on the right hand side of (23) can be written with negligible error as

$$\sum_{\alpha_2} p(\alpha_3, \alpha_2, \alpha_1) \approx \sum_{\alpha'_2 \alpha_2} \left\langle \Psi | P^1_{\alpha_1} P^2_{\alpha'_2} P^3_{\alpha_3} P^3_{\alpha_3} P^2_{\alpha_2} P^1_{\alpha_1} | \Psi \right\rangle \ . \tag{24}$$

the extra terms in the sum being vanishingly small. But now, applying the first of (15) we see

$$\sum_{\alpha_2} p(\alpha_3, \alpha_2, \alpha_1) \approx \left\langle \Psi | P^1_{\alpha_1} P^3_{\alpha_3} P^3_{\alpha_3} P^1_{\alpha_1} | \Psi \right\rangle = p(\alpha_3, \alpha_1) \tag{25}$$

so that the sum rule (22) is satisfied.

Given an initial state $|\Psi\rangle$ and a Hamiltonian H, one could, in principle, identify all possible sets of decohering histories. Among these will be the exactly decohering sets where the orthogonality of the branches is exact. Indeed, trivial examples can be supplied by resolving $|\Psi\rangle$ into a sum of orthogonal vectors at time t_1, resolving those vectors into sums of further vectors such that the whole set is orthogonal at time t_2, and so on. However, such sets of exactly decohering histories will not, in general, have a simple description in terms of fundamental fields nor any connection, for example, with the quasiclassical domain of familiar experience. For this reason sets of histories that approximately decohere are of interest. As we will argue in the next two Sections, realistic mechanisms lead to the decoherence of histories constituting a quasiclassical domain to an excellent approximation. When the decoherence condition (20) is approximately enforced, the probability sum rules such as (22) will only be approximately obeyed. However, as discussed earlier, these probabilities for single systems are meaningful up to the standard they are used. Approximate probabilities for which the sum rules are satisfied to a comparable standard may therefore also be employed in the process of prediction. When we speak of approximate decoherence and approximate probabilities we mean decoherence achieved and probability sum rules satisfied beyond any standard that might be conceivably contemplated for the accuracy of prediction and the comparison of theory with experiment.

5 THE ORIGINS OF DECOHERENCE IN OUR UNIVERSE

What are the features of coarse-grained sets of histories that decohere in our universe? In seeking to answer this question it is important to keep in mind the basic aspects of the theoretical framework on which decoherence depends. Decoherence of a set of alternative histories is not a property of their operators *alone*. It depends

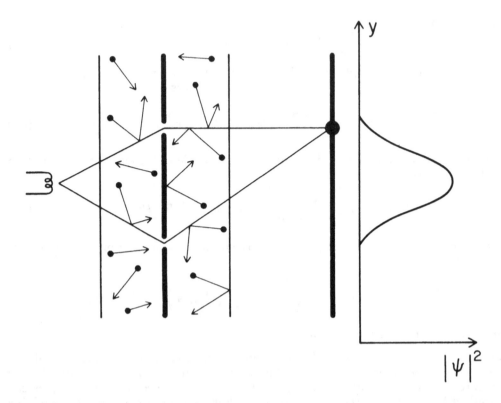

Fig. 4: *The two slit experiment with an interacting gas. Near the slits light particles of a gas collide with the electrons. Even if the collisions do not affect the trajectories of the electrons very much they can still carry away the phase correlations between the histories in which the electron arrived at point y on the screen by passing through the upper slit and that in which it arrived at the same point by passing through the lower slit. A coarse graining that described only of these two alternative histories of the electron would approximately decohere as a consequence of the interactions with the gas given adequate density, cross-section, etc. Interference is destroyed and probabilities can be assigned to these alternative histories of the electron in a way that they could not be if the gas were not present (cf. Fig. 1). The lost phase information is still available in correlations between states of the gas and states of the electron. The alternative histories of the electron would not decohere in a coarse graining that included both the histories of the electron and operators that were sensitive to the correlations between the electrons and the gas.*

This model illustrates a widely occuring mechanism by which certain types of coarse-grained sets of alternative histories decohere in the universe.

on the relations of those operators to the initial state $|\Psi\rangle$, the Hamiltonian H, and the fundamental fields. Given these, we could, in principle, *compute* which sets of alternative histories decohere.

We are not likely to carry out a computation of all decohering sets of alternative histories for the universe, described in terms of the fundamental fields, anytime in the near future, if ever. It is therefore important to investigate specific mechanisms by which decoherence occurs. Let us begin with a very simple model due, in its essential features to Joos and Zeh (1985). We consider the two-slit example again, but this time suppose that in the neighborhood of the slits there is a gas of photons or other light particles colliding with the electrons (Figure 4). Physically it is easy to see what happens; the random uncorrelated collisions can carry away delicate phase correlations between the beams even if the trajectories of the electrons are not affected much. The interference pattern will then be destroyed and it will be possible to assign probabilities to whether the electron went through the upper slit or the lower slit.

Let us see how this picture in words is given precise meaning in mathematics. Initially, suppose the state of the entire system is a state of the electron $|\psi>$ and N distinguishable "photons" in states $|\varphi_1\rangle$, $|\varphi_2\rangle$, etc., viz.

$$|\Psi\rangle = |\psi\rangle|\varphi_1\rangle|\varphi_2 > \cdots |\varphi_N\rangle . \tag{26}$$

Suppose further that $|\psi\rangle$ is a coherent superposition of a state in which the electron passes through the upper slit $|U\rangle$ and the lower slit $|L\rangle$. Explicitly:

$$|\psi\rangle = \alpha|U\rangle + \beta|L\rangle . \tag{27}$$

Both states are wave packets in x, so that position in x recapitulates history in time. We now ask whether the history where the electron passes through the upper slit and arrives at a detector at point y on the screen, decoheres from that in which it passes through the lower slit and arrives at point y as a consequence of the initial condition of this "universe". That is, as in Section 4, we ask whether the two branches

$$P_y(t_2)P_U(t_1)|\Psi\rangle \quad , \quad P_y(t_2)P_L(t_1)|\Psi\rangle \tag{28}$$

are nearly orthogonal, the times of the projections being those for the nearly classical motion in x. We work this out in the Schrödinger picture where the initial state evolves, and the projections on the electron's position are applied to it at the appropriate times.

Collisions occur, but the states $|U\rangle$ and $|L\rangle$ are left more or less undisturbed. The states of the "photons", of course, are significantly affected. If the photons are dilute enough to be scattered once by the electron in its time to traverse the gas the two branches (28) will be approximately

$$\alpha P_y|U\rangle S_U|\varphi_1\rangle S_U|\varphi_2\rangle \cdots S_U|\varphi_N\rangle , \tag{29a}$$

and

$$\beta \, P_y|L\rangle S_L|\varphi_1\rangle S_L|\varphi_2\rangle \cdots S_L|\varphi_N\rangle . \tag{29b}$$

Here, S_U and S_L are the scattering matrices from an electron in the vicinity of the upper slit and the lower slit respectively. The two branches in (29) decohere because the states of the "photons" are nearly orthogonal. The overlap of the branches is proportional to

$$\langle\varphi_1|S_U^\dagger S_L|\varphi_1\rangle\langle\varphi_2|S_U^\dagger S_L|\varphi_2\rangle \cdots \langle\varphi_N|S_U^\dagger S_L \,|\varphi_N\rangle . \tag{30}$$

Now, the S-matrices for scattering off the upper position or the lower position can be connected to that of an electron at the orgin by a translation

$$S_U = \exp(-i\mathbf{k} \cdot \mathbf{x}_U)S \, \exp(+i\mathbf{k} \cdot \mathbf{x}_U) \, , \tag{31a}$$

$$S_L = \exp(-i\mathbf{k} \cdot \mathbf{x}_L)S \, \exp(+i\mathbf{k} \cdot \mathbf{x}_L) \, . \tag{31b}$$

Here, $\hbar\mathbf{k}$ is the momentum of a photon, $\mathbf{x}_U$ and $\mathbf{x}_L$ are the positions of the slits and S is the scattering matrix from an electron at the origin.

$$\langle \mathbf{k}'|S|\mathbf{k}\rangle = \delta^{(3)}(\mathbf{k} - \mathbf{k}') + \frac{i}{2\pi\omega_{\mathbf{k}}}f(\mathbf{k},\mathbf{k}')\delta(\omega_k - \omega'_k) \, , \tag{32}$$

where f is the scattering amplitude and $\omega_k = |\vec{k}|$.

Consider the case where initially all the photons are in plane wave states in an interaction volume V, all having the same energy $\hbar\omega$, but with random orientations for their momenta. Suppose further that the energy is low so that the electron is not much disturbed by a scattering and low enough so the wavelength is much longer than the separation between the slits, $k|\mathbf{x}_U - \mathbf{x}_L| \ll 1$. It is then possible to work out the overlap. The answer according to Joos and Zeh (1985) is

$$\left(1 - \frac{(k|\mathbf{x}_U - \mathbf{x}_L|)^2}{8\pi^2 V^{2/3}}\sigma\right)^N \tag{33}$$

where σ is the effective scattering cross section and the individual terms have been averaged over incoming directions. Even if σ is small, as N becomes large this tends to zero. In this way decoherence becomes a quantitative phenomenon.

What such models convincingly show is that decoherence is frequent and widespread in the universe for histories of certain kinds of variables. Joos and Zeh calculate that a superposition of two positions of a grain of dust, 1mm apart, is decohered simply by the scattering of the cosmic background radiation on the timescale of a nanosecond. The existence of such mechanisms means that the only realistic isolated systems are of cosmological dimensions. So widespread is this kind of phenomena with the initial condition and dynamics of our universe, that we may meaningfully speak of habitually decohering variables such as the center of mass positions of massive bodies.

6 THE COPENHAGEN APPROXIMATION

What is the relation of the familiar Copenhagen quantum mechanics described in Section 3 to the more general "post-Everett" quantum mechanics of closed systems described in Sections 4 and 5? Copenhagen quantum mechanics predicts the probabilities of the histories of measured subsystems. Measurement situations may be described in a closed system that contains both measured subsystem and measuring apparatus. In a typical measurement situation the values of a variable not normally decohering become correlated with alternatives of the apparatus that decohere because of *its* interactions with the rest of the closed system. The correlation means that the measured alternatives decohere because the alternatives of the apparatus with which they are correlated decohere.

The recovery of the Copenhagen rule for when probabilities may be assigned is immediate. Measured quantities are correlated with decohering histories. Decohering histories can be assigned probabilities. Thus in the two-slit experiment (Figure 1), when the electron interacts with an apparatus that determines which slit it passed through, it is the decoherence of the alternative configurations of the apparatus that enables probabilities to be assigned for the electron.

There is nothing incorrect about Copenhagen quantum mechanics. Neither is it, in any sense, opposed to the post-Everett formulation of the quantum mechanics of closed systems. It is an *approximation* to the more general framework appropriate in the special cases of measurement situations and when the decoherence of alternative configurations of the apparatus may be idealized as exact and instantaneous. However, while measurement situations imply decoherence, they are only special cases of decohering histories. Probabilities may be assigned to alternative positions of the moon and to alternative values of density fluctuations near the big bang in a universe in which these alternatives decohere, whether or not they were participants in a measurement situation and certainly whether or not there was an observer registering their values.

7 QUASICLASSICAL DOMAINS

As observers of the universe, we deal with coarse-grained histories that reflect our own limited sensory perceptions, extended by instruments, communication and records but in the end characterized by a large amount of ignorance. Yet, we have the impression that the universe exhibits a much finer-grained set of histories, independent of us, defining an always decohering "quasiclassical domain", to which our senses are adapted, but deal with only a small part of it. If we are preparing for a journey into a yet unseen part of the universe, we do not believe that we need to equip ourselves with spacesuits having detectors sensitive, say, to coherent superpositions of position or other unfamiliar quantum variables. We expect that the familiar quasiclassical variables will decohere and be approximately correlated in time by classical deterministic laws in any new part of the universe we may visit just as they are here and now.

Since the post-Everett quantum mechanics of closed systems does not posit a quasiclassical domain, it must provide an explanation of this manifest fact of everyday experience. No such explanation can be provided from the dynamics of quantum theory alone. Rather, like decoherence, the existence of a quasiclassical domain in the universe must be a consequence of both initial condition of the universe and the Hamiltonian describing evolution.

Roughly speaking, a quasiclassical domain should be a set of alternative histories that decoheres according to a realistic principle of decoherence, that is maximally refined consistent with that notion of decoherence, and whose individual histories are described largely by alternative values of a limited set of quasiclassical variables at different moments of time that exhibit as much as possible patterns of classical correlation in time. To make the question of the existence of one or more quasiclassical domains into a *calculable* question in quantum cosmology we need measures of how close a set of histories comes to constituting a "quasiclassical domain". A quasiclassical domain cannot be a *completely* fine-grained description for then it would not decohere. It cannot consist *entirely* of a few "quasiclassical variables" repeated over and over because sometimes we may measure something highly quantum mechanical. Quasiclassical variables cannot be *always* correlated in time by classical laws because sometimes quantum mechanical phenomena cause deviations from classical physics. We need measures for maximality and classicality (Gell-Mann and Hartle, 1990a).

It is possible to give crude arguments for the type of habitually decohering operators we expect to occur over and over again in a set of histories defining a quasiclassical domain (Gell-Mann and Hartle, 1990a). Such habitually decohering operators are called "quasiclassical operators". In the earliest instants of the universe the operators defining spacetime on scales well above the Planck scale emerge from the quantum fog as quasiclassical. Any theory of the initial condition that does not imply this is simply inconsistent with observation in a manifest way. A background spacetime is thus defined and conservation laws arising from its symmetries have meaning. Then, where there are suitable conditions of low temperature, density, etc., various sorts of hydrodynamic variables may emerge as quasiclassical operators. These are integrals over suitably small volumes of densities of conserved or nearly conserved quantities. Examples are densities of energy, momentum, baryon number, and, in later epochs, nuclei, and even chemical species. The sizes of the volumes are limited above by maximality and are limited below by classicality because they require sufficient "inertia" resulting from their approximate conservation to enable them to resist deviations from predictability caused by their interactions with one another, by quantum spreading, and by the quantum and statistical fluctuations resulting from interactions with the rest of the universe that accomplish decoherence (Gell-Mann and Hartle, 1993). Suitable integrals of densities of approximately conserved quantities are thus candidates for habitually decohering quasiclassical operators. These "hydrodynamic variables" *are* among the principal variables of classical physics.

It would be in such ways that the classical domain of familiar experience could be an emergent property of the fundamental description of the universe, not generally in quantum mechanics, but as a consequence of our specific initial condition and the Hamiltonian describing evolution. Whether the universe exhibits a quasiclassical domain, and, indeed, whether it exhibits more than one essentially inequivalent domain, thus become calculable questions in the quantum mechanics of closed systems.

8 CONCLUSION

Quantum mechanics is best and most fundamentally understood in the context of quantum mechanics of closed systems, most generally the universe as a whole. The founders of quantum mechanics were right in pointing out that something external to the framework of wave function and the Schrödinger equation *is* needed to interpret the theory. But it is not a postulated classical domain to which quantum mechanics does not apply. Rather it is the initial condition of the universe that, together with the action function of the elementary particles and the throws of the quantum dice since the beginning, is the likely origin of quasiclassical domain(s) within quantum theory itself.

ACKNOWLEDGMENTS

The formulation of quantum mechanics described in this paper is a result of joint work with Murray Gell-Mann. It is a pleasure to thank him for the many conversations over the years and for permission to summarize aspects of our work here. Preparation of the report was supported in part by the National Science Foundation under grant PHY90-08502.

9 REFERENCES

Bohr, N. (1958) *Atomic Physics and Human Knowledge*, Science Editions, New York.

Coleman, S. *Nucl. Phys.* **B310**, 643, (1988).

DeWitt, B. and Graham, R.N. (1973) eds. *The Many Worlds Interpretation of Quantum Mechanics*, Princeton University Press, Princeton.

Everett, H. *Rev. Mod. Phys.* **29**, 454, (1957).

Gell-Mann, M. and Hartle, J.B. (1990a) in *Complexity, Entropy, and the Physics of Information, SFI Studies in the Sciences of Complexity*, Vol. VIII, ed. by W. Zurek, Addison Wesley, Reading or in *Proceedings of the 3rd International Symposium on the Foundations of Quantum Mechanics in the Light of New Technology* ed. by S. Kobayashi, H. Ezawa, Y. Murayama, and S. Nomura, Physical Society of Japan, Tokyo.

Gell-Mann, M. and Hartle, J.B. (1990b) in the *Proceedings of the 25th International Conference on High Energy Physics, Singapore, August, 2-8, 1990*, ed. by K.K. Phua and Y. Yamaguchi (South East Asia Theoretical Physics Association and Physical Society of Japan) distributed by World Scientific, Singapore.

Gell-Mann, M. and Hartle, J.B. (1993) Classical Equations for Quantum Systems. Preprint UCSBTH-91-15, to be published in *Physical Review*, **D15**.

Giddings, S. and Strominger, A. *Nucl. Phys.* **B307**, 854, (1988).

Griffiths, R. *J. Stat. Phys.* **36**, 219, (1984).

Groenewold, H.J. (1952) *Proc. Akad. van Wetenschappen*, Amsterdam, Ser. B, **55**, 219.

Halliwell, J. (1991) in *Quantum Cosmology and Baby Universes: Proceedings of the 1989 Jerusalem Winter School for Theoretical Physics*, eds. S. Coleman, J.B. Hartle, T. Piran, and S. Weinberg, World Scientific, Singapore, pp. 65-157.

Hartle, J.B. (1991) *The Quantum Mechanics of Cosmology*, in *Quantum Cosmology and Baby Universes: Proceedings of the 1989 Jerusalem Winter School for Theoretical Physics*, eds. S. Coleman, J.B. Hartle, T. Piran, and S. Weinberg, World Scientific, Singapore, pp. 65-157.

Hawking, S.W. *Phys. Lett.* **B196**, 337, (1983).

Joos, E. and Zeh, H.D. *Zeit. Phys.* **B59**, 223, (1985).

Landau, L. and Lifshitz, E. (1958) *Quantum Mechanics*, Pergamon, London.

London, F. and Bauer, E. (1939) *La théorie de l'observation en mécanique quantique*, Hermann, Paris.

Misner, C.W. *Phys. Rev.* **186**, 1328, (1969).

———— (1972) in *Magic Without Magic: John Archibald Wheeler*, ed. by J.R. Klauder, W.H. Freeman, San Francisco.

Omnès, R. *Rev. Mod. Phys.* **64**, 339, (1992).

Wigner, E. *Am. J. Phys.* **31**, 6, (1963).

Zeh, H.D. *Found. Phys.* **1**, 69, (1971).

Zurek, W. *Phys. Rev.* D **24**, 1516, (1981).

———— *Phys. Rev.* D **26**, 1862, (1982).

———— *Physics Today* **44**, 36, (1991).

Cosmological Vacuum Phase Transitions

WILLIAM A. HISCOCK and DAVID A. SAMUEL

Department of Physics
Montana State University
Bozeman, MT 59717

1. Introduction

Developments in physics on the smallest and largest scales (elementary particle theory and cosmology) over the past several decades have provided strong evidence for the hot Big Bang theory of the early Universe and for gauge field models of elementary particle interactions, particularly as embodied in the Standard Model of elementary particle physics.

In adopting these two theories it is very difficult to avoid the phenomenon of cosmological vacuum phase transitions. Such phase transitions arise because of the temperature dependence of the Higgs[1] field potential. This potential plays several roles within gauge field theories, from providing a mechanism for spontaneous symmetry breaking to providing a possible vacuum energy density. The Higgs field potential (which we shall henceforth label as $U(\phi)$, where ϕ represents the field) at high temperature possesses a single vacuum state, which we may arbitrarily choose to be located at $\phi = 0$. The Higgs potential undergoes a continuous evolution as the temperature is changed and at low temperatures the potential may possess multiple vacuum states (*i.e.*, multiple minima). The nature of the transition from the high temperature vacuum state to the low temperature vacuum state will depend upon the shape of the potential during the course of its evolution. It is possible that the Higgs field will undergo a classical "rolling" evolution between the initial and final vacuum states. However, it is also possible that the evolution and shape of the potential will be such that a potential barrier will prevent the classical evolution of the field to the true vacuum state (*i.e.*, the vacuum state with the lowest energy), and thus the field will find itself trapped in a false vacuum state. Quantum tunneling will offer an eventual escape from such a false vacuum state.

In the standard hot Big Bang cosmology, matter in the early universe was hot and dense. Initially, then, the non-zero temperature quantum field theory corrections to the effective Higgs potential would place the early universe in the high-energy, symmetric vacuum state. The subsequent expansion and cooling of the universe may then have led to a number of vacuum phase transitions [see Figure (1)]: within the Standard Model of particle physics, there are at least two: the electroweak phase transition, in which the intermediate vector bosons became massive, at a temperature of around 300 GeV, and the quark-hadron transition, at around 100-300 MeV, below which temperature the quarks are confined. Experiments now taking place in relativistic heavy ion accelerators are seeking to re-create the quark-hadron phase transition in the laboratory today. In addition to these two transitions which are part of the current, accepted, experimentally supported Standard Model, almost every theoretical effort at extending particle physics beyond the Standard Model contains further vacuum phase transitions: Grand Unification, supersymmetry, superstrings, etc.

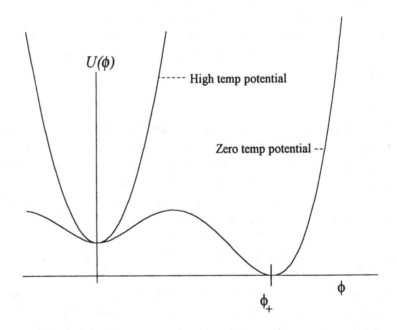

Figure (1): A typical Higgs potential illustrated for high temperature and zero temperature. In the early universe, at high temperature, the universe would necessarily have been in the $\phi = 0$ vacuum state. As the universe expands and cools, that state becomes a false vacuum, while the $\phi = \phi_+$ state is the true vacuum state, by virtue of its lower energy. If there is a potential barrier between the vacua (as in the figure), the vacuum phase transition will be first-order, proceeding via nucleation and expansion of bubbles of the true vacuum within the false.

Such vacuum phase transitions in the early universe may play a crucial role in determining many features of our universe: the original (and many derivative) forms of the inflationary universe utilize a supercooled false vacuum state to cause an exponential expansion of the universe, solving the horizon and flatness problems. Topological defects such as cosmic strings and textures which may be created in cosmological vacuum phase transitions are often invoked as providing the key physics which explains the formation of the observed structure in the cosmos.

In the later universe (*i.e.*, today), we may also address the disquieting possibility that we are presently living within a false vacuum state. This possibility is, amazingly, less speculative than ideas such as Grand Unification and inflation; the Weinberg-Salam-Glashow model of the electroweak interaction admits the possibility that our present, cherished, vacuum state is merely a local minimum of the Higgs potential, doomed to eventual decay.

In this review we will concentrate on the fundamental physics of vacuum phase transitions in cosmological settings, rather than focus on the already well-reviewed applications such as inflation and topological defect physics[2-5].

2. First Order Vacuum Phase Transitions and the "Thin-Wall" Approximation

The lifetime of a false vacuum state can be determined by calculating the nucleation rate of bubbles of true vacuum within the medium of false vacuum. The pioneering semiclassical analysis by Coleman[6] has shown that the number of nucleating bubbles per unit four volume is given by $\Gamma = A\exp(-B)$, where B is the difference between the Euclidean actions for the spacetime with and without the nucleating bubble. (We shall sometimes loosely refer to B as the Euclidean action of the bubble, though by this we strictly mean the difference between the Euclidean actions for the spacetime with and without the nucleating bubble). The coefficient A is a functional determinant whose order of magnitude is characterized by the field mass to the fourth power.

The Euclidean action of a nucleating bubble is evaluated from a scalar field profile which is the solution to the Euclideanized scalar field equation with appropriate boundary conditions. Those solutions which have the smallest Euclidean actions will be the dominant modes of vacuum decay. In the absence of gravity, Coleman, Glazer and Martin have proven that $O(4)$ symmetric bubbles have the least action[7]. The Euclideanized scalar field equation, in the absence of gravity, assuming $O(4)$ symmetry, takes the form:

$$\frac{d^2\phi}{d\rho^2} + \frac{3}{\rho}\frac{d\phi}{d\rho} = \frac{dU}{d\phi} \tag{1}$$

where ρ is the $O(4)$ radial coordinate (*i.e.*, $\rho^2 = r^2 + \tau^2$). The required boundary conditions are given by $\phi \to 0$ (the false vacuum value) as $\rho \to \infty$, and $d\phi/d\rho = 0$ at $\rho = 0$. The Euclidean action for the nucleating bubble is given by the integral,

$$S_E = 2\pi^2 \int_0^\infty \rho^3 \left[\frac{1}{2} \left(\frac{\partial \phi}{\partial \rho} \right)^2 + U(\phi) \right] d\rho. \tag{2}$$

The "thin-wall" approximation developed by Coleman[6] has proven to be a popular method for explicitly calculating false vacuum decay rates [i.e., solutions to Eq. (2) for a given field theory]. The "thin-wall" approximation assumes that a nucleating bubble has a well defined core of true vacuum, a thin wall in which the Higgs field makes a rapid transition from its true vacuum value to the false vacuum value, and a bubble exterior of false vacuum. Real bubbles may approach this description when the energy density difference between the true and false vacuum states is small. Within the "thin-wall" analysis the Euclidean action for the bubble is described in terms of two parameters: the difference between the true and false vacuum energy densities, ϵ, and the bubble wall energy density, σ. In the thin-wall approximation, the Euclidean action for the bubble is then found to be:

$$B_0 = S_{tw} = 27\pi^2 \sigma^4 / 2\epsilon^3, \tag{3}$$

while the radius of the bubble is $\rho_0 = 3\sigma/\epsilon$.

The "thin-wall" approximation was extended by Coleman and De Luccia[8] to include the effect of the self gravity of the Higgs field, specifically for the most astrophysically interesting transitions, from de Sitter to Minkowski space (positive $\to$ zero vacuum energy density) and Minkowski to anti-de Sitter space (zero $\to$ negative vacuum energy density). This required the approximate solution of the coupled Euclideanized scalar field and Einstein equations. In this case their analysis showed that the effect of gravity was to increase the decay rate from de Sitter to Minkowski space, but to impede the decay rate from Minkowski to anti-de Sitter space. They also found that certain decays from Minkowski to anti-de Sitter space which are allowed (and indeed are inevitable) in flat space are forbidden when gravitational effects are taken into account. The forbidden decays correspond to the case where the negative energy density of the anti-de Sitter space is small compared to the bubble wall energy density.

3. Exact Results for False Vacuum Decay Rates

The validity of the "thin-wall" approximation has subsequently been examined by Samuel and Hiscock[9,10] both with and without gravitational effects, for the same scenarios considered by Coleman and De Luccia[8]. The scalar field and Einstein field equations were solved numerically for a model theory that utilized a polynomial potential $U(\phi) = m^2\phi^2 - \eta\phi^3 + \lambda\phi^4$ (which shall be referred to as a ϕ^{2-3-4} potential).

The generality of the results obtained was verified by testing with other, highly non-polynomial, potentials[9,11].

The ϕ^{2-3-4} potential was considered because it is the simplest polynomial potential in which we may independently vary the relative energy density between the true and false vacuum states and also the field separation between the two respective vacuum states. The potential may be rewritten in the useful (dimensionless) form:

$$U = \left\{ \Psi^2 - 2(2\omega + 1)\Psi^3 + (3\omega + 1)\Psi^4 \right\} \psi_+^2, \tag{4}$$

where $\Psi = \psi/\psi_+$, $\omega = \epsilon/\psi_+^2$, ϵ is the (dimensionless) difference in energy density between the false and true vacuum states, and ψ_+ is the (dimensionless) field separation of true and false vacuum states. Conventional dimensions may be restored by multiplying these variables by appropriate powers of the field mass m.

The exact numerical analysis revealed that the validity of the "thin-wall" approximation was questionable in most physically interesting scenarios, where one might expect the dimensionless parameter ω to be of order unity. The results are illustrated in Figure (2), where we notice that the "thin-wall" approximation provides reliable results for the Euclidean action of the bubble only for a very small range of ω close to zero.

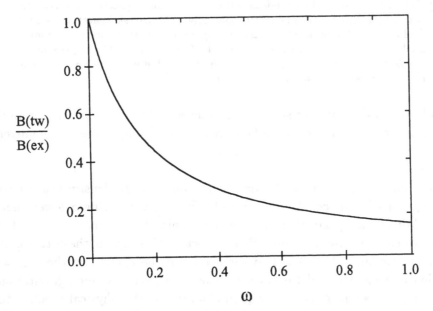

Figure (2): A comparison of the "thin-wall" approximation and exact results for the Euclidean action of nucleating bubbles formed with a ϕ^{2-3-4} potential. The curve represents the ratio of the "thin-wall" Euclidean action to the exact Euclidean action as a function of ω.

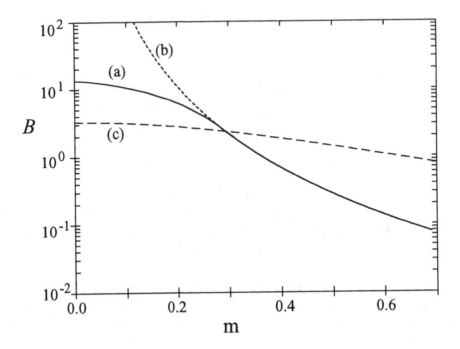

Figure (3): Difference between the Euclidean action for the bubble spacetime and the empty spacetime (*i.e.,B*) for the decay from de Sitter spacetime to Minkowski spacetime as a function of the field mass 'm' (in units of the Planck mass), where the other potential parameters ω and ψ_+ are kept at the constant values of 0.5 and 1.0, respectively. Curve (a) shows the exact value of B obtained numerically, curve (b) show the "thin-wall" approximation, and curve (c) shows the Hawking-Moss tunneling mode.

Figure (4) illustrates the different scalar field profiles during this transition from the nucleation of an isolated bubble to the homogenous tunneling of the entire universe at once.

Gravity has a surprising effect on the decay of a zero energy density space into a negative energy density space. Despite anti-de Sitter space having a lower energy density than Minkowski space, there is a region within the parameter space of the potential in which the false vacuum will not decay. Gravity, in these cases, stabilizes the false vacuum. This was first discovered in the context of the thin-wall approximation by Coleman and De Luccia[8]. The possibility that such gravitational stabilization might play an important role in determining the physical ground state in supergravity theories was pointed out by Weinberg[13]. The mechanism by which gravity stabilizes the higher energy density state may be easily understood within the thin-wall approximation. The volume contained within a bubble of a given proper

We also notice that the "thin-wall" approximation always underestimates the Euclidean action and hence overestimates the false-vacuum decay rate. When this analysis was repeated with a highly non-polynomial potential to test the robustness of this result, it was found that there was a scale change on the w-axis, but the overall functional profile remained quite similar to that shown for the ϕ^{2-3-4} potential.

When considering the effects of gravity on false vacuum decay, there are two special cases of particular interest. These are the decays from de Sitter to Minkowski space, which provides a reasonable model of what may have occurred in the early Universe, and the decay from Minkowski to anti-de Sitter space, which might provide a reasonable model for the possible future decay of our present vacuum state. These spacetimes are chosen (as opposed to more general Robertson-Walker cosmologies) because of the $O(4)$ symmetry they possess, which considerably simplifies the form the Euclideanized field equations; the solutions to which are required in order to evaluate the false vacuum lifetimes. The line element for these spacetimes may be written quite generally as $ds^2 = d\xi^2 + \rho(\xi)^2 d\Omega^2$, where $d\Omega^2$ is the metric of the three-sphere. The solution to the coupled Euclideanized scalar field and Einstein equations now provides not only the scalar field profile $\phi(\xi)$, but also the metric function $\rho(\xi)$. The scalar field and Einstein equations are then given by,

$$\frac{d^2\phi}{d\xi^2} + \frac{3}{\rho}\frac{d\rho}{d\xi}\frac{d\phi}{d\xi} = \frac{dU}{d\phi},\tag{5}$$

and,

$$\left(\frac{d\rho}{d\xi}\right)^2 = 1 + \frac{8\pi G}{3}\rho^2 T_{\xi\xi},\tag{6}$$

where

$$T_{\xi\xi} = \frac{1}{2}\left(\frac{d\phi}{d\xi}\right)^2 - U.\tag{7}$$

The initial condition for the Einstein equation [Eq. (6)] is simply $\rho = 0$ at $\xi = 0$. The boundary conditions for the Euclideanized scalar field equation depend on the decay scenario. For the decay from Minkowski space to anti-de Sitter space they are $\phi \to 0$ (the false vacuum value) as $\rho \to \infty$, and $d\phi/d\xi = 0$ at $\xi = 0$. The boundary conditions are somewhat different for the decay from de Sitter space to Minkowski space due to the closed topology of Euclideanized de Sitter space (i.e., a 4-sphere). We still require that $d\phi/d\xi = 0$ at $\xi = 0$; the other boundary condition must be changed however, as the $O(4)$ radial coordinate ρ has a finite upper bound in Euclideanized de Sitter space. We thus require that $d\phi/d\xi = 0$ at $\xi = \xi_{max}$ where ξ_{max} is the maximum value of ξ within the Euclideanized spacetime (i.e., at the antipodal point of the de Sitter space).

The results obtained for the Euclidean action by Coleman and De Luccia[8] for the thin-wall approximation including the effects of gravity may be summarized by

$$B = \frac{B_0}{\left[1 \pm \left(\frac{\rho_0}{2\Lambda}\right)^2\right]^2}, \tag{8}$$

where the $(+)$ sign represents the decay from de Sitter to Minkowski space, and the $(-)$ sign represents the decay from Minkowski to anti-de Sitter space; the subscript zero represents the associated value in the absence of gravity [see Eq. (3)], and $\Lambda = (3/8\pi|\epsilon|)^{1/2}$ is the radius of curvature of the de Sitter or anti-de Sitter spacetime, where ϵ is the energy density of the de Sitter or anti-de Sitter spacetime.

An alternative to the formation and subsequent expansion of a bubble (often referred to as the "Coleman-De Luccia tunneling mode") has been pointed out by Hawking and Moss[12] for the decay from de Sitter to Minkowski space. This involves the field tunneling in a homogenous fashion throughout the de Sitter space, 'everywhere at once', from the false vacuum state to the top of the barrier separating the two vacuum states. Since the de Sitter space has a finite volume, there is a non-zero probability of such uniform tunneling actually occurring. There would then be a subsequent classical evolution of the field to the true vacuum state. This alternative tunneling mode is also an exact solution to the Euclideanized field equations [Eqs. (5) and (6)], with appropriate boundary conditions (i.e., $\phi = \phi_{top}, d\phi/d\xi = 0$, where ϕ_{top} is the ϕ value corresponding to the top of the potential barrier).

The exact numerical solution of Eqs. (5) and (6), for the ϕ^{2-3-4} potential, provides an interesting insight into the effects of gravity on false vacuum decay. A transition is observed to occur between the Hawking-Moss tunneling mode and the Coleman-De Luccia tunneling mode for the decay from de Sitter space to Minkowski space. This is illustrated in Figure (3), which shows the evolution of B (i.e., the difference between the Euclidean action for the spacetime with and without the nucleating bubble) as a function of the Higgs field mass.

Notice that for small field masses the exact tunneling mechanism favors the Coleman-De Luccia tunneling mode. The observed discrepancy in B is merely the usual failure of the "thin-wall" approximation to yield valid values for B, and may be reduced by decreasing the size of ω. As the mass of the field is increased the Euclidean action of the Hawking-Moss mode crosses the Coleman-De Luccia curve and yields a smaller action. The exact tunneling mode is found to favor the mode that takes the smaller value for B, and thus at higher field masses the exact tunneling mechanism is well described by the Hawking-Moss mode.

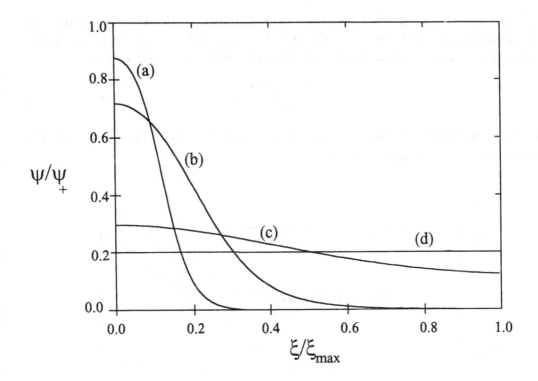

Figure (4): Scalar field profile associated with the decay from de Sitter space to Minkowski space for several values of the field mass. The potential parameters ω and ψ_+ are kept at the constant values of 0.5 and 1.0 respectively. The masses associated with the field profiles are: (a) $m=0.1$, (b) $m=0.2$, (c) $m=0.3$, and $m=0.4$, in natural units. All bubble profiles with $m > 0.4$ are well described by the Hawking-Moss tunneling mode.

surface area is smaller for anti-de Sitter space than for Minkowski space. Any vacuum bubble is required to have zero total energy (or, more precisely, exactly the same energy as the three-volume of false vacuum it has displaced) at all times. At the moment of nucleation, the bubble is at rest, and its radius is determined by setting the sum of the (effectively) negative energy in the interior and the positive bubble wall surface energy equal to zero; e.g., in flat space, ignoring gravity,

$$E = 0 = 4\pi R^2 \sigma - \frac{4}{3}\pi R^3 \epsilon, \tag{9}$$

where σ is the positive bubble wall energy density and $-\epsilon$ is the (negative) true vacuum energy density. In curved spacetime, however, the volume interior to a two-sphere of area $4\pi R^2$ is not given by the usual flat space formula. In particular, in anti-de Sitter space, the volume within a 2-sphere increases only as r^2 for large r; it is then possible to have a situation (with appropriate values of σ and ϵ) in which no bubble, no matter how large, ever succeeds in having zero energy. Thus no bubble can nucleate, and the vacuum decay in question is forbidden. This is illustrated for the "thin-wall" case by the divergence of the action in Eq. (8) when $\rho_0 = 2\Lambda$; for the ϕ^{2-3-4} potential, the divergence will occur when the scalar field mass reaches the critical value

$$m_c = \frac{1}{\psi_+^2} \left(\frac{3\epsilon}{\pi}\right)^{1/2} m_{Planck}. \tag{10}$$

Decay of the Minkowski vacuum with a scalar field mass larger than m_c is not possible within the thin-wall approximation.

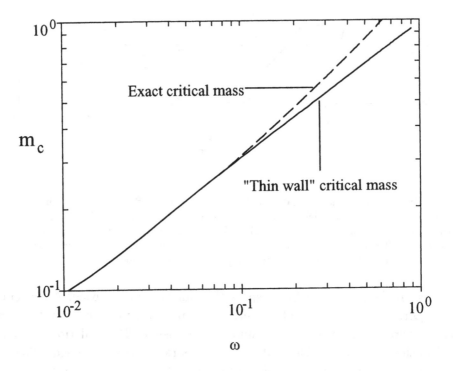

Figure (5): Critical mass associated with the decay from Minkowski to anti-de Sitter space, above which no $O(4)$-symmetric decays are permitted. In the "thin-wall" regime (*i.e.*, small ω) the exact critical mass asymptotically approaches the value given by the "thin-wall" approximation, and there is good agreement between the exact and approximate results. Away from the "thin-wall" regime, as ω becomes larger, the exact critical line diverges and lies above the approximate critical line, resulting in a smaller forbidden region.

The forbidden region is also found to occur within the exact solution to equations to Eqs. (5) and (6). Figure (5) illustrates both the exact and thin-wall forbidden regions for the ϕ^{2-3-4} potential. The thin-wall approximation is found to predict the forbidden area for small field masses very well. However, discrepancies arise as the field mass is increased, and the exact forbidden region is found to be somewhat smaller than that predicted by the thin-wall approximation. Thus, there are false vacuum decays possible which would be predicted as forbidden within the Coleman-De Luccia model.

4. Nucleation Sites for Cosmological Vacuum Phase Transitions

The rate of a first-order phase transition in ordinary matter may be greatly enhanced by the presence of nucleation sites. Examples are the formation of raindrops around atmospheric dust particles in the atmosphere, or bubbles of CO_2 on rough spots in the interior of a glass of beer. In the case of vacuum phase transitions, an interesting possibility is that the curved spacetime associated with gravitationally compact objects (such as black holes, neutron stars, boson stars, or very massive elementary particles) can act as a nucleation site for the phase transition. It is possible to illustrate in a simple model the mechanism by which gravitationally compact objects may enhance the vacuum decay rate. The Euclidean action for an $O(4)$-symmetric "thin-wall" nucleating bubble may be written in the form,

$$B = \sigma S - \epsilon V \qquad (\sigma, \epsilon > 0) \qquad (11)$$

where S is the surface area of the bubble, and V the volume (remembering that these are three-areas and four-volumes). In flat spacetime, if we fix the surface area of the bubble, then the volume is also fixed; however, for a curved spacetime the surface area does not *a priori* determine the contained volume. The curvature of spacetime, caused by the presence of a gravitationally compact objects within the bubble, can provide us with a 'surplus' volume within a given surface area as compared to the volume contained within in a flat spacetime. Increasing V in Eq. (11), while holding ϵ, σ, and S constant, will decrease B and thus increase the bubble nucleation rate.

The fact that the Euclidean action may be lower for bubbles nucleated around black holes or other compact objects does not guarantee that, in the large, the decay rate of the false vacuum will be enhanced significantly. One must take into account the number density of nucleation sites available in a particular cosmological scenario to determine the *quantitative* importance of $O(3)$ nucleation, as will be shown.

Hiscock[14] provided a first analysis which showed that within the "thin-wall" approximation black holes may act as nucleation sites for first order vacuum phase transitions. This analysis utilized the Israel[15] boundary layer formalism in general relativity to model the thin-wall bubbles. Thus, for example, when considering the decay from Schwarzschild to Schwarzschild-anti-de Sitter spacetime (analogous to the "nucleated" decay from Minkowski to anti-de Sitter spacetime) one would patch together a "core" of true-vacuum Schwarzschild-anti-de Sitter to an false-vacuum exterior of Schwarzschild spacetime. The Euclidean actions for $O(3)$ bubbles forming around Schwarzschild black holes were found to be significantly smaller than the $O(4)$ Euclidean actions; reductions in the action of up to 40% were found to occur in some cases. This analysis was extended to nucleation around material compact objects (neutron stars, elementary particles) by Mendell and Hiscock[16]; additional $O(3)$ symmetric black hole nucleating solutions have been found by Berezin et al.[17].

A recent study by Samuel and Hiscock[18] has also considered bubble nucleation around gravitationally compact objects, but without the restrictive assumption of the thin-wall approximation. The analysis utilized a perturbative approach about the $O(4)$ bubble solutions; the spatial curvature around the compact object is taken to be a perturbation on the $O(4)$ Euclidean space. The results of this study are thus trust-worthy as long as the nucleation center is not overly gravitationally compact.

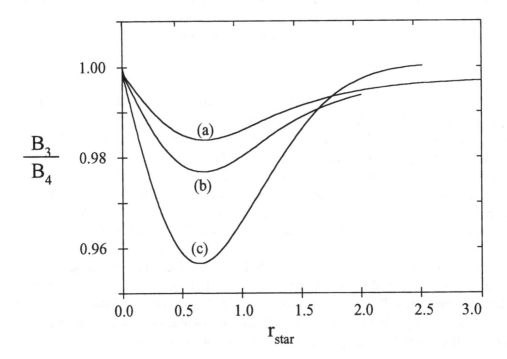

Figure (6): Ratio of Euclidean action for bubble nucleating around a gravitationally compact object to the Euclidean action for the bubble nucleating in a flat spacetime as a function of the ratio of the "radius" of the compact object to the "radius" of the bubble. The three curves represent different degrees of gravitational compactness; from (a) to (c) the ratio M/r_{star} is increased.

The perturbative analysis started with an exact $O(4)$-symmetric bubble, evaluated numerically, which corresponds to bubble nucleation in flat space. The scalar field configuration corresponding to this bubble was then placed in the fixed background spacetime of the gravitationally compact object. The bubble was assumed to possess an $O(3)$-symmetry within this spacetime. The bubble was then subjected to oblate and prolate spheroidal perturbations until the field configuration satisfied two constraints: (i) the energy of the nucleating bubble was zero, (ii) the field configuration was an extremum of the Euclidean action.

The perturbative analysis confirmed the results of the "thin-wall" analysis, and reductions in B were observed. The analysis went further, to show that gravitationally compact objects only play a significant role as nucleation sites when their size is roughly the same as the characteristic size of the nucleating bubble. Also, the reduction in B increased with an increase in the 'compactness' of the compact object as one might expect from the crude volume surplus argument presented earlier. Figure (6) shows the ratio of the $O(3)$ to $O(4)$ Euclidean actions as a function of R_{star}/R_{bubble} of several values of 'stellar compactness'.

Exact $O(3)$ bubble nucleation rate calculations are needed to fill the gap between the thin-wall analysis and the perturbative analysis. Unfortunately, such calculations have not yet been completed, owing to their difficulty. To accurately determine the Euclidean action for $O(3)$ bubble nucleation around a black hole or other compact object with a realistic (non-thin-wall) Higgs potential, one will need to solve coupled elliptic (in the Euclidean sector) partial differential equations. This problem is well within current numerical capabilities; it has, however, not attracted the necessary attention to date.

A reduction in 'B' due to the nucleation of a bubble around a gravitationally compact object does not however necessarily result in a reduced false vacuum lifetime. It is also necessary to take into account the number of gravitationally compact objects which can act as nucleation sites within the spacetime.

For example, consider a (fiducial) box with sides of length L, and volume L^3. The characteristic time that one would have to wait for an $O(4)$ bubble to appear within the box is given by,

$$T_4 = \frac{\exp(B_4)}{L^3 m^4}, \tag{12}$$

where B_4 is the Euclidean action for an $O(4)$-symmetric bubble, and m is mass scale of the field undergoing the phase transition.

The nucleating bubble may appear anywhere within the box, there is no preferential location for nucleation. However, when we consider the effect of a gravitationally compact object on false vacuum decay we make the assumption that the bubble will be forming around the compact object when calculating the associated Euclidean action. Let us now consider a box containing N similar compact objects which can act as nucleation sites. Of course a bubble may appear anywhere within the box and does not have to form around a compact object. If we assume a dilute gas approximation for the compact objects, then Eq. (12) provides a good estimate of the characteristic time for a bubble to nucleate in the space between the compact objects. The characteristic time for a single bubble to form around one of the compact objects is

$$T_3 = \frac{\exp(B_3)}{Nm} \tag{13}$$

where B_3 is the Euclidean action for a bubble nucleating about one of the compact objects. The relative importance of the gravitationally compact objects to false vacuum decay may be expressed by the ratio of T_3 to T_4,

$$\frac{T_3}{T_4} = \frac{m^3}{n}\exp(-\alpha B_4) \tag{14}$$

where $n = N/L^3$, and α is the fractional reduction in the Euclidean action due to the compact object: $\alpha = 1 - B_3/B_4$. If the ratio T_3/T_4 is significantly less than unity, then nucleation around compact objects will play an important role in a cosmological phase transition. If, on the other hand, the ratio is much greater than one, then the enhanced nucleation rate around compact objects will be but a minor perturbation on the lifetime of the false vacuum. Inspection of Eq. (14) shows that the crucial parameters in determining the importance of gravitational nucleation sites are the number density of nucleation sites, n, and the fractional reduction in the action, α.

As an example, we may ask whether nucleated false-vacuum decay is likely to play a dominant, or important, role in a hypothetical false-vacuum decay at the electroweak energy scale at the present epoch (under the assumption that we are currently living in a false-vacuum state).

We shall assume an optimal scenario where the entire mass density within the Universe is in the form of "optimal" nucleation sites, possessing the mass and size which yields the maximal decrease in the Euclidean action for the phase transition in question. We shall assume that the nucleation sites are black holes, with a Schwarzschild radius equal to the characteristic length scale of the field undergoing the phase transition (*i.e.*, the electroweak scale). The gravitationally compact object will then be comparable in size to the nucleating bubble, which is required for it to act efficiently as a nucleation site. Thus with a characteristic length scale $\approx 10^{-16}cm$, we would then have a characteristic mass for the nucleation sites $M_{nuc} \approx 10^{12}g$.

The maximum allowed mass density of the Universe, within observational constraints, is $\rho_{max} \approx 10^{-28} g/cm^3$, which gives as an observational bound on the maximum number density of nucleation sites, $n_{max} < 10^{-40} cm^{-3}$. $O(3)$ nucleation processes will begin to dominate over $O(4)$ spontaneous creation of bubbles in free space when the ratio T_3/T_4 is equal to unity. This implies a minimum number density of nucleation sites given by $n_{req} > m^3\exp(-\alpha B_4)$. We next insert $m \approx 10^{16} cm^{-1}$ and $\alpha = 0.25$ (the maximal decrease in the $O(4)$ action found in this case in the "thin-wall" analysis of Hiscock[14]). We then set $B_4 \approx 400$, obtained by setting the $O(4)$ decay lifetime of the false vacuum in this case equal to the present age of the universe. We then find $n_{req} \approx 10^3 cm^{-3}$, which is much greater than the observationally allowed value of $10^{-40} cm^{-3}$.

Thus we conclude that nucleated electroweak-scale false vacuum decay at this current epoch within the Universe is far less likely than spontaneous false-vacuum decay. It should be pointed out that this sort of conclusion is very strongly dependent upon the particular value of α. If we had assumed an $O(3)$ action decrease corresponding to $\alpha = 0.5$ were possible in this case, then the limit would be much weaker: $n_{req} > 10^{-39} cm^{-3}$. This strong dependence illustrates the need for further exact $O(3)$ Euclidean calculations (without the use of the "thin-wall" or any other approximation) for physically interesting vacuum phase transitions.

5. Stability of our Present Vacuum State

A question of more than academic interest is whether we might be living today within a false vacuum state, the ultimate "fool's paradise". Remarkably, the standard model of particle physics admits this possibility, without speculative extensions such as Grand Unification or supersymmetry. The masses of the electroweak gauge bosons $[\gamma, \{W^+, W^-, Z^0\}]$ and the fermions $[(\{e, \nu_e\}, \{\mu, \nu_\mu\}, \{\tau, \nu_\tau\}), (\{u, d\}, \{c, s\}, \{t, b\})]$ are generated via the Higgs mechanism within the Glashow[19]-Weinberg[20]-Salam[21] (GWS) theory. The theory does not put any bounds upon the fermionic masses, _per se_, but it has been found that very heavy fermionic masses will de-stabilize the broken symmetry vacuum state (<u>our</u> vacuum state!) within the theory[22-26].

The 1-loop, renormalized, effective potential for the Higgs field within the GWS theory is given by,

$$V(\phi) = -\frac{1}{2}\mu^2\phi^2 + \frac{1}{4}\lambda\phi^4 + \Xi\phi^4\ln\left(\frac{\phi^2}{M^2}\right), \tag{15}$$

where $\frac{1}{2}\phi^2 = \Phi^+\Phi$ ($\Phi = Higgs\ doublet$), M is the renormalization mass scale, and $\frac{1}{2}\mu^2$ and $\frac{1}{4}\lambda$ are the quadratic and quartic coefficients of the tree level potential, respectively. The coefficient Ξ is given by,

$$\Xi = \frac{1}{64\pi^2}\left[\sum_{bosons} 3g_i^4 - \sum_{fermions} f_i^4\right], \tag{16}$$

where g_i, and f_i are the respective coupling constants.

If the coefficient Ξ is positive semi-definite then the 1-loop effective potential is bounded below. However, if Ξ becomes negative then the 1-loop effective potential has no lower bound for large values of the Higgs field. This is generally regarded as being unphysical and may illustrate a breakdown in the theory, or a forbidden region within the coupling constant parameter space.

From Eq. (16) we observe that if there is a sufficiently massive fermion within the theory (remembering that the fermion masses are proportional to the couplings f_i) then it is possible that Ξ may become negative. The known fermions do not have sufficient mass to make Ξ negative; however, there is no definitive experimental bound on the mass of the top-quark at this time.

We may place an initial theoretical upper bound upon the top-quark mass by demanding that it not have sufficient mass to render Ξ negative[22,23] (there are of course other phenomenological arguments for placing rough bounds upon the top-quark mass). However, an interesting development occurs when one considers the two-loop corrections to the effective potential[24,25]. The full two-loop effective potential is a complex object and cannot be written in closed form, though it may be studied in an approximate form via the use of running coupling constants within a renormalization group prescription. This analysis reveals that there is a range of one-loop potentials which are unbounded below which become bounded at the two loop level.

The set of unbounded one-loop potentials which become bounded at the two-loop level have an additional vacuum state lying beyond the usual vacuum state of the one loop potential. If the actual potential within the Standard Model is described by one of these new potentials then our current vacuum state would be a false-vacuum state, and unstable to eventual quantum decay.

We may then place theoretical bounds upon the mass of the top-quark within the two-loop analysis as follows: (1) If the top-quark is sufficiently massive then it will render the two-loop effective potential unbounded; we therefore demand that the mass be insufficient for this to occur. (2) This is a range of top-quark masses for which our present vacuum state is a false vacuum state, and is therefore unstable to quantum decay. The lifetime of the false vacuum state will be a function of the two-loop effective potential which in turn is a function of the mass of the top quark. We demand that the lifetime of the false vacuum state be at least 10^{10} years, corresponding to the age of such a vacuum state. Sher and his collaborators have done a number of detailed studies, within the thin-wall approximation, of the upper bounds which may be placed on the top quark mass in this manner[24-26].

6. The Bubble Wall

One of the primary differences between the "thin-wall" approximation and the exact solution to the Euclideanized scalar field equation is the nature of the scalar field transition from the false vacuum state to the true vacuum state at the bubble wall. In the "thin-wall" approximation this transition is assumed to occur very rapidly (*i.e.*, within either a small spatial or time extent). However, the exact scalar field profile does not usually conform to this assumption, especially if the energy density separation between the true and false vacuum states is anything other than very small. For example, Figure (7) illustrates both the exact and "thin-wall" scalar field profile for a nucleating bubble within the ϕ^{2-3-4} potential model, with $\omega = 1$.

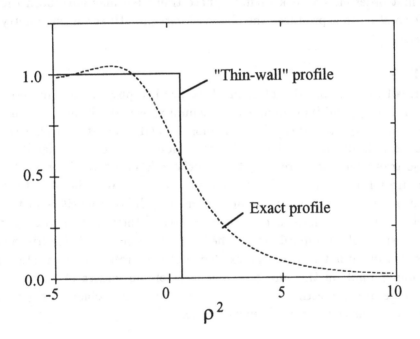

Figure (7): The exact profile of the scalar field, determined numerically, for an $O(4)$ symmetric bubble in the model theory with ϕ^{2-3-4} potential and $\omega = 1$, compared to the thin-wall approximate profile. In the Lorentzian sector, $\rho^2 = r^2 - t^2$.

We also note that there is significant evolution of the scalar field within the interior of the lightcone (Euclidean "radius" ρ^2 negative), which is not present within the thin-wall approximation. This evolution, when field interactions are added, is the source of "reheating" after a supercooled phase transition.

The nature of the transition from the false vacuum state to the true vacuum state is of importance because the vacuum state of the field is ill defined during this transition. As the usual procedure for building up the phenomenology of particle physics (*e.g.*, determining which quantum states describe particles) begins with the vacuum state, there is clearly the possibility of some rather new and exotic physics in this domain. If, for example, we consider the electroweak phase transition in the early universe, and hypothesize that it was first-order, then we note that particles such as fermions will be massless outside the bubble, but will be massive inside the bubble. This occurs because their mass is generated via a coupling to the scalar field and is proportional to the expectation value of the scalar field. More exciting possibilities have recently been pointed out by Dine *et al.*[27] and McLerran *et al.*[28], who have suggested that CP and T violation effects associated with the ambiguous vacuum state of the bubble wall in a first order electroweak vacuum phase transition may have been the source of particle creation that produced the observed baryon-antibaryon asymmetry within the Universe.

7. Conclusion

Essentially all modern theories of elementary particle physics contain spontaneous symmetry breaking, and its consequence, vacuum phase transitions, as a fundamental cornerstone of the theories. The cosmological and gravitational physics of such transitions is rich and as yet, incompletely explored. Effects associated with vacuum phase transitions are sure to play a crucial role in future theories of inflation and structure formation in the early universe; they may also play a critical role in the formation of the baryon-antibaryon asymmetry. Inhomogeneities in the quark-hadron phase transition have been studied to see if they could render primordial nucleosynthesis results compatible with the inflation-enforced flatness criterion, $\Omega = 1$ (they apparently cannot). The very existence of our present vacuum state implies a strong upper limit on the sum of the squares of the masses of all undiscovered quarks. There are interesting cosmological and physical problems associated with vacuum phase transitions at every energy scale.

This work was supported in part by NSF grants PHY88-03234 and PHY92-07903.

REFERENCES

1. Higgs, P. W. , Phys. Rev. Lett. **12**, 132 , (1964).

2. Linde, A. D. , Rep. Prog. Phys. **42**, 389 (1979); Rep. Prog. Phys. **47**, 925 (1984).

3. Guth, A. H. , and Weinberg, E. J. , Nucl.Phys. **B212**, 321 (1983).

4. Brandenberger, R. H. , Rev. Mod. Phys. **57**, 1 (1985).

5. Vilenkin, A. , Phys. Rep. **121**, 265 (1985).

6. Coleman, S. , Phys. Rev. **D15**, 2929, (1977) See also: Callan C. G. , and Coleman S. , Phys. Rev. **D16**, 1762, (1977).

7. Coleman, S. ; Glazer, V. ; and Martin, A. , Commun. Math. Phys. **58**, 211, (1978).

8. Coleman, S. ; and De Luccia, F. , Phys. Rev. **D21**, 3305, (1980).

9. Samuel, D. A. ; and Hiscock, W. A. , Phys. Lett. **B261**, 251 , (1991).

10. Samuel, D. A. ; and Hiscock, W. A. , Phys. Rev. **D44**, 3052 , (1991).

11. Samuel, D. A. , Ph. D. dissertation, Montana State University, 1991, unpublished.

12. Hawking S. W. , and Moss I. G. , Phys. Lett. **B110**, 35 , (1982).

13. Weinberg, S. , Phys. Rev. Lett. **48**, 1776 (1982).

14. Hiscock W. A. , Phys. Rev. **D35**, 1161 , (1987).

15. Israel, W. , Nuovo Cimento **B44**, 1 (1966) and Nuovo Cimento **B48**, 463 (1967).

16. Mendell, G. and Hiscock, W. A. , Phys. Rev. **D39**, 1537 (1989).

17. Berezin, V. A. ; Kuzmin, V. A. ; and Tkachev, I. I. , Phys. Rev. **D43**, R3112 (1991).

18. Samuel, D. A. and Hiscock, W. A. , Phys. Rev. **D45**, 4411, (1992).

19. Glashow S. L. , Nucl. Phys. **22**, 597, (1961).

20. Weinberg S. , Phys. Rev. Lett. **19**, 1264, (1967).

21. Salam A. , Elementary Particle Theory, Ed. N. Swartholm, p.367 (Almqvist and Wiksell, Stockholm, 1968).

22. Krive, I. V. , and Linde, A. D. , Nucl. Phys. **B117**, 265 (1976).

23. Politzer, H. D. , and Wolfram, S. , Phys. Lett. **B82**, 242 (1979); **B83**, 421 (1979) (E).

24. Flores, R. A. , and Sher, M. , Phys. Rev. **D27**, 1679 (1983).

25. Duncan, M. J. , Philippe, R. , and Sher, M. , Phys. Lett. **B153**, 165 (1985).

26. for a thorough review, see Sher, M. , Phys. Rep. **179**, 273 (1989).

27. Dine M. , Huet P. , Singleton R. , and Susskin L. , Phys. Lett. **B257**, 351, (1991).

28. Mc Lerran L. , Shaposhnikov M. , Turok N. , Voloshin M. , Phys. Lett. **B256**, 451 (1991).

Minisuperspace as a Quantum Open System

B. L. Hu *   *Juan Pablo Paz* [†]   *Sukanya Sinha* [‡]

Abstract

We trace the development of ideas on dissipative processes in chaotic cosmology and on minisuperspace quantum cosmology from the time Misner proposed them to current research. We show
1) how the effect of quantum processes like particle creation in the early universe can address the issues of the isotropy and homogeneity of the observed universe,
2) how viewing minisuperspace as a quantum open system can address the issue of the validity of such approximations customarily adopted in quantum cosmology, and
3) how invoking statistical processes like decoherence and correlation when considered together can help to establish a theory of quantum fields in curved spacetime as the semiclassical limit of quantum gravity.

Dedicated to Professor Misner on the occasion of his sixtieth birthday, June 1992.

1.   Introduction

In the five years between 1967 and 1972, Charlie Misner made an idelible mark in relativistic cosmology in three aspects.

First he introduced the idea of chaotic cosmology. In contrast to the reigning standard model of Friedmann-Lemaitre-Robertson-Walker universe where isotropy and homogeneity are 'put in by hand' from the beginning, chaotic cosmology assumes that the universe can have arbitrary irregularities initially. This is perhaps a more general and philosophically pleasing assumption. To reconcile an irregular early universe with the observed large scale smoothness of the present universe, one has to introduce physical mechanisms to dissipate away the anisotropies and inhomogeneities. This is why dissipative processes are essential to the implementation of the chaotic cosmology

---
*Department of Physics, University of Maryland, College Park, MD 20742, USA

[†] T6 Theoretical Astrophysics, ms 288, LANL, Los Alamos, NM 87545, USA

[‡] Instituto de Ciencias Nucleares, UNAM, Circuito Exterior, Apartado Postal 70-543, CU Mexico, DF 04510, Mexico

program. Misner (1968) was the first to try out this program in a Bianchi type-I universe with the neutrino viscosity at work in the lepton era. Though this specific process was found to be too weak to damp away the shear, the idea remains a very attractive one. As we will see, it can indeed be accomplished with a more power-ful process, that of vacuum particle production at the Planck time. The philosophy of chaotic cosmology is similar to that behind the inflationary cosmology of Guth (1981), where the initial conditions are rendered insignificant by the evolutionary process, which, in this case is inflation.

Second, the chaotic cosmology program ushered in serious studies of the dynamics of Bianchi universes. This was exemplified by Misner's elegant work on the mixmaster universe (1969a). (That was when one of us first entered the scene.) The ingeneous use of pictorial representation of the curvature potentials made the complicated dy-namics of these models easy to follow and opened the way to a systematic analaysis of this important class of cosmology. (See Ryan and Shepley 1975.) This develop-ment complements and furthers the ongoing program of Lifshitz, Khalatnikov and Belinsky (1963, 1970) where, in seeking the most general cosmological solutions of Einstein's equations using rigorous applied mathematics techniques (see also Eard-ley, Liang and Sachs 1971), they found the inhomogeneous mixmaster universe (the 'generalized' Kasner solution) as representing a generic behavior near the cosmolog-ical singularity. This also paralleled the work of Ellis and MacCallum (1969) who, following Taub (1951) and Heckmann and Schucking (1962), gave a detailed analysis of the group-theoretical structure of the Bianchi cosmologies. Misner's chaotic cos-mology program provided the physical rationale for these studies. It also prepared the ground for minisuperspace quantum cosmology (1969b, 1972).

Third, the establishment of the quantum cosmology program, which includes the ap-plication of the ADM quantization (1962) to cosmological spacetimes, and the adap-tation of the Wheeler- DeWitt superspace formulation of quantum gravity (Wheeler 1968, De Witt 1967) to minisuperspace cosmology. The physical motivation was to understand 'the issue of the initial state' (Wheeler 1964), and, in particular, the quantum effects of gravity on the cosmological singularity. It was Misner and his associates who started the first wave of activity in quantum cosmology (Ryan, 1972; see also Kuchar 1971, Berger 1974, 1975). The second wave, as we know, came with the work of Hartle and Hawking (1983), and Vilenkin (1983) who, while formulating the problem in Euclidean path integral terms, opened the question on the boundary conditions of the universe.

These three directions in cosmology initiated by Misner and his collaborators in the early seventies –chaotic cosmology, dissipative processes and minisuperspace quantum cosmology– have developed in major proportions in the past two decades. A lot of current research work in these areas still carry the clear imprint of his influence. The present authors have had the good fortune to be influenced by his way of thinking. We want to trace out some major developments since Misner wrote these seminal papers and describe the current issues in these areas, with more emphasis on their interconnections from early universe quantum processes to quantum cosmology.

1.1.   Particle Creation as a Quantum Dissipative Process

Misner (1968) and Matzner and Misner (1972) showed that neutrino viscosity in the lepton era ($\approx 1sec$ from the Big Bang) is in general not strong enough to dissipate away the shear in the Bianchi universes. It was Zel'dovich (1970) who first suggested that vacuum particle creation from the changing gravitational field in anisotropic spacetimes may act as a powerful dissipative source. This created a real possibility for the idea of chaotic cosmology to work, now relying on quantum field processes effective near the Planck time (10^{-43} sec from the Big Bang). One needs a new theory — the theory of quantum fields in curved spacetime. Actually this theory was just beginning to take shape through the work of Parker (1966, 1969) and Sexl and Urbantke (1968) at about the same time when Misner was working on the isotropy of the universe problem (1968). It was later realized that the reason why the Bianchi I universe and not the Robertson-Walker universe produces a copious amount of particles is because it breaks the conformal invariance of the theory (Parker 1973). Zel'dovich used a simple dimensional analysis to explain why the produced particles can strongly influence the dynamics of spacetime at the Planck time. The details are actually more involved than this, because one needs to remove the divergences in the energy-momentum tensor of the matter field before it can be used as the source of the Einstein equation to solve for the metric functions. This is what has been called the 'backreaction' problem. Zel'dovich and Starobinsky (1971) were the first to attempt such a calculation for conformal scalar fields in a Bianchi I universe. (See also Lukash and Starobinsky 1974). That was before the basic issues of particle creation processes were fully understood (e.g., ambiguities in the choice of the vacuum in curved space, see, Fulling 1973) and the regularization techniques were perfected. But they managed to show the viability of such processes. It took several years (1974-77) before the common ground between the different ways of regularization (e.g., adiabatic, point-splitting, dimensional, zeta-function) was reached and a firmer foundation for this new theory of semiclassical gravity was established. Hawking's discovery (1974) of black hole radiance, and the confirmed 'legality' of trace anomaly (Capper and Duff 1975, Deser, Duff and Isham 1976) certainly invigorated this endeavour. The theoretical importance of the subject and the infusion of talents from general relativity, field theory, and particle physics have made this field an important component of contemporary theoretical physics (Birrell and Davies 1982). It is, as we shall see, also an intermediate step (the low energy or adiabatic limit) towards a theory of quantum gravity and quantum cosmology.

Vigorous calculation of the backreaction of particle creation with regularized energy momentum sources began with the work of Hu and Parker (1977,1978) using adiabatic regularization methods, and Hartle (1977), Fischetti et al (1979) and Hartle and Hu (1979, 1980), using effective action methods. Their results on the effect of particle creation on anisotropy dissipation confirmed the qualitative estimate of Zel'dovich, and lent strong support for the viability of Misner's chaotic cosmology program.

1.2.   Entropy of Quantum Fields and Spacetime?

These efforts in quantum field theory in curved spacetime and dissipative processes in the early universe contained the germs for two interesting directions of later development. Both involve quantum field theory and statistical mechanics applied to cosmological problems. They are

i) Dissipation in quantum fields as a statistical thermodynamic process:
Can one associate a viscosity with vacuum particle creation (Zel'dovich 1979, Hu 1984) like that of neutrino in the lepton era, or leptoquark boson decays in the GUT era? Can one define an entropy of quantum fields? (Hu and Pavon 1986, Hu and Kandrup 1987, Kandrup 1988)? An entropy of spacetime? (Penrose 1979, Hu 1983, Davies, Ford and Page 1986) Can one rephrase the chaotic cosmology idea in terms of some analog of second law of thermodynamics involving both fields and geometry (Hu 1983)? Can one generalize the fluctuation-dissipation relation (FDR) for quantum processes in spacetimes with event horizons (Candelas and Sciama 1977, Mottola 1986) to general dynamical spacetimes? Indeed can one view the backreaction process as a manifestation of a generalized FDR for dynamical systems like quantum fields in cosmological spacetimes (Hu 1989, Hu and Sinha 1993)?

This first direction of research has been pursued in the eighties, with the infusion of ideas and techniques in statistical and thermal field theory. It is intended to be a generalization of thermodynamic ideas successfully applied to black hole mechanics (Davies 1978) to non-equilibrium quantum field systems characteristic of cosmological problems. This was an offshoot of Misner's interest in dissipative processes in the early universe, only that one now takes on the challenge of doing everything with quantum fields in non-equilibrium conditions. This exercise probes into many deeper issues of thermodynamics and cosmology, such as the statistical nature of gravitational systems as different from ordinary matter, the thermodynamic nature of spacetimes (with or without event horizons), the nature of irreversibility and the cosmological origin of time-asymmetry. We will not belabor this direction of research here, but refer the reader to some recent reviews (Hu 1989, 1991, 1993a, 1993b). As we will see, inquiries into the statistical nature of particle creation and backreaction processes actually have a role to play in the second direction, i.e.,

ii) Quantum field theory as the semiclassical limit of quantum gravity. How good is modified Einstein's theory with backreaction from quantum matter fields an approximation to quantum gravity? Under what conditions can the quantum dynamics of spacetime described by a wave function make the transition to the classical picture described by trajectories in superspace? We will use the paradigms of quantum open systems applied to Misner's minisuperspace models to illustrate these ideas.

1.3.   From Dissipation to Quantum Open Systems

Speaking from the path traversed by one of us, one can identify a few critical nodes in the search for a pathway which connects i) particle creation as a dissipative processes to ii) backreaction and semiclassical gravity to iii) the statistical aspects in the issue of quantum to classical transition which underlies the relation of quantum gravity and semiclassical gravity. (See the Ph.D. thesis of Sinha 1991 for an overview of other major developments and references in these areas.)   The (in-out) effective action formalism (Schwinger 1951, DeWitt 1964, 1975) was a correct start, because it gives the rate of particle production and the backreaction equations in a self-consistent manner (Hartle and Hu 1979).   However the conditions for calculating the in-out vacuum persistence amplitude are unsuitable for the consideration of backreactions, as the effective equations are acausal and complex. The second major step was the recognition by DeWitt (1986), Jordan (1986), and Calzetta and Hu (1987) that the in-in or closed-time-path formalism of Schwinger (1961) and Keldysh (1964) is the right way to go. Calzetta and Hu (1987) derived the backreaction equations for conformal scalar fields in Bianchi I universe, which provided a beachhead for later discussion of the semiclassical limit of quantum cosmology. They also recognized the capacity of this framework in treating both the quantum and statistical aspects of dynamical fields and indeed used it for a reformulation of non-equilibrium statistical field theory (Calzetta and Hu 1988). The third step in making the connection with statistical mechanics is the construction of the so-called coarse-grained effective action by Hu and Zhang (1990), similar in spirit to the well-known projection operator method in statistical mechanics (Zwanzig 1961).

This is where the connection with 'quantum open systems' first enters. The general idea behind the study of such systems is that one starts with a closed system (universe) , splits it into a "relevant" part (system) and an "irrelevant" (environment) part according to the specific physical conditions of the problem (for example, the split between light and heavy modes or slow and fast variables), and follows the behavior only of the relevant variables (now comprising an open system) by integrating out the irrelevant variables. (See Hu 1993a and 1993b for a discussion of the meaningfulness of such splits in the general physical and cosmological context respectively). Thus the coarse-grained effective action generalizes the background field/ fluctuation-field splitting usually assumed when carrying out a quantum loop expansion in the ordinary effective action calculation to that between a system field and an environment field. It takes into account the averaged effect of the environment variables on the system and produces an effective equation of motion for the system variables. This method was used by Hu and Zhang (1990) (see also Hu 1991) to study the effect of coarse-graining the higher modes in stochastic inflation, and by Sinha and Hu (1991) to study the validity of minisuperspace approximations in quantum cosmology. The last step in completing this tour is the recognition by Hu, Paz and Zhang (1992) that the coarse-grained closed-time-path effective action formalism is equivalent to the Feynman-Vernon (1963) influence functional which was popularized by Caldeira and Leggett (1983) in their study of quantum macroscopic dissipative processes. Paz and Sinha (1991, 1992) have used this paradigm for a detailed analysis of the relation of quantum cosmology with semiclassical gravity. We like to illustrate this development with two problems, initially proposed and studied by Misner. One is on the validity of minisuperspace approximation, the other is on the classical limit of quantum cosmology.

2. Validity of the Minisuperspace Approximation

In quantum cosmology, the principal object of interest is the wave function of the universe $\Psi(h_{ij}, \phi)$, which is a functional on superspace. It obeys the Wheeler DeWitt equation, which is an infinite-dimensional partial differential equation. Its solution is a formidable task in the general case. To make the problem more tractable, Misner suggested looking at only a finite number of degrees of freedom (usually obtained by imposing a symmetry requirement), and coined the word minisuperspace (Misner 1972) quantization. We note that such a tremendous simplification has its preconditions. Since in the process of this transition one is truncating an infinite number of modes, nonlinear interactions of the modes among themselves and with the minisuperspace degrees of freedom are being ignored. It is also well-known that this truncation violates the uncertainty principle, since it implies setting the amplitudes and momenta of the inhomogeneous modes simultaneously to zero. It is therefore important to understand under what conditions it is reasonable to consider an autonomous evolution of the minisuperspace wave function ignoring the truncated degrees of freedom. Misner was keenly aware of the problems when he introduced this approximation. The first attempt to actually assess the validity of the minisuperspace approximation was made only recently by Kuchar and Ryan (1986, 1989).

Two of us (Sinha and Hu 1991) have tried to address this question in the context of interacting field theory with the help of the coarse-grained effective action method mentioned above. The model considered is that of a self interacting ($\lambda\phi^4$) scalar field coupled to a closed Robertson Walker background spacetime. The "system"(minisuperspace) consists of the scale factor a and the lowest mode of the scalar field, while the "environment" consists of the inhomogeneous modes. [1] In this model

---

[1] In this example of minisuperspace the scalar field should not be thought of as providing a matter source for the Robertson-Walker background metric, since in that case varying the action with respect to the scale factor will not give the full set of Einstein equations. In particular, the

calculation the scalar field is mimicking the inhomogeneous gravitational degrees of freedom in superspace. We are motivated to do this because linearized gravitational perturbations (gravitons) in a special gauge can be shown to be equivalent to a pair of minimally coupled scalar fields (Lifshitz 1946). Quantum cosmology of similar models of gravitational perturbations and scalar fields have been studied by several authors (Halliwell and Hawking 1985, Kiefer 1987, Vachaspati and Vilenkin 1988).

Our basic strategy will be to try to derive an "effective" Wheeler-DeWitt equation for the minisuperspace sector which contains the "averaged" effect of the higher modes as a backreaction term. We will then explicitly calculate the backreaction term using the effective action and consequently present a criterion for the validity of the minisuperspace approximation.

2.1. Effective Wheeler DeWitt Equation

The gravitational and matter actions in our model are given by

$$S_g = \frac{1}{2} \int d\eta \, a^2 (1 - \frac{a'^2}{a^2}) \tag{1}$$

$$S_m = -\frac{1}{2} \int \sqrt{-g} d^4 x [\Phi \Box \Phi + m^2 \Phi^2 + \frac{2\lambda}{4!} \Phi^4 + \frac{R}{6} \Phi^2] \tag{2}$$

where a is the scale factor of a closed Robertson-Walker universe, a factor $l_p^2 = 2/(3\pi m_p^2)$ is included in the metric for simplification of computations, and η is the conformal time. $\Box$ is the Laplace Beltrami operator on the metric $g_{\mu\nu}$, and m is the mass of the conformally coupled scalar field. Defining a conformally related field $\chi = (al_p)\Phi$ and expanding χ in scalar spherical harmonics $Q_{lm}^k(x)$ on the S^3 spatial sections,

$$\chi = \frac{\chi_0(t)}{(2\pi^2)^{\frac{1}{2}}} + \sum_{klm} f_{klm} Q_{lm}^k(x) \tag{3}$$

where $k = 2, 3, \ldots \infty, l = 0, 1, \ldots k - 1, m = -l, -l+1 \ldots l - 1, l$ (henceforth we will use k to denote the set $\{klm\}$). We will make the further assumption that the interactions of orders higher than quadratic of the lowest(minisuperspace) mode χ_0 with the higher modes f_n's ($\sim \chi_0 f_k f_l f_m$) as well as the quartic self interaction of the higher modes ($\sim f_m f_n f_k f_l$) are small and can be neglected. With this assumption and with the following redefinitions $m^2 \to m^2/l_p^2, \lambda \to \lambda/2\pi^2$, the matter action can be written as

$$S_m = \int d\eta \left\{ \frac{1}{2} \left[\chi_0'^2 - m^2 a^2 \chi_0^2 \right] - \frac{\lambda}{4!} \chi_0^4 - \frac{1}{2} \sum_k f_k \left[\frac{d^2}{d\eta^2} + k^2 \right] f_k - \frac{1}{2} \sum_k m^2 a^2 f_k^2 \right.$$

$$\left. - \frac{\lambda}{4} \sum_k \chi_0^2 f_k^2 \right\} \tag{4}$$

$G_{ij} = 8\pi G T_{ij}$ equations that constrain the energy momentum tensor via $T_{ij} = 0$ for $i \neq j$ will be missing.

The Hamiltonian constructed from the action $S = S_g + S_m$ is given by

$$H = -\frac{1}{2}\pi_a{}^2 + \frac{1}{2}\pi_{\chi_0}{}^2 + \frac{1}{2}\sum_n \pi_{f_n}{}^2 + V_0(a, \chi_0) + V(a, \chi_0, f_n) \tag{5}$$

where

$$V_0(a, \chi_0) = -\frac{1}{2}a^2 + \frac{1}{2}m^2 a^2 \chi_0{}^2 + \frac{\lambda}{4!}\chi_0{}^4 \tag{6}$$

and

$$V(a, \chi_0, f_n) = \frac{\lambda}{4}\sum_k \chi_0{}^2 f_k{}^2 + \frac{1}{2}\sum_k (k^2 + m^2 a^2) f_k{}^2 \tag{7}$$

where $\pi_a, \pi_{\chi_0}, \pi_{f_n}$ are the momenta canonically conjugate to a, χ_0, f_n respectively. The Wheeler-DeWitt equation for the wave function of the Universe Ψ is obtained from the Hamiltonian constraint by replacing the momenta by operators in the standard way, and is given by

$$\left[\frac{1}{2}\frac{\partial^2}{\partial a^2} - \frac{1}{2}\frac{\partial^2}{\partial \chi_0{}^2} - \frac{1}{2}\sum_n \frac{\partial^2}{\partial f_n{}^2} + V_0 + V\right]\Psi(a, \chi_0, f_n) = 0 \tag{8}$$

where we choose a factor ordering such that the kinetic term appears as the Laplace-Beltrami operator on superspace.

Writing

$$\Psi(a, \chi_0, f_n) = \Psi_0(a, \chi_0)\Pi_n\Psi_n(a, \chi_0, f_n) \tag{9}$$

we would like to obtain from (2.8) an effective Wheeler-DeWitt equation of the form

$$(H_0 + \Delta H)\Psi_0(a, \chi_0) = 0 \tag{10}$$

where H_0 is the part of the Hamiltonian operator in (2.8) independent of f_n and ΔH represents the influence of the higher modes. By making the assumption that Ψ_n varies slowly with the minisuperspace variables (see Sinha and Hu 1991 for details of this approximation) one can identify

$$\Delta H = -\sum_n < H_n > \tag{11}$$

where the expectation value is taken with respect to Ψ_n. It is evident that the examination of this term will enable us to comment on the validity of the minisuperspace description. We will then make a further assumption that Ψ_0 has a WKB form, i.e, $\Psi_0 = e^{iS(a,\chi_0)}$ which can be used in regions of superspace where the wavefunction oscillates rapidly, such that using eqn. (2.10), S satisfies

$$\frac{1}{2}(\nabla S)^2 + V_0 = -\sum_n < H_n > \tag{12}$$

where ∇ is the gradient operator on minisuperspace. The above equation can be regarded as a Hamilton-Jacobi equation with backreaction. It can be also shown that the Ψ_n's satisfy a Schrödinger - like equation with respect to the WKB time. This approximation can therefore be roughly thought of as the semiclassical limit where the minisuperspace modes behave classically, but the higher modes behave quantum mechanically (for further subtleties regarding the semiclassical limit see Sec. 3). Identifying $\frac{\partial S}{\partial a} = \pi_a$ and $\frac{\partial S}{\partial \chi_0} = \pi_{\chi_0}$, and substituting for the canonical momenta in terms of "velocities" a' and χ_0' Eq. (12) reduces to

$$\frac{1}{2}a'^2 - \frac{1}{2}\chi_0'^2 + V_0(a,\chi_0) = -\sum_n <H_n> \qquad (13)$$

This is the effective Wheeler- De Witt equation in the WKB limit, which we will compare with the backreaction equation derived in the next section, in order to calculate the term on the right hand side.

2.2. Backreaction of the Inhomogeneous Modes

We would now like to calculate the backreaction term (2.11) explicitly. Since we are making a split of the system from the environment based on the mode decomposition, we should use the coarse-grained effective action where the coarse graining consists of functionally integrating out the higher modes. As we would like to generate vacuum expectation values from the effective action rather than the matrix elements one should use the in-in or closed-time-path (CTP) version of the effective action (Calzetta and Hu 1987) rather than the in-out version. The CTP coarse-grained effective action in our case is given by (for details, see Sinha and Hu 1991)

$$e^{iS_{eff}(a^+,\chi_0^+,a^-,\chi_0^-)} = \int \mathcal{D}f_k^+ \mathcal{D}f_k^- \, e^{i\left(S(a^+,\chi_0^+,f_k^+)-S^*(a^-,\chi_0^-,f_k^-)\right)} \qquad (14)$$

S^* indicates that in this functional integral, m^2 carries an $i\epsilon$ term. $a^\pm, \chi_0^\pm, f_k^\pm$ are the fields in the positive (negative) time branch running from $\eta = \overline{+} \infty$ to $\pm\infty$. The path integral is over field configurations that coincide at $t = \infty$. $\mathcal{D}f_k$ symbolizes the functional integration measure over the amplitudes of the higher modes of the scalar field.

We have derived the one loop renormalized coarse-grained effective action given as follows (omitting terms that involve $-$ fields only)

$$\begin{aligned}
S_{eff} = \; & S_{g+} + \frac{1}{2}\int d\eta \left\{\chi_0^{+\prime 2} - \tilde{m}^{+2}\chi_0^{+2}\right\} \\
& - \frac{\lambda}{4!}\int d\eta \chi_0^{+4} + \frac{13\lambda}{48}\int d\eta M^{+2} + \frac{1}{16}\int d\eta M^{+4} \\
& + \frac{1}{32}\int d\eta_1 d\eta_2 M^{+2}(\eta_1)K(\eta_1-\eta_2)M^{+2}(\eta_2) \\
& + \frac{1}{32}\int d\eta_1 d\eta_2 M^{+2}(\eta_1)\bar{K}(\eta_1-\eta_2)M^{-2}(\eta_2)
\end{aligned} \qquad (15)$$

where the coupling constants have their renormalized values. S_{g+} represents the classical gravitational part of the action. $M^{\pm^2} = \tilde{m}^{\pm^2} + \frac{1}{2}\lambda\chi_0^{\pm^2}$, and K and $\bar{K}$ are complex nonlocal kernels with explicitly known forms. The effective equations of motion are obtained from this effective action via [2]

$$\frac{\delta S_{eff}}{\delta a^+}\bigg|_{\substack{a^+=a^-=a \\ \chi_0^+=\chi_0^-=\chi_0}} = 0 \quad \text{and} \quad \frac{\delta S_{eff}}{\delta\chi_0^+}\bigg|_{\substack{a^+=a^-=a \\ \chi_0^+=\chi_0^-=\chi_0}} = 0 \tag{16}$$

Since we are interested in comparing with (2.13) which is equivalent to the G_{00} Einstein equation with backreaction, we need the first integral form of (2.16), which can be derived as

$$\frac{1}{2}a'^2 - \frac{1}{2}\chi_0'^2 + \frac{1}{2}m^2a^2\chi_0^2 - \frac{1}{2}a^2 + \frac{\lambda}{4!}\chi_0^4 - \frac{13\lambda}{96}\chi_0^2 - \frac{13}{48}m^2a^2 - \frac{1}{16}M^2 ln\mu a$$

$$+ \frac{1}{32}\int d\eta_1 M^2(\eta)\mathcal{K}(\eta-\eta_1)M^2(\eta_1) = 0 \tag{17}$$

with the assumption of having no quanta of the higher modes in the initial state. $\mathcal{K} = K + \bar{K}$ and is real and hence the above equation is also manifestly real. Equation (2.17) is then equivalent to the effective G_{00} Einstein equation or the Einstein- Hamilton-Jacobi equation plus backreaction. This in turn can be identified with (2.13), the effective Wheeler- De Witt equation in the WKB limit. Therefore, comparing (2.17) and (2.13) we can identify the backreaction piece in (2.13) as

$$\sum_n <H_n> = -\frac{13\lambda}{96}\chi_0^2 - \frac{13\lambda}{48}m^2a^2 - \frac{1}{16}M^2 ln\mu a$$

$$-\frac{1}{32}\int d\eta_1 M^2(\eta)\mathcal{K}(\eta-\eta_1)M^2(\eta_1) \tag{18}$$

when the boundary conditions on the wave function are appropriate for the Ψ_n's to be in a conformal "in" vacuum state. [3]

Since equation (2.17) is the "effective" Wheeler- DeWitt equation for the minisuperspace sector within our approximation scheme, the condition for validity of this approximation can be stated as

$$\sum_n <H_n> \ll V_0 \tag{19}$$

where by the left hand side we mean the regularized value given by (2.18). It was shown that the term in equation (2.18) involving the nonlocal kernel is related to

---

[2] we need to compute only those terms in the effective action that involve the + fields. Terms containing only − fields will not contribute to the equations of motion.

[3] Since $\mathcal{K}$ is real the backreaction term given above is real and represents a genuine expectation value in the "in" vacuum state rather than an in-out matrix element generated using the in-out effective action.

dissipative behavior in closely related models (Calzetta and Hu 1989). This dissipative behavior in turn has been related to particle production by the dynamical background geometry in semiclassical gravity models (Hu, 1989). In our case this can be interpreted as scalar particles in the higher modes being produced as a result of the dynamical evolution of the minisuperspace degrees of freedom generating a back-reaction that modifies the minisuperspace evolution. We can therefore think of this term as introducing dissipation in the minisuperspace sector due to interaction with the higher modes that are integrated out. One can justifiably think of autonomous minisuperspace evolution only when this dissipation is small. Since we have used the scalar field modes to simulate higher gravitational modes these considerations can also be directly extended to include gravitons. [4]

This is an example of how ideas of open systems can be useful in understanding dissipation and backreaction, even in quantum cosmology. One can also use this paradigm to address the problem of quantum to classical transition, specifically the relation of semiclassical gravity with quantum gravity, which we will now address. To do this the formalism will need to be elevated to the level of density matrices.

3.   Semiclassical Limit of Quantum Cosmology

We now report on the result of some recent work by two of us (Paz and Sinha 1991, 1992) on this problem. Quantum cosmology rests on the rather bold hypothesis that the entire universe can be described quantum mechanically. This pushes us to question the usual Copenhagen interpretational scheme that relies on the existence of an *a priori* classical external observer/apparatus. Since in the case of quantum cosmology this cannot be assumed, the theory needs to predict the "emergence" of a quasiclassical domain starting from a fundamentally quantum mechanical description. Recently, there has been a lot of interest in this subject, and two basic criteria for classicality have emerged (see Halliwell 1991 and references therein) from these endeavors. The first is *decoherence* – which requires that quantum interference between distinct alternatives must be suppressed. The second is that of *correlation*, which requires that the wave function or some distribution constructed from it (e.g. Wigner 1932) predicts correlations between coordinates and momenta in accordance with classical equations of motion. We will study the emergence of a semiclassical limit with the help of a quantum open system paradigm applied in the context of quantum cosmology. The basic mechanism for achieving decoherence and the appearance of correlations is by coarse-graining certain variables acting as an environment coupled to the system of interest.

---

[4]A similar idea has been discussed by Padmanabhan and Singh (1990) in a linearized gravity model where they claim that in order that the minisuperspace approximation be valid, the rate of production of gravitons should be small.

3.1.   Reduced Density Matrix

We will consider a D-dimensional minisuperspace with coordinates r^m $(m = 1, \ldots, D)$ as our system for which the Hamiltonian can be written as:

$$H_g = \frac{1}{2M} G^{mm'} p_m p_{m'} + M V(r^m) \tag{1}$$

where $G^{mm'}$ is the (super)metric of the minisuperspace and M is related to the square of the Planck mass. The minisuperspace modes are coupled to other degrees of freedom Φ, such as the modes of a scalar field, or the gravitational wave modes with a Hamiltonian $H_\Phi(\Phi, \pi_\Phi, r^m, p_m)$. These constitute the environment, or the 'irrelevant' part in our model. Thus, the wave function of the Universe is a function $\Psi = \Psi(r^m, \Phi)$ which satisfies the Wheeler-De Witt equation (Wheeler 1968, DeWitt 1967) :

$$H\Psi = (H_g + H_\Phi)\Psi = 0 \tag{2}$$

where the momenta are now replaced by operators and H_Φ. We use the same factor ordering as before. If one is interested in making predictions only about the behavior of the minisuperspace variables, a suitable quantity for such a coarse-grained description is the reduced density matrix defined as:

$$\rho_{red}(r_2, r_1) = \int d\Phi \Psi^*(r_1, \Phi)\Psi(r_2, \Phi) \tag{3}$$

We will consider an ansatz for the wave function of the following form:

$$\Psi(r, \Phi) = \sum_n e^{iM S_{(n)}(r)} C_{(n)}(r) \chi_{(n)}(r, \Phi). \tag{4}$$

where S_n is a solution of a Hamilton-Jacobi equation with respect to the r coordinate which is the same as (2.12) without the backreaction term and with the gradient given by $G^{mm'} \frac{\partial}{\partial r^{m'}}$. The prefactor $C_{(n)}$ is determined through the equation $2G^{mm'} \partial_m C_{(n)} \partial_{m'} S_{(n)} + \Box_G S_0 = 0$, and $\chi_{(n)}$ is a solution of a Schrödinger equation with respect to the WKB time $\frac{d}{dt} = G^{mm'} \frac{\partial S_0}{\partial r^{m'}} \frac{\partial}{\partial r^m}$ and with Hamiltonian H_Φ. These equations are derived by the well -known order by order expansion in M^{-1}, where M acts as a parameter analogous to the mass of the heavy modes (in our case the minisuperspace ones) in the Born-Oppenheimer approximation. The Hamilton-Jacobi equation satisfied by $S_{(n)}$ will have a $D-1$ parameter family of solutions, the subindex (n) labeling such parameters characterizes a particular solution and thus a specific WKB branch.

The reduced density matrix associated with the wave function (3.4) is then given by:

$$\rho_{red}(r_2, r_1) = \sum_{n, n'} e^{iM[S_{(n)}(r_1) - S_{(n')}(r_2)]} C_{(n)}(r_1) C_{(n')}(r_2) I_{n, n'}(r_2, r_1) \tag{5}$$

where we call

$$I_{n,n'}(r_2, r_1) = \int \chi^*_{(n')}(r_2, \Phi)\chi_{(n)}(r_1, \Phi)d\Phi \qquad (6)$$

the influence functional. The justification for this name comes from the fact that it has been shown (Paz and Sinha 1991, Kiefer 1991) that it is exactly analogous to the Feynman-Vernon (1963) influence functional in the case in which the environment is initially in a pure state. For models with $r > 1$ it can be shown that it is indeed a functional of two histories, which is in turn related to the Gell-Mann Hartle decoherence functional (Gell-Mann and Hartle 1990, Griffith 1984, Omnes 1988). It is the central object of our consideration.

3.2.   Decoherence and Correlation

We will now study the emergence of classical behavior in the minisuperspace variables as a consequence of their interaction with the environment. As we will show the two basic characteristics of classicality, the issues of correlations and decoherence, are indeed interrelated.

i) **Decoherence:** Decoherence occurs when there is no interference effect between alternative histories. If each of the diagonal terms in the sum of (3.4) corresponds to a nearly classical set of histories, the interference between them is contained in the non–diagonal terms ($n \neq n'$) of that sum. Thus, the system decoheres if influence functional $I_{n,n'} \propto \delta_{n,n'}$ approximately. [5] As a consequence the density matrix (3.4) can be considered to describe a "mixture" of non–interfering WKB branches representing distinct Universes.

ii) **Correlations:** The second criterion for classical behavior is the existence of correlations between the coordinates and momenta which approximately obey the classical equations of motion. To analyze this aspect we will compute the Wigner function associated with the reduced density matrix and examine whether the Wigner function has a peak about a definite set of trajectories in phase space corresponding to the above correlations (Halliwell 1987, Padmandabhan and Singh 1990). Once we have established the decoherence between the WKB branches using criterion i), we look for correlations using the Wigner function *within* a "decohered WKB branch", i.e, that associated with ($n = n'$). In this sense the two criteria are interrelated, i.e, we *need* decoherence between the distinct WKB branches to be able to meaningfully predict correlations.

---

[5] for a suggestive argument to justify the choice of $I_{n,n'}(r, r)$ as an indicator of the degree of decoherence between the WKB branches, see Paz and Sinha (1992) Sec. V

The Wigner function associated with one of the diagonal terms is given by

$$W_{(n)}(r, P) = \int\limits_{-\infty}^{+\infty} d\xi^m C_{(n)}(r_1) C_{(n)}(r_2) e^{-2iP_m\xi^m} e^{iM[S_{(n)}(r_1) - S_{(n)}(r_2)]} I_{n,n}(r_2, r_1) \quad (7)$$

where $r_{1,2}^m = r^m \pm \frac{\xi^m}{M}$.

The functional $I_{n,n}(r_2, r_1)$ plays the dual role of producing the diagonalization of $\rho_{(n)}$ and affecting the correlations (the phase affects the correlations and the absolute value the diagonalization). [6] In the language of measurement theory (Wheeler and Zurek 1986) one can say that the environment is "continuously measuring" the minisuperspace variables (Zurek 1981, 1991, Zeh 1986, Kiefer 1987) and that this interaction not only suppresses the $n \neq n'$ terms in the sum of equation (3.11), but also generates a "localization" of the r variables inside each WKB branch. This localization effect is essential in order to obtain a peak in the Wigner function. The form of the peak and the precise location of its center are determined by the form of $I_{n,n}(r_2, r_1)$. To illustrate this better let us assume that the state of the environment is such that the influence functional can be written in the form

$$I_{(n,n)}(r) \simeq e^{i\beta_m(r)\xi^m - \sigma^2\xi_m\xi^m} \quad (8)$$

where β and σ are real and $r = \frac{r_1 + r_2}{2}$, and we notice that σ is related to the degree of diagonalization of the density matrix. The Wigner function for this is computed to be

$$W_{(n)}(r, P) \simeq C_{(n)}^2(r) \sqrt{\frac{\pi}{\sigma^2}} e^{-(P_m - M\frac{\partial S}{\partial r^m} - \frac{1}{2}\beta_m)^2} \quad (9)$$

It is a Gaussian peaked about the classical trajectory ($P_m - M\frac{\partial S}{\partial r^m} = 0$) shifted by a backreaction term $\frac{1}{2}\beta_m$, but with a spread characterized by σ, both of these arising from coarse graining the environment. We also notice a competition between sharp correlations (related to the sharpness of the above peak) and the diagonalization of the density matrix, and this can be formalized (see Paz and Sinha 1991, 1992) in a set of criteria for the emergence of classical behavior through a compromise between decoherence and correlations. It can also be shown that when the influence functional is of the form (3.8) the correlations predicted are exactly those given by the semiclassical Einstein equations. Thus the above arguments can be tied to the emergence of the semiclassical limit of quantum fields in curved space time starting from quantum cosmology. This has been illustrated in Paz and Sinha (1991, 1992) in the context of various specific cosmological models.

---

[6]The diagonalization produced by $I_{n,n}(r_2, r_1)$ has been studied by various authors (Kiefer 1987, Halliwell 1989, Padmandabhan 1989, Laflamme and Louko 1991) and has been identified with decoherence. However, as noted by Paz and Sinha (1992), this term applies more properly to the lack of interference between WKB branches which, as emphasized by Gell-Mann and Hartle (1990), is the more relevant effect. However, it is clear that the diagonalization is an effect that accompanies the former one and has the same origin.

4.   Statistical Mechanics and Quantum Cosmology

The above two examples give a good illustration of how models of quantum open systems can be used to understand some basic issues of quantum cosmology. The advantage of using statistical mechanical concepts for the study of issues in quantum cosmology and semiclassical gravity has been discussed in general terms by Hu (1991) who emphasized the importance of the interconnection between processes of decoherence, correlation, dissipation, noise and fluctuation, particle creation and backreaction. (see also Hu, Paz and Zhang 1992, 1993, and Gell-Mann and Hartle 1993). In our view, these processes are actually different manifestations of the effect of the environment on the different attributes of the system (the phase information, energy distribution, entropy content, etc). Decoherence is related to particle creation as they both can be related to the Bogolubov coefficients, as is apparent in the example of Calzetta and Mazzitelli (1991) and Paz and Sinha (1992). It is also important to consider the nature of noise and fluctuation for any given environment. In this way their overall effect on the system can be captured more succintly and effectively by general categorical relations like the fluctuation-dissipation relations (see Hu 1989, Hu and Sinha 1993). This also sheds light on how to address questions like defining gravitational entropy and related questions mentioned in the Introduction (see Hu 1993b).

So far one has only succeeded in finding a pathway to show how semiclassical gravity can be deduced from quantum cosmology. One important approximation which makes this transition possible is the assumption of a WKB wave function. To see the true colors of quantum gravity, which is nonlinear and is likely to be also nonlocal, one needs to avoid such simplifications. One should incorporate dynamical fluctuations both in the fields and in the geometry without any background field separation, and deal with nonadiabatic and nonlinear conditions directly. This is a difficult but necessary task. Calzetta (1991) has tackled the anisotropy dissipation problem in quantum cosmology without such an approximation. Recently Calzetta and Hu (1993) have proposed an alternative approach to address the quantum to classical transition issue in terms of correlations between histories. It uses the BBGKY hierarchy truncation scheme to provide a more natural coarse-graining measure which brings about the decoherence of correlation histories. This scheme goes beyond a simple system-environment separation and enables one to deal with nonadiabaticity and nonlinearity directly as in quantum kinetic theory.

Acknowledgement

This essay highlights the work done by Calzetta and the three of us in the years 1987-1992, a good twenty years after Charlie Misner's seminal works in relativistic cosmology. It is clear from this coarse sampling how much our work is indebted to Misner intellectually. It is also our good fortune to have had personal interactions with Charlie as colleague and friend. We wish him all the best on his sixtieth birthday and look forward to his continuing inspiration for a few more generations of relativists. This work is supported in part by the National Science Foundation under grant PHY91-19726.

References

A. Anderson, Phys. Rev. **D42**, 585 (1990)

R. Arnowitt, S. Deser, and C. W. Misner, "The Dynamics of General Relativity." in Gravitation: An Introduction to Current Research, ed. L. Witten, pp. 227-65 (Wiley, New York, 1962)

V. A. Belinskii, E. M. Lifschitz and I. M. Khalatnikov, Adv. Phys. **19**, 525 (1970)

B. K. Berger, Phys. Rev. **D11**, 2770 (1975)

B. K. Berger, Ann. Phys. (N. Y.) **83**, 203 (1974)

L. Bianchi, *Mem. di. Mat. Soc. Ital. Sci.*, **11**, 267 (1897)

N. D. Birrell and P. C. W. Davies, *Quantum Fields in Curved Space,* (Cambridge University Press, Cambridge, 1982).

A. O. Caldeira and A. J. Leggett, Physica **121A**, 587 (1983)

E. Calzetta, Class. Quant. Grav. **6**, L227 (1989)

E. Calzetta, Phys. Rev. **D43**, 2498 (1991)

E. Calzetta and B. L. Hu, Phys. Rev. **D35**, 495 (1987)

E. Calzetta and B. L. Hu, Phys. Rev. **D37**, 2838 (1988)

E. Calzetta and B. L. Hu, Phys. Rev. **D40**, 380 (1989)

E. Calzetta and B. L. Hu, Phys. Rev. **D40**, 656 (1989)

E. Calzetta and B. L. Hu, "Decoherence of Correlation Histories" in *Directions in General Relativity* Vol 2 (Brill Festschrift), eds. B. L. Hu, M. P. Ryan and C. V. Vishveshwara (Cambridge Univ. Cambridge, 1993)

E. Calzetta and F. Mazzitelli, Phys. Rev. **D42**, 4066 (1991)

P. Candelas and D. W. Sciama, Phys. Rev. Lett. **38**, 1372 (1977)

D. M. Capper and M. J. Duff, Nuovo Cimento **A23**, 173 (1975)

S. Deser, M. J. Duff and C. J. Isham, Nucl. Phys. **B111**, 45 (1976)

P. C. W. Davies, Rep. Prog. Phys. **41**, 1313 (1978)

P. C. W. Davies, L. H. Ford and D. N. Page Phys. Rev. **D34**, 1700 (1986)

B. S. DeWitt, Phys. Rev. **160**, 1113 (1967)

B. S. DeWitt, Phys. Rep. **19C**, 297 (1975)

B. S. DeWitt, in *Quantum Concepts in Space and Time*, ed. R. Penrose and C. J. Isham (Claredon Press, Oxford, 1986)

H. F. Dowker and J. J. Halliwell, Phys. Rev. **D46**, 1580 (1992)

D. Eardley, E. Liang and R. K. Sachs, J. Math. Phys. **13**, 99 (1971)

G. F. R. Ellis and M. A. H. MacCallum, Commun. Math. Phys. **12**, 108 (1969)

R. P. Feynman and F. L. Vernon, Ann. Phys. **24**, 118 (1963)

M. V. Fischetti, J. B. Hartle and B. L. Hu, Phys. Rev. **D20**, 1757 (1979)

S. A. Fulling, Phys. Rev. **D7**, 2850 (1973)

M. Gell-Mann and J. B. Hartle, in *Complexity, Entropy and the Physics of Information*, ed. W. Zurek, Vol. IX (Addison-Wesley, Reading, 1990)

M. Gell-Mann and J. B. Hartle, Phys. Rev. **D47** (1993)

R. Griffiths, J. Stat. Phys. **36**, 219 (1984)

A. H. Guth, Phys. Rev. **D23**, 347 (1991)

S. Habib , Phys. Rev. **D42**, 2566 (1990)

S. Habib and R. Laflamme, Phys. Rev. **D42**, 4056 (1990)

J. J. Halliwell, Phys. Rev. **D36**, 3627 (1987)

J. J. Halliwell, Phys. Rev. **D39**, 2912 (1989)

J. J. Halliwell, Lectures at the 1989 Jerusalem Winter School, in *Quantum Mechanics and Baby Universes*, eds. S. Coleman, J. Hartle, T. Piran and S. Weinberg (World Scietific, Singapore, 1991)

J. J. Halliwell and S. W. Hawking, Phys. Rev. **D31**, 1777 (1985)

J. B. Hartle, Phys. Rev. Lett. **39**, 1373 (1977)

J. B. Hartle, in *Gravitation in Astrophysics*, NATO Advanced Summer Institute, Cargese, 1986 ed. B. Carter and J. Hartle (NATO ASI Series B: Physics Vol. 156, Plenum, N.Y 1987).

J. B. Hartle and S. W. Hawking, Phys. Rev. **D28**, 1960 (1983)

J. B. Hartle and B. L. Hu, Phys. Rev. **D20**, 1772 (1979)

J. B. Hartle and B. L. Hu, Phys. Rev. **D21**, 2756 (1980)

S. W. Hawking, *Nature* (London),**248**, 30 (1974)

O. Heckman and E. L. Schucking, In *Gravitation: An Introduction to Current Research*, ed. L. Witten (Wiley, New York, 1962)

B. L. Hu, Phys. Lett. **A90**, 375 (1982)

B. L. Hu, Phys. Lett. **A97**, 368 (1983)

B. L. Hu, in *Cosmology of the Early Universe*, ed. L. Z. Fang and R. Ruffini (World Scientific, Singapore, 1984)

B. L. Hu, Physica **A158**, 399 (1989)

B. L. Hu "Quantum and Statistical Effects in Superspace Cosmology" in *Quantum Mechanics in Curved Spacetime*, ed. J. Audretsch and V. de Sabbata (Plenum, London 1990)

B. L. Hu, in *Relativity and Gravitation: Classical and Quantum*, Proc. SILARG VII, Cocoyoc, Mexico 1990, eds. J. C. D' Olivo et al (World Scientific, Singapore 1991)

B. L. Hu, "Statistical Mechanics and Quantum Cosmology", in *Proc. Second International Workshop on Thermal Fields and Their Applications*, eds. H. Ezawa et al (North-Holland, Amsterdam, 1991)

B. L. Hu, "Fluctuation, Dissipation and Irreversibility" in *The Physical Origin of Time-Asymmetry*, Huelva, Spain, 1991, eds. J. J. Halliwell, J. Perez-Mercader and W. H. Zurek (Cambridge University Press, 1993)

B. L. Hu, "Quantum Statistical Processes in the Early Universe" in *Quantum Physics and the Universe*, Proc. Waseda Conference, Aug. 1992 ed. Namiki, K. Maeda, et al (Pergamon Press, Tokyo, 1993)

B. L. Hu and H. E. Kandrup, Phys. Rev. **D35**, 1776 (1987)

B. L. Hu and L. Parker, Phys. Lett. **63A**, 217 (1977)

B. L. Hu and L. Parker, Phys. Rev. **D17**, 933 (1978)

B. L. Hu and D. Pavon, Phys. Lett. **B180**, 329 (1986)

B. L. Hu, J. P. Paz and Y. Zhang, Phys. Rev. **D45**, 2843 (1992)

B. L. Hu, J. P. Paz and Y. Zhang, Phys. Rev. **D47** (1993)

B. L. Hu, J. P. Paz and Y. Zhang, "Stochastic Dynamics of Interacting Quantum Fields" Phys. Rev. D (1993)

B. L. Hu and S. Sinha, "Fluctuation-Dissipation Relation in Cosmology" Univ. Maryland preprint (1993)

B. L. Hu and Y. Zhang, "Coarse-Graining, Scaling, and Inflation" Univ. Maryland Preprint 90-186 (1990)

E. Joos and H. D. Zeh, Z. Phys. **B59**, 223 (1985)

R. D. Jordan, Phys. Rev. **D33**, 44 (1986)

H. E. Kandrup, Phys. Rev. **D37**, 3505 (1988)

L. V. Keldysh, Zh. Eksp. Teor. Fiz. **47**, 1515 (1964) [Sov. Phys. JEPT **20**, 1018 (1965)]

C. Kiefer, Class. Quant. Grav. **4**, 1369 (1987)

C. Kiefer, Class. Quant. Grav. **8**, 379 (1991)

K. Kuchar, Phys. Rev. **D4**, 955 (1971)

K. Kuchar and M. P. Ryan Jr., in *Proc. of Yamada Conference XIV* ed. H. Sato and T. Nakamura (World Scientific, 1986)

K. Kuchar and M. P. Ryan Jr., Phys. Rev. **D40**, 3982 (1989)

R. Laflamme and J. Luoko, Phys. Rev. **D43**, 3317(1991)

E. M. Lifschitz and I. M. Khalatnikov, Adv. Phys. **12**, 185 (1963)

V. N. Lukash and A. A. Starobinsky, JETP **39**, 742 (1974)

R. A. Matzner and C. W. Misner, Ap. J. **171**, 415 (1972)

C. W. Misner, Ap. J. **151**, 431 (1968)

C. W. Misner, Phys. Rev. Lett. **22**, 1071 (1969)

C. W. Misner, Phys. Rev. **186**, 1319 (1969)

C. W. Misner, in *Magic Without Magic*, ed. J. Klauder (Freeman, San Francisco, 1972)

E. Mottola, Phys. Rev. **D33**, 2136 (1986)

R. Omnes, J. Stat. Phys. **53**, 893, 933, 957 (1988); Ann. Phys. (NY) **201**, 354 (1990): Rev. Mod. Phys. **64**, 339 (1992)

T. Padmanabhan, Phys. Rev. **D39**, 2924 (1989)

T. Padmanabhan and T. P. Singh, Class. Quan. Grav. **7**, 411 (1990)

L. Parker, "The Creation of Particles in an Expanding Universe," Ph.D. Thesis, Harvard University (unpublished, 1966).

L. Parker, Phys. Rev. **183**, 1057 (1969)

L. Parker, Phys. Rev. **D7**, 976 (1973)

J. P. Paz, Phys. Rev. **D41**, 1054 (1990)

J. P. Paz, Phys. Rev. **D42**, 529 (1990)

J. P. Paz and S. Sinha, Phys. Rev. **D44**, 1038 (1991)

J. P. Paz and S. Sinha, Phys. Rev. **D45**, 2823 (1992)

J. P. Paz and W. H. Zurek, in preparation (1993)

R. Penrose, "Singularities and Time-Asymmetry" in *General Relativity: an Einstein Centenary Survey*, eds. S.W. Hawking and W. Israel (Cambridge University Press, Cambridge, 1979)

M. P. Ryan, *Hamiltonian Cosmology* (Springer, Berlin, 1972)

M. P. Ryan, Jr. and L. C. Shepley, *Homogeneous Relativistic Cosmologies* (Princeton University Press, Princeton, 1975)

J. Schwinger, Phys. Rev., **82**, 664 (1951)

J. S. Schwinger, J. Math. Phys. **2** 407 (1961).

R. U. Sexl and H. K. Urbantke, Phys. Rev., **179**, 1247 (1969)

S. Sinha, Ph. D. Thesis, University of Maryland (unpublished 1991)

S. Sinha and B. L. Hu, Phys. Rev. **D44**, 1028 (1991).

A. H. Taub, Ann. Math., **53**, 472 (1951)

T. Vachaspati and A. Vilenkin, Phys. Rev. **D37**, 898 (1988)

A. Vilenkin, Phys. Rev. **D27**, 2848 (1983), **D30**, 509 (1984); Phys. Lett. **117B**, 25 (1985)

J. A. Wheeler, in *Relativity, Groups and Topology*, ed. B. S. De Witt and C. De Witt (Gordon and Breach, New York, 1964)

J. A. Wheeler, in *Battelle Rencontres* ed. C. DeWitt and J.A Wheeler (Benjamin, N. Y. 1968)

J. A. Wheeler and W. H. Zurek, *Quantum Theory and Measurement* (Princeton University Press, Princeton 1986)

E. P. Wigner, Phys. Rev. **40**, 749 (1932)

H. D. Zeh, Phys. Lett. **A116**, 9 (1986).

Ya. B. Zel'dovich, Pis'ma Zh. Eksp. Teor. Fiz, **12** ,443 (1970) [JETP Lett. **12**, 307(1970)]

Ya. B. Zel'dovich, in *Physics of the Expanding Universe*, ed. M. Diemianski (Springer, Berlin, 1979)

Ya. B. Zel'dovich and A. A. Starobinsky, Zh. Teor. Eksp. Fiz. **61** (1971) 2161 [JEPT **34**, 1159 (1972)]

W. H. Zurek, Phys. Rev. **D24**, 1516 (1981); **D26**, 1862 (1982); in *Frontiers of Nonequilibrium Statistical Physics*, ed. G. T. Moore and M. O. Scully (Plenum, N. Y., 1986); Physics Today **44**, 36 (1991)

R. Zwanzig, in *Lectures in Theoretical Physics*, ed. W. E. Britten, B. W. Downes and J. Downes (Interscience, New York, 1961)

Ricci Flow on Minisuperspaces and the Geometry-Topology Problem

James Isenberg *   *Martin Jackson* †

Abstract

We review recent work which studies the behavior of the Ricci flow of locally homogeneous Riemannian 3-geometries on closed 3-manifolds. We discuss the role this work plays in Hamilton's program for the study of Thurston's geometrization conjecture.

1.   Introduction

Ever since Einstein first proposed the field equations for his general relativistic theory of gravity, researchers searching for solutions have used the technique of assuming spacetime isometries to simplify this system of equations. Virtually all of the known exact solutions of Einstein's equations have been found in this way.

It was Misner's insight to extend this technique to the qualitative study of the behavior of solutions. By assuming spatial homogeneity, Misner and his coworkers [1] were able to reduce the daunting dynamical analysis of the full collection of solutions to the quite manageable study of a finite dimensional system. Much has been learned [2], [3] about Einstein's theory and its solutions by applying qualitative dynamical systems ideas to this finite dimensional "minisuperspace," in which each point is a homogeneous 3-geometry, and in which the dynamically determined trajectories correspond to spatially homogeneous cosmological spacetimes.

In recent work [4], we have used this same approach to gain some understanding of how Ricci flows behave. In the study of Ricci flow, one follows the evolution of Riemannian geometries as prescribed by the nonlinear parabolic Ricci flow equation

$$\frac{\partial}{\partial t}\gamma_{ij}(t,x) = -2R_{ij}[\gamma(t,x)]; \tag{1}$$

---

*Department of Mathematics and Institute of Theoretical Science, University of Oregon,Eugene, OR 97403. This work has been partially supported by the National Science Foundation under grant PHY-902301 at the University of Oregon.

†Department of Mathematics and Computer Science, University of Puget Sound, Tacoma, WA 98416 .

here $\gamma_{ij}(t,x)$ is a t-parametrized family of Riemannian metrics, while R_{ij} is the Ricci tensor corresponding to $\gamma_{ij}(t,x)$. This heat-like evolution, first proposed for study by Bourgignon [5], has been vigorously analyzed by Hamilton [6], [7], [8]. Besides proving a number of very nice mathematical results concerning the properties of solutions of (1), Hamilton has outlined a program for using these solutions to study the relationship between low dimensional manifolds and the geometries which they admit (the "geometry-topology problem"). This program depends not on obtaining exact solutions to equation (1), but rather on obtaining a qualitative understanding of the collective behavior of solutions. Hence, a qualitative dynamical analysis is appropriate. Since the Ricci flow equation is, like Einstein's equation, a complicated nonlinear PDE system, even a rough qualitative picture of the behavior of the Ricci flow for general geometries is presently far beyond reach. Thus, as a model study, we have examined Ricci flow on minisuperspace.

There is a big difference between the Einstein evolution and the Ricci flow evolution on minisuperspace: While the Einstein evolution (determined by a nonlinear hyperbolic system) is more or less chaotic,[1] the Ricci flow (determined by a nonlinear parabolic system) is more or less asymptotically stable. Thus, as we shall see, we are able to obtain a complete understanding of the asymptotic behavior of Ricci flow in minisuperspace. We describe our results in Section 3.

While a minisuperspace analysis of Ricci flow is only a small step towards an understanding of the general behavior of Ricci flows, such an analysis does play an important role in Hamilton's program. We describe this role in Section 2, after discussing some of the important results and conjectures of the geometry-topology problem (including the Thurston geometrization conjecture), and remarking on how Ricci flow may serve as a tool for studying this problem. In Section 3, we set up our analysis of Ricci flow in minisuperspace, and discuss the results. Finally in Section 4 we conclude with remarks concerning other work–e.g., Ricci flow on various "midisuperspaces"–which may help in the study of the geometry-topology problem.

2. The Geometry-Topology Problem and Ricci Flow

Nearly one hundred years ago, analysts [9] completed work on a remarkable result which relates the topology of two-dimensional closed manifolds to the constant curvature Riemannian geometries admissable on such manifolds. Recall that a two-dimensional closed (compact without boundary) orientable manifold M^2 is completely classified by a non-negative integer g, called the "genus" of the manifold. Roughly speaking, g counts the number of holes in M^2: a sphere has $g = 0$, a torus has $g = 1$, and so on.

Recall, too, that in two dimensions the scalar curvature determines the full curva-

---

[1]in practice, if not always by official definition

ture of a Riemannian metric, and there are exactly three types of constant curvature geometries: the positive curvature spherical geometries, the zero curvature flat geometries, and the negative curvature hyperbolic geometries.

The remarkable result, connecting the topology and the geometry of two-dimensional closed manifolds, says the following:

Two Dimensional Uniformization Theorem: *Every closed, orientable, two-dimensional manifold M^2 admits one and only one type of constant curvature geometry, with the type admitted being determined by the genus of M^2:*

	spherical	*if and only if $g = 0$;*
	flat	*if and only if $g = 1$;*
and	*hyperbolic*	*if and only if $g \geq 2$.*

While the original proof of this theorem involves techniques from complex analysis and the study of Riemann surfaces, recent work by Hamilton [8] and Chow [10] shows that one can also prove the Uniformization Theorem using two dimensional Ricci flow. Here is the idea:

Fix M^2 and choose an arbitrary Riemannian metric $\gamma_0(x)$ on M^2. Now evolve $\gamma_0(x)$ into a one-parameter family of metrics $\gamma(t, x)$ via the Ricci flow equation in its volume-preserving form

$$\frac{\partial}{\partial t}\gamma_{ij}(t, x) = -2R_{ij}[\gamma(t, x)] + \langle R[\gamma(t, x)]\rangle \gamma_{ij}(t, x); \qquad (2)$$

here $\langle R[\gamma(t, x)]\rangle$ is the spatial average of the scalar curvature over M^2. The only fixed points of the Ricci flow in two dimensions are the constant curvature geometries,[2] so if γ_0 does not have constant curvature (and if the Ricci flow initial value problem is well-posed), then $\gamma(t, x)$ evolves. If one can show that, for any initial choice of γ_0 on any M^2, the Ricci flow exists for $t \geq 0$ and converges for $t \to \infty$, then it follows that M^2 admits a constant curvature geometry since the flow can only converge to a fixed point.

Hamilton and Chow have proven that indeed two dimensional Ricci flow always converges. Hence, as outlined, the two-dimensional Uniformization Theorem follows.

The behavior of three-dimensional Ricci flow is considerably more complicated than that of its two-dimensional counterpart. Indeed, one easily verifies that in three dimensions (unlike in two) Ricci flows may become singular. The geometry consisting of a round two-sphere crossed (simple product) with a circle provides a good (easy) analytic example of this: the two sphere shrinks to zero volume and the circle expands without bound as the (volume preserving) Ricci flow of this geometry proceeds [see Figure 1].

---

[2] $\frac{\partial}{\partial t}\gamma_{ij} = 0$ iff $R_{ij} = \frac{1}{2}\langle R\rangle\gamma_{ij}$, which implies constant curvature.

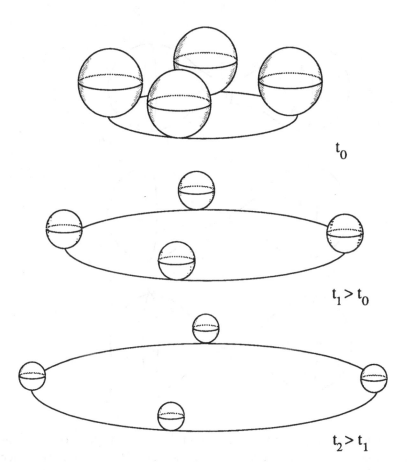

t_0

$t_1 > t_0$

$t_2 > t_1$

Figure 1. Schematic representation of Ricci flow for a round two-sphere crossed with a circle

As another example, one verifies numerically that a sufficiently tightly-corseted three-sphere has a singular Ricci flow as the waist pinches down in finite time [see Figure 2]. Note that the Ricci flow of any tightly-corseted two-sphere, by contrast, always springs back to a round S^2.

Since three-dimensional Ricci flows can become singular, one cannot hope to use it to prove a three-dimensional uniformization theorem. Fittingly, there is no such theorem – note, for example, that the manifold $S^2 \times S^1$ admits no constant curvature geometry. Thurston [11],[12], however, has conjectured that while closed three-manifolds do not always admit constant curvature geometries, the following holds:

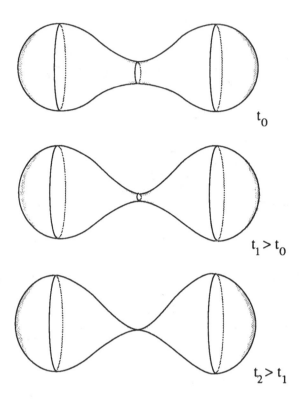

t_0

$t_1 > t_0$

$t_2 > t_1$

Figure 2. Schematic representation of Ricci flow for a tightly corseted three-sphere. The three-sphere is pictured here as a two-sphere. (Note that the Ricci flow of a two-sphere, tightly corseted or not, always converges to a round two sphere rather than pinching down.)

Thurston's Three-Dimensional Geometrization Conjecture: *For any closed, orientable three-manifold M^3, there is a canonical decomposition[3] of M^3 into a set of three-manifolds $\{M_\alpha^3\}$ such that each M_α^3 admits a locally homogeneous geometry.*

---

[3]This decomposition is done in two steps. The first step decomposes M^3 into a set of "prime manifolds" $\{M_i^3\}$ such that $M^3 = \#_i M_i^3$. A manifold N^3 is "prime" if any expression $N^3 = N_1^3 \# N_2^3$ implies that either N_1^3 or N_2^3 is homeomorphic to S^3. The operator "#" denotes a connected sum. One splits N^3 into a connected sum $N_1^3 \# N_2^3$ by cutting N^3 along an embedded two-sphere, and then patching a three-ball onto the boundary S^2 on each piece. N_1 and N_2 result. Work by Kneser [13] and Milnor [14] shows that the decomposition $M^3 = \#_i M_i^3$ can always be done, with $\{M_i^3\}$ a unique set.

The second step decomposes the M_i^3 further. Jaco and Shalen [15] show that in each M_i^3, there is a finite (possibly empty) collection of disjoint incompressible two-sided tori which split M_i^3 so that the manifold on each side is either a Seifert fiber space (with S^1 action) or it admits no further embedded incompressible torus except possibly one which cuts parallel to the boundary. One cuts the M_i^3 along these tori to obtain the $\{M_\alpha^3\}$ which make up the decomposition. Note that generally the M_α^3 are either closed or have $T^2 \times \Re^1$ ends.

What is a locally homogeneous geometry? By definition, a Riemannian metric γ on M^3 is locally homogeneous if, for every pair of points $x, y \in M^3$, there exist neighborhoods U_x of x and V_y of y such that there is an isometry Ψ mapping $(U_x, \gamma \mid U_x)$ to $(V_y, \gamma \mid V_y)$ with $\Psi(x) = y$. Generally, these local isometries do not extend to isometries of the whole space (M^3, γ). Any one of the compact hyperbolic Riemannian geometries (M^3, γ_{hyp}) such as that of Löbel [16] or the Seifert-Weber dodecahedral space [11] provides an example of this. If the isometries *do* extend to all of M^3 – e.g., for the hyperbolic metric γ_{hyp} on $M^3 = \Re^3$ – then the geometry is homogeneous. That is, (M^3, γ) is homogeneous if for every pair of points $x, y \in M^3$ there exists an isometry Ψ from (M, γ) to itself which takes x to y.

The homogeneous geometries are familiar to relativists as essentially the Bianchi types. While the set of locally homogeneous geometries sounds like a much bigger class, a result of Singer [18] shows that, for our purposes here, it is not really any bigger. Singer proves that for every locally homogeneous geometry (M^3, γ), the universal cover $(\tilde{M}^3, \tilde{\gamma})$ is homogeneous. We will exploit this result below in our study of the Ricci flow of locally homogeneous geometries.

Thurston's geometrization conjecture is a very strong one. One finds, for example, that if it proves to be true, then the Poincare conjecture would be an immediate corollary.[4] It is, however, far from proven and some mathematicians believe it is false. How could Ricci flow be useful in studying it? Hamilton suggests the following scenario:

Fix M^3, and choose an arbitrary Riemannian metric $\gamma_0(x)$ on M^3. Now evolve $\gamma_0(x)$ into the family $\gamma(t, x)$ via the three-dimensional (volume preserving) Ricci flow

$$\frac{\partial}{\partial t} \gamma_{ij}(t, x) = -2R_{ij}[\gamma(t, x)] + \frac{2}{3} < R[\gamma(t, x)] > \gamma_{ij}(t, x) \tag{3}$$

Singularities may occur. Those which occur in finite time should be of the "S^2 neck pinching" form [see Figure 3a]. When a singularity of this type occurs, cut the neck and patch each side with a three-ball, and then continue the Ricci flow in each of the two resulting geometries. As t goes to ∞, "T^2 necks" may also form [see Figure 3b]. Cut them, and leave the resulting $T^2 \times \Re$ cylinders as ends on the resulting manifolds.

After all the S^2 neck cutting and patching as well as the T^2 neck cutting is over, M^3 has been decomposed into a set of manifolds $\{M_\alpha^3\}$. Now allow the Ricci flow on each

---

[4]The Poincare conjecture says that every manifold M^3 which is homotopic to S^3 must be homeomorphic to S^3. This follows from Thurston's conjecture as follows:

If M^3 is homotopic to S^3, then the canonical topological decomposition is trivial – one gets $\{M_\alpha^3\} = \{M^3\}$. The Thurston conjecture (if true) then requires that M^3 admit a locally homogeneous geometry. But the only *locally homogeneous* geometries which a manifold M^3 that is homotopic to S^3 admits are the SU(2) symmmetric ones – i.e., Bianchi IX. Since every three-manifold which admits SU(2) symmetric geometries is homeomorphic to either S^3 or one of the lens spaces, and since the lens spaces are not homotopic to S^3, it follows that M^3 is homeomorphic to S^3.

piece M_α^3 to proceed, and try to show that it either converges to a locally homogeneous geometry, or asymptotically approaches the Ricci flow of a family of such geometries.

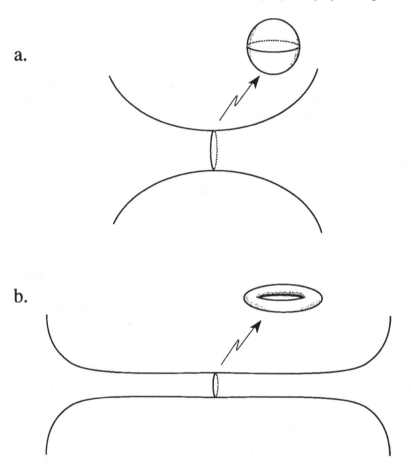

a.

b.

Figure 3. Schematic representation of three-dimensional Ricci flow singularities. a) A two-sphere neck pinch, which can occur in finite time. b) A two-torus neck pinch, which can occur as t goes to infinity

This is a fairly complex, rigid, scenario, and some find it to be a somewhat far-fetched approach for attempting to prove Thurston's conjecture. It is certainly an interesting one, however, which we believe is well worth pursuing.

As noted, according to Hamilton's scenario, after all the cutting and patching of M^3 has been completed, the Ricci flow on each M_α^3 is supposed to approach that of some family of locally homogeneous geometries. To understand what this means, it is clearly necessary to explore how Ricci flows of locally homogeneous geometries behave. Some locally homogeneous geometries, like the round three-sphere geometry, the Löbel hyperbolic geometry, etc., are Einstein geometries (with $R_{ij}[\gamma] = \Lambda\gamma_{ij}$ for some constant Λ) and hence fixed points of Ricci flow. These can be the limiting geometries of some of the Ricci flows on the manifolds M_α^3, according to Hamilton's scenario. But most locally homogeneous geometries are Einstein geometries, and further most of these cannot be smoothly deformed into Einstein geometries. For these, it is crucial to study how the Ricci flow behaves asymptotically if we want to make sense of, and carefully test, Hamilton's scenario. We discuss the Ricci flows of locally homogeneous geometries in the next section.

3. The Ricci Flow of Locally Homogeneous Geometries

The analysis which reduces the Einstein equations for spatially homogeneous space-times into a system of ordinary differential equations is a familiar one to general relativists. First, one chooses the three (or four) dimensional Lie group which characterizes the spatial homogeneity – either one of the Bianchi types, or $SO(3) \times S^1$ for the Kantowski-Sachs spacetimes. Then, one writes the spatial metric in the form

$$\gamma = \gamma_{ab}(t)\theta^a \otimes \theta^b \tag{4}$$

where $\{\theta^a\}$ is a fixed triad which is invariant under the group action, and γ_{ab} is a spatially constant time-dependent matrix. One similarly expands the extrinsic curvature tensor in this basis

$$K = K_{cd}(t)\theta^c \otimes \theta^d \tag{5}$$

The constraint equations now become a set of algebraic equations relating the matrices γ_{ab} and K_{cd}. The Einstein evolution equations (with a suitable choice of lapse function to control the time slicing) become an ordinary differential equation system for the time dependent $\gamma_{ab}(t)$ and $K_{cd}(t)$. The exact form of these constraint and evolution equations for $\gamma_{ab}(t)$ and $K_{cd}(t)$ depend upon the Lie group chosen (see, e.g., Ryan and Shepley [2]).

The same sort of analysis works for the study of Ricci flow of homogeneous geometries. There are, however, two important differences between the formulation of Ricci flow as an initial value problem and that of the Einstein equations: First, since Ricci flow is parabolic rather than hyperbolic, one does not need the extrinsic curvature K to be included in the initial data or among the dynamic variables. Second, the Ricci flow system has no constraints. So one obtains, for each choice of homogeneity group, an ordinary differential equation system (and unconstrained initial value problem) for the time dependent matrix γ_{ab}.

We are interested in the Ricci flow of locally homogeneous geometries, not just homogeneous geometries. However, as we noted in Section 2, every locally homogeneous geometry (M^3, γ) has, for its simply connected cover, a homogeneous geometry $(\tilde{M}^3, \tilde{\gamma})$. Now locally γ and $\tilde{\gamma}$ are represented by exactly the same matrix γ_{ab}.[5] As well, the

---

[5]More specifically, we have $\gamma = \gamma_{ab}\theta^a \otimes \theta^b$ and $\tilde{\gamma} = \gamma_{ab}\tilde{\theta}^a \otimes \tilde{\theta}^b$, where $\tilde{\theta}^a = \Psi^*\theta^a$ for the covering map Ψ.

Ricci flow equations for $\tilde{\gamma}$ and γ reduce to exactly the same ordinary differential equation system for the matrix γ_{ab}. Hence the study of the Ricci flow for any family of locally homogeneous geometries is identical to that of a corresponding family of homogeneous geometries - -i.e., Ricci flow on a minisuperspace.

While it may ultimately be necessary to understand the asymptotic behavior of Ricci flow for all families of locally homogeneous geometries, we have chosen for the present to focus on those families which live on closed three-manifolds. In Hamilton's program for the study of Thurston's conjecture, one starts with a closed three-manifold M^3. If only S^2 neck pinches occur, then the daughter manifolds M_a^3 are also closed. More generally, (i.e., if T^2 necks occur) the daughter manifolds M_α^3 may be nonclosed, so it will probably be necessary to learn about a wider class of locally homogeneous Ricci flows. For now, however, we restrict our attention to geometries on closed manifolds.

Locally homogeneous geometries on closed three-manifolds fit into nine classes. These classes are each characterized by a particular Lie group G of dimension 3, 4 or 6. A geometry (M^3, γ) is contained in the class C_G if (a) G acts transitively and isometrically on the covering geometry $(\tilde{M}^3, \tilde{\gamma})$ of (M^3, γ), and (b) G is the smallest Lie group for which this is true. For example, G = SU(2) labels one of these classes of locally homogeneous geometries. The group SU(2) is topologically the three sphere (simply connected) so all geometries (M^3, γ) in $C_{SU(2)}$ have $\tilde{M}^3 = S^3$ as the simply-connected cover of M^3 (hence M^3 is either S^3 or one of the lens spaces) and all have γ taking the local form $\gamma = \gamma_{ab}\theta^a \otimes \theta^b$, where $\tilde{\theta}^a = \Psi^*\theta^a$ for $\{\tilde{\theta}^a\}$ a set of linearly-independent, left-invariant one-forms on SU(2) = S^3, and where $\Psi : S^3 \rightarrow M^3$ is the covering map. Included in $C_{SU(2)}$ are the round sphere geometries, each with the six-dimensional isometry group $SU(2) \times SO(3)$. Also included in $C_{SU(2)}$ are geometries with the four-dimensional isometry group $SU(2) \times S^1$ (these are called "locally rotationally symmetric" by relativists), and geometries with just the three-dimensional isometry group SU(2). The geometries of the class $C_{SU(2)}$ are parametrized by the (three-dimensional) set of distinct inner products on $\Re^3$ (labeled by symmetric positive definite matrices), with the round spheres labeled by the isotropic inner products $(\gamma_{ab} = \gamma I_{ab})$. The class $C_{SU(2)}$, of course, corresponds to the Bianchi IX geometries which were studied by Misner and his co-workers as part of the "mixmaster program."

As a quite different example of these classes of locally homogeneous geometries we note the class $C_{SO(3) \times \Re^1}$. The group $SO(3) \times \Re^1$ is a four dimensional Lie group; it acts transitively (but not simply transitively) on $S^2 \times \Re^1$. All the geometries (M^3, γ) in $C_{SO(3) \times R^1}$ have covering manifolds $\tilde{M}^3 = S^2 \times \Re^1$, and the metrics γ all take the simple product form

$$\gamma = \alpha\gamma_{S^2} + \beta\gamma_{\Re^1}$$

where α and β are spatial constants, γ_{S^2} is the standard round metric on the two-sphere, and $\gamma_{\Re^1}$ is the standard flat metric line on the line. For all of the geometries in this class, the isometry group is exactly $SO(3) \times \Re^1$. To relativists, this class is familiar in the study of Kantowski-Sachs cosmological models.

There are, as noted, seven other classes. We list them all in Table 1, according to their defining group G. For each class, we note (in the table) the corresponding Bianchi type (if any), we list the dimension of the isometry groups which the covering geometries of members of the class may have, and we list the number of parameters needed to label all the geometries in the class. Note that if G is three-dimensional and hence isotopic to the covering manifold $\tilde{M}^3$ of any geometry (M^3, γ) in C_G, then M^3 can be closed *only* if G is unimodular. Hence the restriction to *closed* locally homogeneous geometries eliminates Bianchi types III and IV and most of Bianchi type V, among others.

Table 1. Classes C_G of Locally Homogeneous Geometries on Closed Three-Manifolds

Defining Group	Bianchi Type	Allowed Isometry Groups (dimensions)	Family of Geometries (parameters)
$\mathfrak{R}^3$	I	3,4,6	3
SU(2)	IX	3,4,6	3
SL(2,$\mathfrak{R}$)	VIII	3,4	3
Heisenberg	II	3,4	3
E(1,1)	IV_{-1}	3	3
E(2)	VII_0	3,4,6	3
H(3)	Contained in V	6	1
SO(3) $\times \mathfrak{R}^1$	Kantowski-Sachs	4	2
H(2) $\times \mathfrak{R}^1$	–	4	2

In [4], we examine in detail the volume-preserving Ricci flow for all nine classes of locally homogeneous geometries on closed manifolds, and determine the asymptotic behavior in each class.[6] We will summarize these results below, and relate them to Hamilton's scenario. Before doing that, we wish to work out a couple of the cases to show how the analysis proceeds. Let us start with $C_{SU(2)}$, the mixmaster Ricci flow:

For any geometry (M^3, γ) in $C_{SU(2)}$, there is always a choice of $SU(2)$ invariant frame $\{\theta^a\}$ such that the matrix γ_{ab} is diagonal:

$$\gamma = A\left(\tilde{\theta}^1\right)^2 + B\left(\tilde{\theta}^2\right)^2 + C\left(\tilde{\theta}^3\right)^2. \tag{6}$$

Such a choice is not very useful in the study of Bianchi IX solutions of Einstein's equation because the same basis does not generally diagonalize the extrinsic curvature K; so the diagonal form (6) is not generally preserved. In the case of Ricci flow,

---

[6]Note that Ricci flow preserves isometries and hence the flow stays within a given class.

however, since a diagonal metric γ (in $C_{SU(2)}$) implies a diagonal Ricci tensor, diagonal form is preserved along Ricci flow, and hence diagonal form is useful. We use it here.[7]

In terms of A, B, and C, one easily calculates the curvature for a metric γ in the class $C_{SU(2)}$; the nonzero components of the Ricci tensor are

$$R_{11} = \frac{1}{2}A\left[A^2 - (B - C)^2\right] \tag{7}$$

$$R_{22} = \frac{1}{2}B\left[B^2 - (C - A)^2\right] \tag{8}$$

$$R_{33} = \frac{1}{2}C\left[C^2 - (A - B)^2\right] \tag{9}$$

while the scalar curvature is

$$R = \frac{1}{2}\left[A^2 - (B - C)^2\right] + \left[B^2 - (C - A)^2\right] + \left[C^2 - (A - B)^2\right] \tag{10}$$

From these expressions, we obtain the volume-preserving Ricci flow equations in the form of an ordinary differential equation system for A, B, and C:

$$\frac{d}{dt}A = \frac{2}{3}\left[-A^2(2A - B - C) + A(B - C)^2\right] \tag{11}$$

$$\frac{d}{dt}B = \frac{2}{3}\left[-B^2(2B - C - A) + B(C - A)^2\right] \tag{12}$$

$$\frac{d}{dt}C = \frac{2}{3}\left[-C^2(2C - A - B) + C(A - B)^2\right] \tag{13}$$

We immediately verify from equations (11), (12) and (13) that any geometry γ for which A = B = C is a fixed point of the Ricci flow. Such a geometry is a round sphere, and one easily verifies that the round spheres are the only fixed points for Ricci flow on $C_{SU(2)}$. We shall now show that for any choice of initial data $A(0) = A_0$, $B(0) = B_0$, $C(0) = C_0$, the Ricci flow converges to a round sphere geometry which has the same volume as the initial geometry

$$\gamma = A_0\left(\theta^1\right)^2 + B_0\left(\theta^2\right)^2 + C_0\left(\theta^2\right)^2.$$

Since equations (11), (12) and (13) are symmetric under interchange of A, B, and C, we may without loss of generality choose

$$A_0 \geq B_0 \geq C_0 \tag{14}$$

---

[7]Before trying this diagonal form, we tried to work out $C_{SU(2)}$ Ricci flow using the standard nondiagonal Mixmaster parametrization

$$\gamma = e^{\alpha}e^{\beta ij}\theta_i \otimes \theta_j.$$

The diagonal parametrization makes the analysis quite a bit simpler.

Now from equations (11), (12) and (13) we obtain the evolution equations

$$\frac{d}{dt}(A - B) = \frac{2}{3}\left[-2\left(A^3 - B^3\right) + C\left(A^2 - B^2\right) + C^2\left(A - B\right)\right] \tag{15}$$

$$\frac{d}{dt}(B - C) = \frac{2}{3}\left[-2\left(B^3 - C^3\right) + A\left(B^2 - C^2\right) + A^2(B - C)\right] \tag{16}$$

$$\frac{d}{dt}(C - A) = \frac{2}{3}\left[-2\left(C^3 - A^3\right) + B\left(C^2 - A^2\right) + B^2(C - A)\right] \tag{17}$$

We see from these equations that if $A_0 \geq B_0 \geq C_0$, then it follows that $A(t) \geq B(t) \geq C(t)$ for all values of t. Using this result in equation (13) we see that $\frac{d}{dt}C \geq 0$, and hence C is nondecreasing. We may then derive from eq. (17) the following differential inequality.

$$\frac{d}{dt}(A - C) \leq -2C_0^2(A - C). \tag{18}$$

Since $(A - C) \geq 0$ for all t, we may integrate (18) to obtain

$$A(t) - C(t) \leq (A_0 - C_0)e^{-2C_0^2 t} \tag{19}$$

Hence we see that $A - C$ converges exponentially to zero. Since $B(t)$ is sandwiched between $A(t)$ and $C(t)$ it follows that $A(t) - B(t)$ converges exponentially to zero as well. Since $A(t)B(t)C(t)$ is a constant for this Ricci flow, we have a non-singular round three-sphere limit for Ricci flow starting at any geometry in $C_{SU(2)}$.

Unlike $C_{SU(2)}$, the class $C_{Heisenberg}$ (which corresponds to Bianchi class II) contains no Einstein metric and hence has no fixed point for the Ricci flow. How does Ricci flow on $C_{Heisenberg}$ behave?

Like the geometries of $C_{SU(2)}$, those in $C_{Heisenberg}$ can be diagonalized, and the diagonalization is preserved by the Ricci flow. So we may write the Ricci flow equations on $C_{Heisenberg}$ as a three-dimensional system

$$\frac{d}{dt} A = -\frac{4}{3} A^3 \tag{20}$$

$$\frac{d}{dt} B = \frac{2}{3} A^2 B \tag{21}$$

$$\frac{d}{dt} C = \frac{2}{3} A^2 C \tag{22}$$

These can be integrated directly (starting with 20); we obtain

$$A(t) = A_0 \left(1 + \frac{8}{3} A_0^2 t\right)^{-\frac{1}{2}} \tag{23}$$

$$B(t) = B_0 \left(1 + \frac{8}{3} A_0^2 t\right)^{\frac{1}{4}} \tag{24}$$

$$C(t) = C_0 \left(1 + \frac{8}{3} A_0^2 t\right)^{\frac{1}{4}} \tag{25}$$

Clearly for any choice of the initial data $\{A_0, B_0, C_0\}$, we find that $B(t)$ and $C(t)$ increase together without bound at the rate $t^{1/4}$ while $A(t)$ decreases to zero at the rate $t^{-1/2}$. Calculating the curvature invariants, we find

$$|R(t)| = R(0) \left(1 + \frac{8}{3} A_0^2 t\right)^{-1} \tag{26}$$

and

$$R_{ij}(t) R^{ij}(t) = R_{ij}(0) R^{ij}(0) \left(1 + \frac{8}{3} A_0^2 t\right)^{-2} \tag{27}$$

So as the geometry along the flow shrinks in one dimension and expands in the two others, it becomes increasingly flat. In a sense, the Ricci flow in the minisuper-space of $C_{Heisenberg}$ goes asymptotically to the two-dimensional flat geometries on the "boundary" of $C_{Heisenberg}$. The behavior is much like that of the "pancake singu-larities" which occur in the Kasner solutions of the Einstein equations. We call it "pancake degeneracy."

The two sorts of Ricci flow behavior which we have thus far described – convergence to an Einstein geometry in CSU(2), and flow to a pancake degeneracy in $C_{Heisenberg}$ – oc-cur not just in these two classes, but in most of the others as well. Indeed, the follow-ing general statements of behavior hold for all classes: (1) Whenever the flow can con-verge – i.e., in those classes which contain Einstein metrics ($C_{\Re^3}, C_{SU(2)}, C_{H(3)}, C_{E(2)}$) – it *does* converge. Convergence, when it occurs, is exponential. (2) In all those classes – except $C_{SO(3) \times \Re^1}$ – which do not admit Einstein metrics, the curvature in-variants die along the flow at the rate $\frac{1}{t}$, and the flows approach either a pancake degeneracy or a cigar degeneracy. (3) In $C_{SO(3) \times R^1}$ the curvature diverges along all flows. (4) In all classes except $C_{Heisenberg}$, flows approach those geometries which have maximal isometry group dimension.

We summarize these results in Table 2.

4. Conclusion

While the behavior of solutions of the Einstein equations on minisuperspace is very complicated and not fully explored, we find that we have a fairly complete picture of how Ricci flow behaves on minisuperspace. Our main motivation for studying such Ricci flow, apart from its intrinsic interest, is that according to Hamilton's scenario, after various singularities in the flow have occurred, the Ricci flow in each pinched off piece M_α^3 of M^3 should approach that of locally homogeneous geometries. Is there any evidence for this?

No one has yet managed to find a way to carefully handle the singularities which are expected to occur in a generic Ricci flow. A scheme for flowing past S^2 neck pinches

Table 2. Summary of Ricci Flow Results

Class	Behavior	Rate
$\Re^3$	Convergence to flat space	Trivial flow (all metrics flat)
SU(2)	Convergence to round sphere	Exponential
SL(2,$\Re$)	Pancake Degeneracy	$R^{ij}R_{ij} \sim 1/t^2$
Heisenberg	Pancake Degeneracy	$R^{ij}R_{ij} \sim 1/t^2$
E(1,1)	Cigar Degeneracy	$R^{ij}R_{ij} \sim 1/t^2$
E(2)	Convergence to flat space	Exponential
H(3)	Convergence to hyperbolic space	Trivial (all members hyperbolic)
$SO(3) \times \Re^1$	Curvature Singularity	$R^{ij}R_{ij} \sim \frac{1}{(1-t)^2}$
$H(2) \times \Re^1$	Pancake Degeneracy	$R^{ij}R_{ij} \sim \frac{1}{(1-t)^2}$

has not yet been found. On the other hand, we have studied various families of nonsingular Ricci flows on nonlocally homogeneous geometries, and we have studied locally homogeneous Ricci flows. So far, the expected behavior has been found.

Our most fruitful study of this nature involves certain "midisuperspaces."[8] Motivated by various questions concerning Gowdy spacetimes, we have considered families of metrics taking the form

$$\gamma = e^{2A}d\theta^2 + e^f \left[e^W dx^2 + e^{-W} dy^2 \right] \tag{28}$$

where f is a spatial constant (which is allowed to change under Ricci flow) and where A and W are both functions of θ. We consider metrics of this form on two different manifolds: In the first case, which we call C_I, we let $M^3 = T^3$, with (θ, x, y) being standard coordinates on T^3, and we require that A and W be periodic in θ. So the metrics in C_I all have two commuting orthogonal Killing vector fields. In the second case, which we call C_{II}, we take M^3 be a solv-twisted T^2 bundle over the circle with coordinate θ and we require that A be periodic in θ while W satisfies the aperiodic condition $W(\theta + 2\pi) = W(\theta) + \Lambda$, where Λ is a topological constant characteristic of the bundle $M^3 \to S^1$. The metrics in C_{II}, while locally symmetric under a T^2 action, do *not* admit any true Killing vector fields.

The class $C_{\Re^3}$ of locally homogeneous geometries is a subclass of C_I (all those with $\frac{\partial}{\partial\theta} W(\theta) = 0$ and so one expects all Ricci flows in C_I to approach $C_{\Re^3}$. Indeed, the metrics in $C_{\Re^3}$ are all flat, fixed points of Ricci flow, so one expects all Ricci flows

---

[8]The term "midisuperspace," first used by Kuchar [18], refers to families of geometries which are not (locally) homogeneous, but share a nontrivial isometry group.

in C_I to converge to geometries in $C_{\Re^3}$. In [19], we show (with Carfora) that this occurs.

The solv-twisted geometries of C_{II} do not include flat metrics or any other Einstein metrics, so convergence cannot occur. However, C_{II} intersects $C_{E(1,1)}$ – if W is linear in θ, the metric lies in $C_{E(1,1)}$ – and so one looks for all Ricci flows in C_{II} to approach Ricci flows in $C_{E(1,1)}$, with their asymptotic "cigar degeneracies." In [20], Hamilton and the first author show that for each Ricci flow $\gamma(t)$ in C_{II}, there is a corresponding one $\hat{\gamma}(t)$ in $C_{E(1,1)}$ such that $||\gamma(t) - \hat{\gamma}(t)||_{L^\infty(S^1)} < \frac{1}{t^4}$ for large t. This provides an example of what asymptotic approach to a locally homogeneous Ricci flow can mean.

We have, with D. Dedrickson, studied a few other families of geometries, hoping to test if the flows on these families approach flows on locally homogeneous geometries. These families consist of three-geometries (M^3, γ) which are warped products of two-geometries (M^2, μ) with a circle. So

$$M^3 = M^2 \times S^1 \tag{29}$$

and

$$\gamma = \mu + e^{\phi(x)}d\theta^2 \tag{30}$$

where θ is the coordinate on the circle S^1, and $\phi(x)$ is an arbitrary real-valued function on M^2. If there is no warping – i.e., if $\nabla\phi = 0$ – it follows essentially from the results of Hamilton and Chow that the Ricci flow always approaches that which occurs on locally homogeneous geometries (either $C_{SO(3)\times\Re^1}$, $C_{\Re^3}$, or $C_{H(2)\times\Re^1}$ depending on the genus of M^2). If there is warping, our studies are not yet conclusive; but the indications are that the same behavior holds.

The plausibility of Hamilton's scenario is supported by our results in these special classes of initial data. Obviously there is much work to be done in fully realizing the scenario. It is fair to say that other approaches to Thurston's conjecture are much closer to a full proof. Nonetheless, a continued study of Ricci flow should produce some interesting insights regarding the geometry-topology problem.

References

1. C. Misner (1970), "Classical and Quantum Dynamics of a Closed Universe," in *Relativity* [eds. M. Carmelli, I. Fickler, and L. Witten], Plenum.

2. M. Ryan and L. Shepley (1975), *Homogeneous Relativistic Cosmologies*, Princeton U. Press.

3. R. Jantzen (1986), "Spatially Homogeneous Dynamics: A Unified Picture," in *Gamow Cosmology*, LXXXVI Corso, Soc. Ital. di Fis.,

   Bologna.

4. J. Isenberg and M. Jackson (1992), "Ricci Flow of Locally Homogeneous Geometries on Closed Manifolds," J. Diff. Geom. 35, 723-741.

5. J. Bourgignon (1980), "Deformations des Varietes d'Einstein," Asterisque 80, Soc. Math. Fr.

6. R. Hamilton (1982), "Three-manifolds with Positive Ricci Curvatures," J. Diff. Geom. 17, 255-306.

7. R. Hamilton (1986), "Four-manifolds with Positive Curvature Operator," J. Diff. Geom. 27, 153-179.

8. R. Hamilton (1988), "The Ricci Flow on Surfaces," in *Mathematics and General Relativity* [ed J. Isenberg] Contemp. Math 71, Am. Math. Soc.

9. H. Poincare (1907), "Sur L'Uniformisation des Fonctions Analytiques," Acta. Math. 31, 1.

10. B. Chow (1990), "The Ricci-Hamilton Flow on the 2-Sphere," J. Diff. Geom.

11. W. Thurston (1982), "Three Dimensional Manifolds, Kleinian Groups, and Hyperbolic Geometry," Bull. Am. Math. Soc. 6, 357-381.

12. P. Scott (1983), "The Geometry of 3-Manifolds," Bull. Lond. Math Soc. 15, 401-487.

13. H. Kneser (1929), "Geschlossene Flochen in Dreidimensionale Mannigfattigkeiten,' Jahnesber Deutsch Math-Venein 38, 248-260.

14. J. Milnor (1962), "A Unique Factorization Theorem for 3-Manifolds," Am. J. Math 84, 1-7.

15. W. Jaco and P. Shalen (1979), "Seifert Fibred Spaces in 3-Manifolds," Mem. Am. Math. Soc. 2.

16. F. Löbell (1931), "Beispele Geschlossener Dreidimensionale Clifford-Kleinsche Raume Negativer Kr"ummung," Berl. Verh. Sachs. Akad. Wiss. Leipzig, Math. Phys. K1 83, 167.

17. I. Singer (1960), "Infinitesmally Homogeneous Spaces," Com. P. & Appl. Math XIII, 685.

18. K. Kuchar (1974), private communication.

19. M. Carfora, J. Isenberg, and M. Jackson (1990), "Convergence of the Ricci Flow for Metrics with Indefinite Ricci Curvature," J. Diff. Geom. 31, 249-263.

20. R. Hamilton and J. Isenberg, "Quasi-Convergence of Ricci Flow for a Class of Metrics," Univ. of Oregon preprint.

Classical and Quantum Dynamics of Black Hole Interiors

WERNER ISRAEL

Canadian Institute of Advanced Research Cosmology Program, Theoretical
Physics Institute, University of Alberta, Edmonton, Canada T6G 2J1

1 INTRODUCTION

The internal workings of a black hole constitutes a relatively new field, still in its first
faltering steps. These pages are intended as a brief introduction and progress report
on what I think has been gleaned so far.

As far as external appearances go, black holes are matchlessly simple objects. I have
described elsewhere (Israel 1987; Thorne 1993) how Charles Misner was probably
the first to fully appreciate this. In the wake of a gravitational collapse, quadrupole
and other deformations originally anchored in the star get swallowed by the hole or
carried off by gravitational radiation. The external field and event horizon settle, like
a newly-formed soap bubble, into the simplest configuration that is compatible with
the external constraints—mass, charge and angular momentum. (The soap-bubble
analogy, I believe, stems also from Misner.)

The immaculate exterior of a black hole hides an egregious inner disorder. A solar-
mass black hole holds 10^{20} times as much entropy as a ball of radiation of the same
mass-energy and volume.

The key to this dichotomy is the peculiar causal structure of the hole. The event
horizon not only encloses but *precedes* the inner core where curvatures approach
Planck levels. The surface of a black hole, unlike the surface of a star, is not influenced
by processes in the core. This has allowed us (thanks to the efforts of Christodoulou,
Bekenstein, Hawking and many others (see Novikov and Frolov 1989)) to reach an
understanding of the thermodynamical and quantum properties of the horizon and
the exterior, in blissful ignorance of conditions in the core. And this is why internal
exploration can proceed with some confidence by boring in from the outside, without
serious concern that a future breakthrough in quantum gravity might vitiate what
can now be learned about the outer, sub-Planckian layers.

The dawning of a real understanding of generic black hole interiors dates from a remark of Penrose 1968 that the inner horizon is a surface of infinite blueshift, hence unstable to time-dependent perturbations, and almost certainly singular. This insight was fleshed out in a number of perturbative studies (e.g. Matzner et al 1979, Chandrasekhar and Hartle 1982). A time-dependent disturbance unavoidable in real life is the radiative tail of a generic collapse. Its natural decay — as an inverse power of advanced time (Price 1972) — is swamped by the (exponential) blueshift close to the "Cauchy horizon," i.e., the ingoing sheet of the inner horizon, corresponding to infinite advanced time.

These studies stopped short of actually examining the back-reaction of the blueshifted influx on the geometry near the Cauchy horizon. The back-reaction was first studied by Hiscock 1981 and (allowing for outflow from the collapsing star during its passage through the hole) by Poisson and Israel 1990 and Ori 1991, 1992.

This brought into evidence a curious phenomenon which has been dubbed "mass inflation." Close to the Cauchy horizon, the "Coulomb component" $|\Psi_2|$ of Weyl curvature grows exponentially. (In idealized spherical models, $|\Psi_2|$ is essentially m/r^3 in terms of the local Schwarzschild mass function m. In the generic case, inflation of $|\Psi_2|$ is overlaid with an even stronger divergence of the Weyl component Ψ_0, which represents a free gravitational wave imploding near the Cauchy horizon.)

One can say, in short, that gravitational collapse makes waves which get partially absorbed and blueshifted at the Cauchy horizon to produce an exponential mass-inflation of the black hole's core.

No trace of this is perceptible externally. Outside the hole, the mass function m does not inflate, but remains practically the same as the mass of the original star. News of the drastic change of internal field must propagate with the speed of light (as a gravitational wave), and it can never emerge from the hole.

Classical theory provides no damping mechanism for this growth of curvature near the Cauchy horizon. But as the curvature approaches Planck levels, new, quantum effects will undoubtedly come into play. Should one expect these to have a self-regulatory effect, curbing the growth and holding curvatures below the Planck limit (Frolov, Markov and Mukhanov 1990)?

Before we have a meaningful quantum theory of gravity there is no basis for a definite answer to this question. We cannot peer beyond a preceding epoch when curvatures are still well below Planck levels and a classical description of the spacetime geometry still makes approximate sense. However, there is a much more modest question which can in principle be addressed now. In this earlier, "semi-classical" phase, will vacuum polarization and the cumulative effects of pair creation associated with fields other than gravity act in a direction that tends to restrain the rise of curvature near the Cauchy horizon?

The main part of this review will focus on this question. In later sections, I shall report on a first attack on the problem by Anderson, Brady et al 1992, using a simple spherical model of the black hole interior devised by Ori 1991. But I shall begin with some elementary remarks which may help to demystify the mechanism of classical mass inflation.

2 THE ELEMENTARY MECHANICS OF MASS INFLATION

It is nowadays a trite remark that in general relativity the total mass-energy of an asymptotically flat gravitating system (measured at spatial infinity) is conserved, and (with appropriate restrictions on the local stress-energy of the material) is positive (Witten 1981, Nester 1981). In a closed universe the total mass-energy is reasonably taken as zero (Tryon 1973, Misner, Thorne and Wheeler 1973), the positive energy of the material exactly balanced by the negative contribution of gravitational "potential energy." General relativity, like Newtonian theory, puts no lower bound on potential energy, and it is entirely concordant with what has just been said that arbitrarily large amounts of gravitational energy can be siphoned into the matter during the era of cosmological inflation or in the last stages of collapse. All that is new is that, whereas in Newtonian theory the energy transferred becomes unbounded only for a contraction approaching zero volume, in Einstein's theory the floodgates are already open (in appropriate circumstances) at the Cauchy horizon, or at the past horizon of a white hole (Blau 1989, Barrabès, Brady and Poisson 1992).

Thin spherical shells provide an especially simple model of the mechanism. Consider first a shell of radius R (depending on proper time τ), moving in the field of an interior distribution of gravitational (i.e. Schwarzschild) mass m_-. Both m_- and the exterior Schwarzschild mass m_+ are constant. The proper mass M of the shell satisfies

$$dM + Pd(4\pi R^2) = 0.$$

(It is conserved if the surface pressure $P = 0$.) The relation (Chase 1970)

$$m_+ - m_- = M\left(1 - \frac{2m_-}{R} + \dot{R}^2\right)^{\frac{1}{2}} - \frac{1}{2}\frac{M^2}{R}, \quad \dot{R} \equiv \frac{dR}{d\tau},$$

expresses the total conserved, gravitating mass $m_+ - m_-$ of the shell as a sum of four terms (when the square root is expanded to first order): the rest-mass M, the kinetic energy $\frac{1}{2}M\dot{R}^2$, the mutual potential energy $-Mm_-/R$ and a self-potential

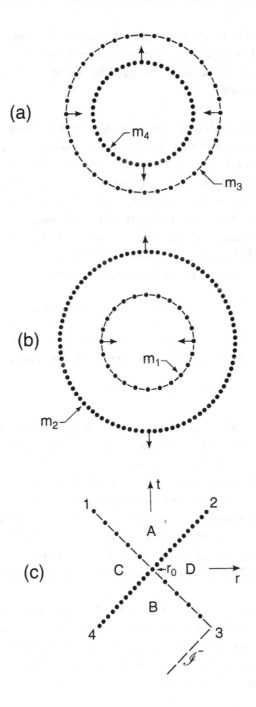

Fig. 1   A pair of transparent, concentric spherical shells, one expanding, the other contracting at the speed of light, encounter and pass through each other. (a) Before, (b) after the encounter; (c) spacetime history of the shells.

energy $-\frac{1}{2}M^2/R$. The point to note is that (at least in this way of formulating the dynamics) the "potential energy" (in general a mutual, unlocalized property shared by a pair of bodies) here contributes to the gravitating mass of the *outer* body: it is a "binding energy" (Blau 1989).

There are circumstances in which the outer body can be released from its gravitational binding, with a consequent increase in its gravitating mass. This is illustrated most graphically in the lightlike limit (Barrabès and Israel 1991). Fig. 1(a) shows a lightlike shell of Schwarzschild mass m_3 imploding onto a concentric interior shell of Schwarzschild mass m_4 which is itself expanding at the speed of light. We may assume that the shells are transparent and simply pass through each other when they cross at a radius r_0. At this instant, their mutual potential energy, of order $-m_3m_4/r_0$, is transferred from the imploding to the exploding shell. Exact calculation shows that their new gravitational masses (see Fig. 1(b)) are, respectively,

$$m_1 = m_3(1 - 2m_4/r_0)^{-1}, \qquad m_2 = m_4(1 - 2m_1/r_0). \qquad (1)$$

(The factors of 2 are peculiar to the lightlike limit and are of relativistic origin.) The total Schwarzschild mass is, of course, conserved as, indeed, follows from (1):

$$m_1 + m_2 = m_3 + m_4.$$

The imploding shell thus gains mass-energy upon being relieved of its gravitational binding at r_0. For a collision just outside a horizon of the interior field, $r_0 = 2m_4 + \epsilon$ ($\epsilon \to 0^+$), the mass increase $m_1 - m_3$ can be made arbitrarily large. (The new Schwarzschild mass m_2 of the outgoing shell would then become negative. This has a simple interpretation. The Schwarzschild mass generally has a potential-energy contribution, i.e., it is the mass-energy calibrated for an observer at infinity, allowing for work done in lifting it against gravity. If $m_2 < 0$ for the "outgoing" shell, this shell is inside a black hole, and actually contracting; its negative Schwarzschild mass just represents the shortfall of energy that would have been needed to reach infinity.)

In referring to the increase $m_1 - m_3$ as "mass inflation," it is important to emphasize that we are talking about real, material mass and not just using a catchphrase for something of merely formal significance. If an imploding shell of photons were absorbed by a blob of charcoal at the centre, it would contribute the full value of its inflated mass m_1 to the mass of the blob.

This elementary example already provides a fair schematic picture of the process of mass inflation inside a black hole. The imploding shell may be imagined to represent the fallout from the radiative tail of the collapse, and the outgoing shell an outflow from the collapsing star after it has dropped inside the hole. In a spacetime diagram

(Fig. 1(c)), future lightlike infinity $\mathcal{I}^+$ and the Cauchy horizon would lie just beyond the history of shell 3 in an extension of sector B, representing the outer spacetime; r_0 is here just a little larger than the initially constant radius of the inner horizon.

Relations equivalent to (1) were first derived by Dray and 'tHooft (1985) and by Redmount (1985). They are special cases of a much more general result. Two concentric lightlike shells, colliding in an arbitrary spherisymmetric background, divide spacetime into four sectors A, B, C, D as shown in Fig. 1(c). Because of the geometrical discontinuities across the shells, the squared gradient $f = g^{\alpha\beta}(\partial_\alpha r)(\partial_\beta r)$ of Schwarzschild's radial coordinate r takes different forms $f_A, \ldots, f_D$ in the four sectors. However, at points O where the shells collide, the values of f are linked by

$$f_A \, f_B|_O = f_C \, f_D|_O. \tag{2}$$

(The relations (1) are recovered as a special case by setting $f_C = 1$, $f_A = 1 - 2m_1/r$ etc.)

The "generalized DTR relations" (2) are not even restricted to spherical symmetry, but hold for the collision of arbitrary lightlike shells (Barrabès et al 1990). The appropriately generalized mass function that is subject to inflation in the nonspherical case is closely related to a quasi-local mass introduced by Hawking 1968. This provides a degree of reassurance that mass inflation is not an artefact of spherical symmetry, as, indeed, seems to be borne out by much more elaborate investigations (Ori 1992, Droz et al 1992). However, for a dissenting opinion see Yurtsever 1992.

3 MASS INFLATION IN BLACK HOLES: ORI'S SPHERICAL MODEL

The simplest way of bringing out the bare bones of the mass inflation phenomenon is to schematize the fallout from the radiative tail of a collapse as a single impulsive burst, as in the thin-shell model of the previous section. But this is too crude to give an idea of how fast the curvature is actually growing close to the Cauchy horizon (CH).

Ori 1991 has devised a model, involving a spherical charged black hole, which captures the essential physics—in particular, the growth rates—yet remains simple enough that the solution can be written explicitly. The influx of gravitational waves is modelled as a radial stream of lightlike material particles. (The large blueshift near CH justifies this representation of the waves as "gravitons" with an "effective material stress-energy" (Isaacson 1968).) Irradiation of CH by outflow from the star (as it passes between outer and inner horizons) plays a merely catalytic role: it focuses the generators of CH and initiates its contraction. It can be treated schematically. In Ori's model, it is represented as a thin, transparent, outgoing lightlike shell Σ (Fig. 2).

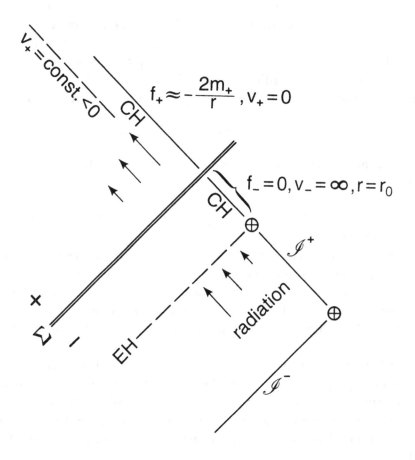

Fig. 2   Ori model. In falling radiation passes through a transparent, "outgoing" lightlike shell Σ inside a charged spherical hole. EH is the event horizon,
and CH the Cauchy horizon.

The metric in each of the domains $\mathcal{V}_-$ and $\mathcal{V}_+$ separated by Σ then has the charged-Vaidya form for pure inflow:

$$(ds^2)_\pm = dv_\pm(2dr - f_\pm\,dv_\pm) + r^2 d\Omega^2, \qquad f_\pm = 1 - \frac{2m_\pm(v_\pm)}{r} + \frac{e^2}{r^2}. \qquad (3)$$

(The charge e is taken to be constant.)

The advanced-time parameters v_+ and v_- are unequal. The functional relation between the two emerges upon noting that the equations of the lightlike 3-space Σ, with respect to the two abutting coordinate systems, are

$$f_+ dv_+ = f_- dv_- = 2dr \quad \text{along } \Sigma. \qquad (4)$$

Continuity of the influx across Σ requires that

$$f_+^{-2} dm_+/dv_+ = f_-^{-2} dm_-/dv_- \qquad (5)$$

(Barrabès and Israel 1991).

From (4) and (5),
$$dm_+/f_+ = dm_-/f_- \quad \text{along } \Sigma, \qquad (6)$$

which clearly shows a divergence of the internal mass function $m_+(v_+)$ as one approaches CH ($f_- \to 0$, $v_- \to \infty$).

The ansatz
$$m_-(v_-) = m_0 - Av_-^{-(p-1)} \quad (A = \text{const.}) \qquad (7)$$

reproduces the correct, power-law decay $dm_-/dv_- \sim v_-^{-p}$ of the externally measured gravitational wave flux (Price 1972). For quadrupole waves, $p = 12$. The solution of $2dr = f_-\,dv_-$ then gives the asymptotic forms of f_- and r along Σ as

$$f_- = -\frac{2pA}{\kappa_0 r_0}\left(1 + \frac{p}{\kappa_0 v_-} + \cdots\right),$$
$$r = r_0 + \frac{A}{\kappa_0 r_0}v_-^{-(p-1)}\left(1 + \frac{p-1}{\kappa_0 v_-} + \cdots\right). \qquad (v_- \to \infty) \qquad (8)$$

Here, $\kappa_0 = (m_0^2 - e^2)^{\frac{1}{2}}/r_0$ is the surface gravity (and r_0 the radius) of the initial, static segment of CH in the outer region $\mathcal{V}_-$.

Substituting (7) and (8) into (5) and (4) now gives the asymptotic growth rate of m_+ (from here on, written simply as m):

$$m(v_+) = m_0^2[\ln(m_0/|v_+|)]^{-p}(-v_+)^{-1} \qquad (v_+ \to 0^-) \qquad (9)$$
$$-v_+ = (\text{const.})\exp(-\kappa_0 v_-) \qquad (v_- \to \infty) \qquad (10)$$

(Ori 1991, Israel 1992). The origin of v_+ has been chosen at CH, and, in (9), I have set equal to unity a dimensionless constant which depends on the luminosity and initial deformation of the collapsing star.

The growth rate (9) agrees with results of a more realistic model (Poisson and Israel 1990), which treats the outflow as continuous.

A charged Vaidya geometry with metric of the form (3) has the Ricci tensor

$$R_{\alpha\beta} = 2\dot{m}(v_+)r^{-2}(\partial_\alpha v_+)(\partial_\beta v_+) + e^2 r^{-4}\Delta_{\alpha\beta}, \tag{11}$$

(the dot indicates differentiation), and

$$-\Psi_2 = \frac{1}{2}C^{\theta\varphi}{}_{\theta\varphi} = [m(v_+) - e^2/r]r^{-3} \tag{12}$$

is the only nonzero Newman-Penrose component of Weyl curvature. In (11),

$$\Delta^\beta_\alpha = \text{diag}\,(-1,-1,1,1), \qquad x^\alpha \equiv (v_+, r, \theta, \varphi).$$

Because $|e| \ll m(v_+)$ near CH, the electrostatic contributions in (11) and (12) are negligible and will often be ignored.

The singularity of this geometry at CH has the notable feature (Ori 1991, 1992; Balbinot et al 1991, Herman and Hiscock 1992, Israel 1992) that the Weyl curvature, though diverging, is an integrable function of v_+, thanks to the logarithmic factor in (9). This leads to a metric that is finite at CH in terms of appropriate coordinates. Ignoring the electric term, and introducing a retarded time U (regular at CH) defined by

$$dU = r\,dr + m_+(v_+)dv_+, \tag{13}$$

the metric in $\mathcal{V}_+$ becomes

$$ds^2 = 2r^{-1}dU\,dv_+ + r^2 d\Omega^2, \tag{14}$$

which is clearly bounded.

4 APPROACHING PLANCKIAN CURVATURE NEAR THE CAUCHY HORIZON

In the remainder of this article I should like to give a brief account of a first attempt (Anderson, Brady et al 1992) to see how quantum corrections affect the classical approach to a singularity at the Cauchy horizon CH.

In the layer just above and before CH, where the Weyl curvature is approaching, but still appreciably below, Planck levels (the "pre-Planck" or "semi-classical" regime), a classical description of the geometry should still be approximately valid. On the other hand, all Compton wavelengths are much larger than the radius of curvature of spacetime, and thus all fields can be treated as effectively massless, and conformally coupled (since the curvature scalar R vanishes for the radiation-dominated geometry (11)).

The goal therefore is to estimate the renormalized expectation value $\langle T_{\mu\nu} \rangle$ (in the Unruh state for an evaporating black hole) of the stress-energy for conformally coupled massless fields on the Vaidya background (3), when the mass function $m(v_+)$ has (classically) the diverging form (9) as v_+ approaches zero from below.

The task of calculating $\langle T_{\mu\nu} \rangle$ is, in general, notoriously difficult. A conjunction of fortunate circumstances makes it relatively tractable here. (i) Only the asymptotic form near CH is of interest. (ii) The singularity is relatively mild, as noted at the end of the previous section. (iii) The essentially lightlike form (11) of $R_{\alpha\beta}$ means that terms nonlinear in $R_{\alpha\beta}$ that are potentially strongly divergent actually vanish. Finally, (iv) the special form (9) of the mass function $m(v_+)$, with its logarithmic factor, implies that, of two terms with the same total number of (m-factors) + (v-derivatives), the term with the smaller number of m-factors is dominant. For example,

$$\ddot{m} \gg m\dot{m}/r^2.$$

The two sides differ "merely" by a logarithmic factor. But this factor does tend to infinity at CH, and for $v_+ = -10^{10}t_p$ ($t_p = 10^{-43}$s is the Planck time), it has already reached a value 10^{23} for a solar-mass black hole!

The remarkable implication of all this is that one is allowed to treat the geometry as a linear perturbation of flat space, even in the pre-Planck regime close to CH. The terms linear in $m(v)$, which are retained, actually dominate the neglected nonlinear terms.

Barvinsky and Vilkovisky 1990 are in the process of building a systematic general formalism applicable in such circumstances, i.e., terms linear (and, more generally, of algebraically lower order) in curvature and derivatives dominant over higher-order terms having the same dimension. For the immediate purposes of this preliminary study, it will suffice to apply a simpler, first-order prescription due to Horowitz 1980.

There is a potential source of worry which should first be addressed. The Weyl curvature approaches Planck levels, $\Psi_2 \to t_p^{-2}$, as $v_+ \to v_p < 0$, where

$$|v_p| = t_p b^{-2} \epsilon [\ln(b/\epsilon^2)]^{-p}, \qquad b \equiv r_0/m_0, \qquad \epsilon \equiv t_p/r_0. \qquad (15)$$

Apparently, it is the advanced time v_p which marks the end of the semi-classical phase and the moment when effects of quantum gravity become important. But according to (11) and (12),

$$|R_{\alpha\beta}/\Psi_2| \approx r\dot{m}/m \sim r_0/|v_+|$$

showing that the Ricci tensor diverges more rapidly than Weyl curvature. This suggests that Ricci curvature should already surpass Planck levels at times much earlier than v_p.

The weak point of this statement is that it is not coordinate-independent. The Ricci curvature $R_{\alpha\beta}u^\alpha u^\beta$ (in effect, the classical radiation density) actually measured by an observer with 4-velocity u^α will be arbitrarily small if he is falling inward at nearly the speed of light, whereas the Weyl curvature scalar $\Psi_2 = -(C_{\alpha\beta\gamma\delta}C^{\alpha\beta\gamma\delta}/48)^{\frac{1}{2}}$ is boost-invariant. Nevertheless, it might be cause for concern that the growth of Ricci curvature in some physically "reasonable" frame (however defined) may invalidate a semi-classical analysis already at times much earlier than v_p.

Fortunately, it is possible to allay such doubts for all practical purposes. Make the conformal transformation

$$ds^2 = (r/r_0)^2 ds_*^2. \qquad (16)$$

The new metric ds_*^2 has a Ricci tensor $R_{\alpha\beta}^*$ free of the worrisome, strongly divergent $\dot{m}$ terms (see (18) below), while Ψ_2 is merely multiplied by a factor $(r/r_0)^2$ of order unity. The strategy will be to first obtain $\langle T_{\mu\nu}\rangle$ in the conformal geometry, and then use Page's (1982) well-known formula to transform $\langle T_{\mu\nu}\rangle$ back to the physical, Vaidya geometry.

It is intuitively helpful to rescale the null coordinate v_+ so that it more nearly represents Planck scales. Set

$$v = \epsilon v_+, \qquad u = 2\epsilon r_0^2/r, \qquad \epsilon \equiv t_p/r_0. \qquad (15)$$

Further, define

$$\bar{\theta} = \epsilon^{-1}\theta, \qquad x + iy = t_p(\epsilon^{-1}\sin\epsilon\bar{\theta})e^{i\varphi}.$$

The upper limit of the range $0 \le \bar{\theta} \le \pi/\epsilon$ is practically infinite, and $r_0^2 d\Omega^2 \approx dx^2 + dy^2$, i.e., the sphere $r = r_0$ (representing the initial static segment of CH) is practically flat on Planck scales. Finally, we may approximate f in (3) by $f \approx -2m(v_+)/r$, since m is much larger than e and r near CH.

The conformal metric now takes a Kerr-Schild (flat+lightlike) form

$$ds_*^2 = -du\,dv + dx^2 + dy^2 + 2L(u,v)dv^2, \qquad (16)$$

$$L = \frac{1}{8}ku^3m(v_+), \qquad k = (t_p r_0^3)^{-1}. \qquad (17)$$

This is manifestly almost flat for $|\Psi_2| \ll t_p^{-2}$. The Ricci curvature derived from (16) is

$$R^*_{\alpha\beta} = 4L_{uu}(2L\ell_\alpha\ell_\beta - \ell_{(\alpha}n_{\beta)}),\qquad(18)$$

where $\ell_\alpha = -\partial_\alpha v$, $n_\alpha = -\partial_\alpha u$ and $L_{uu} = \partial^2 L/\partial u^2$.

5 VACUUM POLARIZATION IN A NEARLY FLAT SPACETIME

The general problem of finding, to linear order, the renormalized in-in vacuum expectation value $\langle T_{\mu\nu}(x)\rangle$ of stress-energy for a massless field in a nearly flat spacetime has been solved by Horowitz 1980. He used general arguments to show that it must be determined by a retarded integral over the past light cone of the point x in the flat background, having the form

$$t_p^{-2}\langle T_{\mu\nu}(x)\rangle = a \int H(x - x')A_{\mu\nu}(x')d^4x' + \alpha A_{\mu\nu}(x) + \beta B_{\mu\nu}(x)\qquad(19)$$

for a conformally coupled field.

Here, a is a positive numerical coefficient whose value is known for different spins:

$$960\pi^2(4\pi a) = \begin{cases} 1 & \text{(conformal scalar)} \\ 3 & \text{(neutrino)} \\ 12 & \text{(Maxwell)} \end{cases}\qquad(20)$$

while α and β are arbitrary numbers, not determinable by the theory. $A_{\mu\nu}$ and $B_{\mu\nu}$ are the (linearized) variational derivatives $\delta/\delta g^{\mu\nu}$ of the actions associated with $C^2_{\alpha\beta\gamma\delta}$ and R^2 respectively. Explicitly,

$$\begin{aligned} A_{\mu\nu} &= -2\Box G_{\mu\nu} - \frac{2}{3}G^\alpha_{\alpha,\mu\nu} + \frac{2}{3}\eta_{\mu\nu}\Box G^\alpha_\alpha, \\ B_{\mu\nu} &= 2\eta_{\mu\nu}\Box G^\alpha_\alpha - 2G^\alpha_{\alpha,\mu\nu}, \end{aligned}\qquad(21)$$

in which $\eta_{\mu\nu}$ and the derivatives of course refer to rectilinear coordinates on the flat background.

The past lightcone distribution H is given by

$$H(x - x') = \delta'(\sigma)\theta(t - t'),\qquad \sigma \equiv \frac{1}{2}(x_\nu - x'_\nu)^2.\qquad(22)$$

As Horowitz shows, (22) is uniquely singled out for the vacuum stress (19) as the only Lorentz-invariant distribution on a flat background having the correct dimension (length)$^{-4}$. Explicitly, for any function f,

$$\int H(x - x')f(x')d^4x' = \int_{4\pi} d\Omega \int_{-\infty}^0 dU \left[\frac{\partial f}{\partial U}\ln\left(-\frac{U}{\lambda}\right) + \frac{1}{2}\frac{\partial f}{\partial V}\right]_{V=0},\qquad(23)$$

where U, V, $d\Omega$ are spherical lightlike coordinates centred on x, so that $V = 0$ is the past lightcone of x. An arbitrary length-scale λ appears in (23) as the result of an integration by parts on the original form (22). The arbitrariness of λ may be considered to reflect the arbitrariness of α in (19), and λ could be adjusted so as to absorb α.

6 VACUUM POLARIZATION NEAR THE CAUCHY HORIZON

In the general in-vacuum expression (19) for $\langle T_{\mu\nu} \rangle$, it is implicit that the geometry is flat in the remote past. For our purposes, it needs to be supplemented by a local, conserved tensor representing initial conditions appropriate to the Unruh state for an evaporating black hole. Inside the hole, this is just the lightlike influx of negative energy that accompanies the thermal outflux to infinity (Fulling 1977). But for black holes heavier than 100 kg this remains negligible up to the moment $v_+ = v_p$ when the classical curvature becomes Planckian (Balbinot et al 1991), and it will therefore be ignored.

According to (21), (18) and (17), the functions to be inserted in (19) are

$$
A_{uu} = B_{uu} = 0, \qquad A_{vv} = \frac{1}{3}B_{vv} = 4kud^2m/dv^2
$$
$$
A_{uv} = \frac{1}{3}B_{uv} = -\frac{1}{2}A_{xx} = \frac{1}{12}B_{xx} = 4kdm/dv
$$
(24)

with m given by (9), and, of course, $A_{yy} = A_{xx}$, $B_{yy} = B_{xx}$.

To evaluate the integrals (19), one chooses the arbitrary point x (having coordinates u, v in the plane lightlike coordinate system of (16)) as origin of a system of spherical lightlike coordinates. Let U, V, Θ, Φ be spherical coordinates of an arbitrary point x', with Θ defined as the angle between the radius vector and the spatial direction of radiative inflow. The relationship between the two coordinate systems is then

$$
2(v - v') = U + V + (V - U)\cos\Theta
$$
$$
2(u - u') = U + V - (V - U)\cos\Theta
$$

For a point (u', v') on the past lightcone of x, $V = 0$. The integrals involved in (23) can now be expressed as

$$
\iint d\Omega\, dU\, \frac{\partial f}{\partial V} = -4\pi \int_0^{a_1} \int_0^{b_1} d\xi\, d\eta\, (\xi + \eta)^{-2} \left(\eta\frac{\partial f}{\partial \xi} + \xi\frac{\partial f}{\partial \eta} \right)
$$
$$
\iint d\Omega\, dU\, \frac{\partial f}{\partial U} \ln\left(-\frac{U}{\lambda}\right) = -4\pi \int_0^{a_1} \int_0^{b_1} d\xi\, d\eta\, (\xi + \eta)^{-2} \ln\left(\frac{\xi + \eta}{\lambda}\right) \left(\xi\frac{\partial f}{\partial \xi} + \eta\frac{\partial f}{\partial \eta} \right),
$$

in which $\xi = u - u'$, $\eta = v - v'$, $a = u - u_1$, $b = v - v_1$, and the initial lightlike surfaces $u = u_1$ and $v = v_1$ represent the outer (event) horizon and an advanced time, just after the collapse, when the radiative influx first becomes appreciable.

It is now straightforward to substitute the expressions (24) for f and evaluate the integrals approximately by Laplace's method (Anderson, Brady et al 1992). The dominant contribution to $\langle T_{\mu\nu}\rangle$ comes from the logarithmic term in (22) with $f = A_{vv}$:

$$t_p^{-2}\langle T_{v_+v_+}\rangle \approx 32\pi a \ r^{-3}\ddot{m}(v_+)\ln(-v_+/\lambda), \qquad (25)$$

with a given by (20). The result quoted is for the physical metric. The conformal transformation (16) affects the leading term of $\langle T_{v_+v_+}\rangle$ only through an effective rescaling of the arbitrary length-scale λ.

Inspection of (24) shows that $\langle T_{uu}\rangle = 0$ in the conformal geometry. This result is not significantly changed on transforming back to the physical metric (3). Page's (1982) conformal transformation formula for $\langle T_{\mu\nu}\rangle$ yields for the physical outflux

$$\langle T^{\mu\nu}\rangle(\partial_\mu v_+)(\partial_\nu v_+) \sim t_p^2 m/r^5 \qquad (26)$$

which is of order $t_p^2\Psi_2$ times the classical outflux from the collapsing star. Thus, up to the time when curvatures reach Planck levels, the classical picture of a CH contracting very slowly (on Planck scales) under irradiation by the collapsing star is not significantly affected by quantum corrections. (Balbinot and Poisson 1992 have arrived at a similar conclusion, using general arguments based on the conservation laws and the trace anomaly.)

7 QUANTUM DAMPING OR ANTI-DAMPING? THE PROBLEM OF SCALE-ARBITRARINESS

We now come to the most interesting question that falls within the scope of our semi-classical approach. Does the leading-order quantum correction (25) tend to counteract or to aggravate the slide towards a singularity at the CH driven by the classical influx

$$(T_{v_+v_+})_{\text{class}} = \dot{m}(v_+)/4\pi r^2$$

(see (11))?

The answer depends on the sign of the logarithmic factor in (25) during the pre-Planck phase (i.e., v_+ approaching $v_p < 0$, given by (15)). There are two essentially different possibilities. If $\lambda \gg |v_p|$, the logarithm is negative for v_+ near v_p, and (25) predicts damping of the classical growth of curvature and mass inflation before the Weyl curvature reaches Planck levels. If, on the other hand, $\lambda \lesssim |v_p|$, quantum effects would (at least initially, in the pre-Planck phase) further destabilize the classical plunge toward a curvature singularity.

Unfortunately, there is no way of resolving this ambiguity within the frame of a theory of conformal massless fields propagating on a fixed classical background. Such a theory has no inherent length scale and provides no information about λ (Adler et al 1977, Wald 1978).

A little more can be said if one is prepared to entertain an arguable hypothesis concerning the source of this incompleteness of the semi-classical theory (Wald 1978).

Quantum effects of gravity itself have been left out of account. A successful (renormalizable or finite) quantum theory of gravity would be expected to have the effect, at moderate curvatures, of modifying the Einstein-Hilbert Lagrangian by terms quadratic in curvature,

$$16\pi L_G = t_p^{-2} R + \alpha_1 C_{\alpha\beta\gamma\delta}^2 + \beta_1 R^2, \tag{27}$$

where α_1 and β_1 are coefficients of order unity. ('tHooft and Veltman 1974, Critchley 1978, Barvinsky and Vilkovisky 1985. Fradkin and Vilkovisky 1978 gave general arguments which suggest that α_1 is negative for pure gravity.) The added terms induced in the effective field equations,

$$G_{\mu\nu} + t_p^2(\alpha_1 A_{\mu\nu} + \beta_1 B_{\mu\nu}) = 8\pi \left\{ T_{\mu\nu}^{\text{class}} + \langle T_{\mu\nu} \rangle \right\} \tag{28}$$

are of the same form as the terms left arbitrary in (19). This suggests that the incompleteness of the semi-classical theory is related to the neglect of quantum gravitational effects.

It is arguable that a quantum theory of massless fields which includes gravity would become a *complete* theory, with the net coefficients of $A_{\mu\nu}$ and $B_{\mu\nu}$ no longer indeterminate.

Suppose this is true. Adjust λ in (23) so that $\alpha = 0$ in (19). Then α_1 in (28) is expected to be of order unity, and the term $\alpha_1 A_{\mu\nu}$ may be interpreted as representing effects associated with gravitational vacuum polarization. Now, it seems reasonable to expect that, once the quantum gravitational degrees of freedom are activated (i.e., when $v_+ \approx v_p$), gravitons will have effects not too dissimilar from photons and other massless fields, i.e., that

$$\langle T_{\mu\nu} \rangle \sim -\alpha_1 t_p^2 A_{\mu\nu} \quad \text{for } v_+ \approx v_p. \tag{29}$$

Comparison of (24) and (25) now suggests that the logarithmic factor is of order unity, i.e., that $\lambda \approx |v_p|$. (The circumstance that this is much shorter than a Planck time t_p has no physical significance, because the scale of the null coordinate v_+ (which λ normalizes) has no intrinsic local meaning.)

If this conclusion is correct, (25) should be interpreted as an intensification (rather than a damping) of the classical influences tending to produce a curvature singularity at the CH, at least for the Ori spherical model considered here. It would indicate strongly that the ultimate, quantum stage of evolution of the hole is inaccessible to semi-classical considerations.

8 CONCLUDING REMARKS

The outcome of this first reconnaissance of quantum effects inside a black hole is the hardly unexpected conclusion that an understanding of the ultimate fate of a gravitational collapse will require nothing less than the full armory of a yet-to-be-discovered quantum theory of gravity. It is nonetheless remarkable (whether or not the tentative picture I have sketched here happens to be near the truth) that the fragmentary grasp of the basic physical laws that we possess now potentially allows us to penetrate to within Planck lengths of the singularity at the Cauchy horizon.

As far as one can now tell, not a single observable prediction can emerge from the internal exploration of black holes. The motivation for such a quest will therefore be cause for head-scratching in some quarters. If so, it is my turn to scratch my head. That the aim of theory is to make observable predictions is a maxim to which all of us at least play lip service. But for me it is hard to imagine how anyone who truly believed this to be the *fundamental* aim should find it worth getting out of bed in the morning. The fundamental aim of all science is to find things out and to understand them, and to that end, to use whatever rational means are at hand.

It is a pleasure to record my indebtedness to the friends and colleagues who have been my collaborators on various aspects of the work reported here: Warren Anderson, Roberto Balbinot, Claude Barrabès, Pat Brady, Serge Droz, Sharon Morsink and Eric Poisson. I am also grateful to many colleagues for helpful discussions and advice, in particular, Andrei Barvinsky, Roberto Camporesi, Valery Frolov, Bill Hiscock and Don Page.

This work was supported by the Canadian Institute for Advanced Research and by the Natural Sciences and Engineering Research Council of Canada.

REFERENCES

Adler, S. L., Lieberman, J. and Yee, J. N. 1977. *Annals of Physics* **106**, 279.

Anderson, W., Brady, P. R., Israel, W. and Morsink, S. M. 1992. Preprint, University of Alberta.

Balbinot, R., Brady, P. R., Israel, W. and Poisson, E. 1991. *Physics Letters* **A161**, 223.

Balbinot, R. and Poisson, E. 1992. Caltech preprint GRP-318.

Barrabès, C., Brady, P. R. and Poisson, E. 1992. Preprint, University of Alberta.

Barrabès, C. and Israel, W. 1991. *Phys. Rev.* **D43**, 1129.

Barrabès, C., Israel, W. and Poisson, E. 1990. *Class. Quantum Gravity* **7**, L273.

Barvinsky, A. O. and Vilkovisky, G. A. 1985. *Physics Reports* **119**, 1.

Barvinsky, A. O. and Vilkovisky, G. A. 1990. *Nucl. Physics* **B333**, 471.

Blau, S. K. 1989. *Phys. Rev.* **D39**, 2901.

Chandrasekhar, S. and Hartle, J. B. 1982. *Proc. Roy. Soc.* **A384**, 301.

Chase, J. E. 1970. *Nuovo Cimento* **67B**, 136.

Critchley, R. 1978. *Phys. Rev.* **D18**, 1849.

Dray, T. and 'tHooft, G. 1985. *Commun. Math. Phys.* **38**, 119.

Droz, S., Israel, W. and Morsink, S. M. 1992. Paper in preparation.

Fradkin, E. S. and Vilkovisky, G. A. 1978. *Physics Letters* **B73**, 209.

Frolov, V. P., Markov, M. A. and Mukhanov, V. F. 1990. *Phys. Rev.* **D41**, 383.

Fulling, S. A. 1977. *J. Phys.* **A10**, 917.

Hawking, S. W. 1968. *J. Math. Phys.* **9**, 598.

Herman, R. and Hiscock, W. A. 1992. *Phys. Rev.* **D46**, 1863.

Hiscock, W. A. 1981. *Physics Letters* **A83**, 110.

Horowitz, G. T. 1980. *Phys. Rev.* **D21**, 1445.

Isaacson, R. A. 1968. *Phys. Rev.* **160**, 1263.

Israel, W. 1987. In "300 Years of Gravitation" (ed. S. W. Hawking and W. Israel), Cambridge Univ. Press, Cambridge.

Israel, W. 1992. In "Black Hole Physics" (ed. V. De Sabbata and Z. Zhang), Kluwer, Boston, pp. 147–183.

Matzner, R. A., Zamorano, N. and Sandberg, V. D. 1979. *Phys. Rev.* **D19**, 2821.

Misner, C. W., Thorne, K. S. and Wheeler, J. A. 1973. "Gravitation," W. H. Freeman, San Francisco, §19.4.

Nester, J. M. 1981. *Physics Letters* **83A**, 241.

Novikov, I. D. and Frolov, V. P. 1989. "Physics of Black Holes," Kluwer, Boston.

Ori, A. 1991. *Phys. Rev. Letters* **67**, 789.

Ori, A. 1992. *Phys. Rev. Letters* **68**, 2117.

Page, D. N. 1982. *Phys. Rev.* **D25**, 1499.

Penrose, R. 1968. In "Battelle Rencontres" (ed. C. M. DeWitt and J. A. Wheeler), W. A. Benjamin, New York, p. 222.

Poisson, E. and Israel, W. 1990. *Phys. Rev.* **D41**, 1796.

Price, R. H. 1972. *Phys. Rev.* **D5**, 2419.

Redmount, I. H. 1985. *Prog. Theor. Phys.* **73**, 401.

'tHooft, G. and Veltman, M. 1974. *Ann. Inst. Henri Poincaré* **20**, 69.

Thorne, K. S. 1993. "Curved Space, Warped Time" (in press).

Tryon, E. P. 1973. *Nature* **246**, 1396.

Wald, R. M. 1978. *Phys. Rev.* **D17**, 1477.

Witten, E. 1981. *Commun. Math. Phys.* **80**, 381.

Yurtsever, U. 1992. Univ. of Calif. (Santa Barbara) preprint.

Matter Time in Canonical Quantum Gravity

Karel V. Kuchař *

Abstract

The functional Schrödinger approach to canonical quantum gravity requires the construction of time and frame variables from the canonical data. I review the difficulties of basing such variables purely on the geometric data. I describe different ways of introducing matter variables for this purpose. I argue that the simplest phenomenological medium, an incoherent dust, does more than one can reasonably expect. The comoving coordinates of the dust particles and the proper time along their worldlines become canonical coordinates in the phase space of the coupled system. The Hamiltonian constraint can easily be solved for the momentum canonically conjugate to the dust time. The ensuing Hamiltonian density has an extraordinary feature not encountered in other systems: it depends only on the geometric variables, not on the dust frame and the dust time. This has three important consequences. Firstly, the functional Schrödinger equation can be solved by separating the dust time from geometric variables. Secondly, the Hamiltonian densities strongly commute and can therefore be simultaneously defined by spectral analysis. Thirdly, the Schrödinger equation can be solved independently of the supermomentum constraint, which is then satisfied by parametrizing this solution.

1. Canonical quantum gravity

One can never pay off old debts; the most one can do is to acknowledge them. Not that I lacked opportunity to acknowledge those which I owe Charlie Misner. He helped to set the entire framework of canonical gravity, and there is hardly a paper which I wrote since coming to the United States in which I did not have the chance – and the pleasure – to refer to his contributions. At this festive libation, I want to relate how his work posed me a problem which has since preoccupied me for years: that of internal time in canonical quantum gravity.

As all of you who work in the field know, in spite of much valuable research which had been done before, we always fall back on the two major outlines of canonical quantum gravity, one by Dirac [1], and the other by Arnowitt, Deser and Misner (ADM) [2].

---

*Department of Physics, University of Utah, Salt Lake City, Utah 84112, U.S.A.

Their starting point is the same. Spacetime is cut into slices whose intrinsic metric and extrinsic curvature become the canonical coordinates and canonical momenta of the gravitational field. Covariance forces on the canonical data at each point of the slice the familiar super-Hamiltonian and supermomentum constraints. The problem then arises of how to take care of those constraints in quantum theory.

There is a subtle difference between the ways in which Dirac and ADM meet that task. Dirac considers all spacelike slices and all reference frames to be equally admissible [3]. States are described by functionals of the canonical data, and constraints are imposed as operator restrictions on such states. In the metric representation, the imposition of the super-Hamiltonian constraint yields the Wheeler-DeWitt equation [4],[5]. This equation is somehow, one does not quite know how, expected to propagate the state from an initial slice to an arbitrary final slice.

Arnowitt, Deser and Misner select a definite foliation and a definite reference frame by coordinate conditions. The leaves of the foliation are labeled by a single time parameter, and the points of the frame by a triplet of coordinates. The constraints are solved for the single momentum canonically conjugate to the time variable. This solution yields the true nonvanishing Hamiltonian depending on two unconstrained pairs of canonical variables, the true gravitational degrees of freedom. States are defined only on the privileged foliation selected by the slicing condition. They depend on two canonical coordinates per point, and on a single time parameter. Quantization proceeds as in an ordinary field theory.

Each approach presents a puzzle. The Wheeler-DeWitt equation is a second-order functional differential equation, and the space of its solutions cannot be easily turned into a Hilbert space [6]. Moreover, the geometric data are not separated into time and frame variables, on the one hand, and dynamical degrees of freedom, on the other. It is thus difficult to say how one initially prepares the state, and what dynamical variables can one finally measure.

None of these difficulties are present in the ADM approach. Time and frame are chosen at the very beginning, dynamical variables are identified, the states are propagated by an ordinary Schrödinger equation, and that equation conserves the standard inner product. However, the approach violates the spirit of general relativity. Why is it that the states can be evolved and the measurements done only on the leaves of a privileged foliation? Why should one set of slicing and coordinate conditions be preferred to any other? Or, when one admits all different sets, how can one compare the ensuing quantum theories [6]?

2.   Internal time

These question intrigued me when I came to Princeton in 1968. I longed for a description of canonical gravity which would bridge the gap between the ADM and

Dirac approaches. I felt that one should not fix the foliation *ab initio*, but I also felt that without separating time and frame variables from the true degrees of freedom one does not know what to measure and when. It was then that I started to explore the idea of internal functional time.

The time variable, I thought, should be able to label all spacelike hypersurfaces, not only those of a privileged foliation. To do that, one needs to know the time $T(\sigma)$ of each point $\sigma \in \Sigma$ of an arbitrary slice. In canonical gravity, such a time should be determined entirely by the geometric canonical data $g(\sigma)$ and $p(\sigma)$ carried by Σ: this is what is meant by the adjective 'internal'. Therefore, $T(\sigma; g, p]$ should be a function of σ and a functional of g and p. In the same spirit, three other functionals $Z^k(\sigma; g, p]$ should specify the frame point at σ. Taken together, the variables (T, Z^k) should carry information about the embedding of Σ in an Einstein spacetime.

What one looks for is a canonical transformation [7]

$$g(\sigma), p(\sigma) \mapsto \begin{pmatrix} T(\sigma), Z^k(\sigma), \psi(\sigma) \\ P(\sigma), P_k(\sigma), \pi(\sigma) \end{pmatrix} \tag{1}$$

which maps the six canonical coordinates $g(\sigma)$ and six canonical momenta $p(\sigma)$ into $1+3+2$ canonical coordinates $T(\sigma)$, $Z^k(\sigma)$ and $\psi(\sigma)$ and $1+3+2$ canonical momenta $P(\sigma)$, $P_k(\sigma)$ and $\pi(\sigma)$. The canonical variables $\psi(\sigma), \pi(\sigma)$ represent the true dynamical degrees of freedom. One then resolves the classical constraints with respect to the momenta $P(\sigma)$ and $P_k(\sigma)$, and replaces them by an equivalent set of constraints

$$\begin{aligned} P(\sigma) + h(\sigma; T, \psi, \pi] &= 0, \\ P_k(\sigma) + h_k(\sigma; T, \psi, \pi] &= 0 \end{aligned} \tag{2}$$

prior to quantization. When one imposes the new constraints as operator restrictions on the state functionals $\Psi[T, Z^k, \psi]$, one obtains a functional Schrödinger equation in the functional time variable $T(\sigma)$. This equation propagates the states between arbitrary hypersurfaces, and it also has a conserved positive-definite inner product. This quantization scheme, so it seemed to me, combined the best features of the ADM approach with those of the Dirac-Wheeler-DeWitt scheme. It reduces to the ADM approach when one restricts the functional variables $T(\sigma), Z^k(\sigma)$ to have constant values on Σ.

I tried the idea first on linearized quantum gravity with the choice of the time and frame variables suggested by the ADM coordinate conditions [8]. It was quite cumbersome. At that point, Charlie Misner came to Princeton for his sabbatical. He brought with him a program which inspired a whole generation of relativity students and, judging from the journals, keeps inspiring the succeeding generations as well: the program of minisuperspace quantization. The idea was to suppress all degrees of freedom but few by symmetry, and to quantize the remaining ones by canonical techniques. I came under its spell like everyone else, and I suspect for the same reason:

It is so difficult to do anything in quantum gravity, and here there was something so easy.

Charlie Misner and his students were involved in minisuperspace quantization of homogeneous cosmological models [9]. This did not give much chance for developing the functional time quantization because homogeneity restricts the slicing to a one-parameter family. Homogeneity also destroys the field-theoretical features of quantum gravity, leaving only a finite number of degrees of freedom. It occurred to me that both of these restrictions can be lifted when one applies the minisuperspace quantization to the Einstein-Rosen cylindrical waves [10]. The functional time approach worked just fine. While the Wheeler-DeWitt equation was unwieldy and unmanageable, I succeeded in finding a non-local transformation (1) on the mini-phase-space of the model which made the problem exactly soluble. Indeed, the quantization of the cylindrical waves turned out to be equivalent to the quantization of a single massless scalar field on a two-dimensional Minkowskian background. I was so cheered by being able to handle an infinite number of degrees of freedom that I felt the technique deserved a linguistic boost. With hemlines dropping and miniskirts disappearing from the Princeton campus, I proudly called my attempt midisuperspace quantization.

I still did not know at that time how to quantize the field on arbitrary (cylindrically symmetric) slices. Only much later did I find how to factor order the constraints so that they do not produce anomalies [11]. In this form, they allow a truly functional evolution. Charles Torre has taken necessary steps toward treating the cylindrical waves in this manner [12].

Under the spell of the minisuperspace charm, it was tempting to believe that the trick could be repeated for full quantum gravity. In a due time I sobered up and learned about the pitfalls of the internal functional time approach. Einstein's theory is not easily cast into that mold. Here are some of the major problems: [6]

1. *Global time problem.* It may happen that the constraint system of vacuum gravity cannot be made globally equivalent to the constraint system (2) by any choice of internal time. This must happen if there does not exist a global choice of the time function on the geometric phase space such that each classical trajectory intersects every hypersurface of constant internal time once and only once. Hájíček analyzed this problem on simple models [13] and Torre showed that it exists in the general theory [14].

2. *Multiple choice problem.* The functional time approach does not restrict the evolution to a privileged foliation, but allows the states to be followed on arbitrary hypersurfaces. This does not mean, however, that quantum theories based on two choices of internal time are equivalent or even comparable [6],[15]. If there is not a geometrically privileged choice, one faces an embarrassment of riches.

3. *Spacetime problem.* In classical gravity, one is asked to reconstruct the time coordinate T of an event x' in an Einstein spacetime $(\mathcal{M}, \gamma)$ from the canonical data $g(\sigma)$, $p(\sigma)$ on an embedding $x = X(\sigma)$ passing through x'. Time should emerge from the canonical formalism as a spacetime scalar which does not depend on the embedding from which it was reconstructed [6]. This requirement is quite strong. No intrinsic time $T(\sigma; g]$ has this property, and the most obvious choices of extrinsic time (as $|g|^{-1/2} p(\sigma)$) are also excluded.

4. *Spectral analysis problem.* The solution of the constraints with respect to the time momentum may not exist for the entire domain of the geometric variables $g(\sigma)$, $p(\sigma)$ and, even if it does, it may be too complicated. To give meaning to such classical expressions typically requires definition of the operators by spectral analysis [16]. This is an involved nonlocal procedure which calls for solving an eigenvalue problem for the operators. More often than not the operators do not commute and the eigenvalue problem makes no sense. The restrictions on the domain mean that the operators do not need to be self-adjoint. At the very best, it is difficult to find the solution in an explicit form [6].

In brief, the implementation of the internal functional time approach to canonical quantum gravity meets quite formidable conceptual and technical difficulties.

3.   Matter time and reference fluids

Many of these can be traced back to the fact that the intrinsic metric and extrinsic geometry enter the constraints in a complicated way. Nothing in the structure of the constraints indicates how the true dynamical degrees of freedom are to be distinguished from the quantities that determine the hypersurface. This makes the task of constructing an internal time from the geometric variables so elusive.

The main role of embedding variables is to provide an identification of spacetime events. In the early days of general relativity, Einstein [17] and Hilbert [18] introduced a conceptual device which does exactly that: the concept of a reference fluid. The particles of the fluid are supposed to identify space points, and clocks carried by them to distinguish instants of time. These fix the reference frame and time foliation. In that frame and on that foliation, not only geometry, but the metric itself, becomes classically measurable.

The reference fluid is traditionally considered as a test system whose back reaction on geometry can be neglected. Instead of deriving its equations of motion from an action, one encodes it in coordinate conditions. These are statements about the metric which hold in the coordinate system of the fluid and are violated in any other system. In canonical theory, this points back back to the ADM approach. To go beyond what has already been done requires turning the reference fluid into a physical system. There are two routes one can follow. The first one is to impose the coordinate conditions *before* variation by adjoining them to the action with Lagrange multipliers. The action is then parametrized by expressing the privileged coordinates as functions of arbitrary coordinates. This process turns the time and frame variables into canonical coordinates on an extended phase space. The additional terms in the action amount to a matter source coupled to gravity. The second route is to forget about the coordinate conditions and to picture the fluid as a realistic material medium described by a reasonable Lagrangian. In either case, it is matter that brings into spacetime the embedding variables $T(\sigma)$, $Z^k(\sigma)$ which are so difficult to construct from pure geometry.

4.   Reference fluids associated with coordinate conditions

The association of a material medium with coordinate conditions stems from the problem of representing spacetime diffeomorphisms in canonical gravity. Chris Isham and I tried to resolve this problem by breaking the invariance of general relativity by Gaussian coordinate conditions and restoring it by parametrization [19]. This procedure leads to modification of the constraints by terms which are linear in the momenta canonically conjugate to the Gaussian coordinates. Hartle discussed the Schrödinger equation obtained by imposing the new constraints as restrictions on the physical states [20]. In the end, he dismissed the new terms as devoid of physical reality. Halliwell and Hartle related the new form of the constraints to the sum-over-histories approach to quantum gravity [21]. Charles Torre and I derived the modified constraints from an action principle by enforcing the Gaussian coordinate condition with Lagrange multipliers, and identified the new terms with energy-momentum densities of the corresponding reference fluid [22]. For the full set of Gaussian conditions, the fluid has the structure of a nonrotating and heat-conducting incoherent dust. The heat term disappears when one enforces merely the Gaussian slicing condition. The method can be applied to various coordinate conditions. Charles Torre, Chris Stone and I extended it first to harmonic coordinate conditions [23]. The harmonic reference fluid is simply a collection of four noninteracting massless scalar fields. The same approach can be used for introducing the conformal, harmonic, and light-cone gauges in the canonical theory of a bosonic string [24] and in the canonical treatment of (induced) two-dimensional gravity [25].

The condition of slicing by hypersurfaces of constant mean extrinsic curvature turned out to destroy the functional time procedure in an interesting fashion [26]. The equations of motion of the reference fluid have a hidden elliptic nature, and the proliferation of second-class constraints forces one in the end from arbitrary hypersurfaces onto the slices of constant mean curvature. The unimodular coordinate condition fixing the spacetime volume element is also exceptional [27]. It is so weak that unimodular coordinates do not label individual embeddings, but only equivalence classes of those embeddings which are separated from each other by zero four-volume. The ensuing Schrödinger equation is not a true functional time equation, and it has interpretational difficulties similar to those of the Wheeler-DeWitt equation.

5. Phenomenological fluids

Phenomenological fluids were introduced as interpretational tools into quantum gravity by DeWitt. In the covariant approach, DeWitt coupled the gravitational field to an elastic medium carrying mechanical clocks [28]. From these objects he constructed idealized apparatuses which, in the spirit of Bohr-Rosenfeld analysis of measurability in quantum electrodynamics, were able to detect appropriate projections of the quantized Riemann curvature tensor. A little later, he used the same device for interpreting the canonical minisuperspace quantization of the Friedmann universe: he introduced a cloud of clocks into the model and studied their correlation with the radius of the universe [5]. DeWitt's approach was recently rediscovered by Rovelli [29]. Rovelli used a square-root Hamiltonian for the clocks. He argued that when the clock's momentum is large in comparison with its mass, the Dirac constraint quantization approximately leads to a Schrödinger equation.

The reference fluid corresponding to the Gaussian slicing condition is actually the simplest of all phenomenological media: it corresponds to incoherent dust. However, that dust is forbidden to rotate. We know forces which practically forbid compression, but we do not know any which would prevent rotation. The condition that the dust cannot rotate is unnatural. Can one introduce time and frame through dust in an arbitrary state of motion?

David Brown devised a canonical description of relativistic fluids in which the time and frame variables play a prominent role [30]. Incoherent dust is a special case of his formalism. He and I set out to find what it implies for canonical quantum gravity [31]. I expected that rotation would merely complicate the neat scheme based on the nonrotating Gaussian reference dust. To my surprise, it made the functional Schrödinger equation remarkably simple. Indeed, it came as close to a phenomenological solution of the time problem in canonical quantum gravity as one can reasonably expect. I am going to explain how the scheme works in detail.

6. Incoherent dust

Fill a globally hyperbolic spacetime $(\mathcal{M} = \mathsf{R} \times \Sigma, \gamma)$ with dust. The dust particles introduce a privileged reference frame into $\mathcal{M}$. Their worldlines z form a congruence $\mathcal{S}$ which can be viewed as an abstract three-dimensional space. As a manifold, $\mathcal{S}$ is isomorphic to Σ. Choose a hypersurface transverse to the worldlines and Lie propagate it by the dust four-velocity U. This generates a privileged time foliation $\mathcal{T}$. A spacetime event can be identified by the worldline $z \in \mathcal{S}$ where it happens and the instant of time $t \in \mathcal{T}$ when it happens. In this way, dust introduces a standard of space and time into the spacetime manifold.

It is often convenient to work with local coordinates $x^\alpha(x)$, $z^k(z)$ and $t(t)$ on the manifolds $\mathcal{M}$, $\mathcal{S}$ and $\mathcal{T}$. I leave the coordinates x^α and z^k arbitrary, but choose the time coordinate $t \in R$ of a leaf t $\in \mathcal{T}$ to be the proper time which separates it from a fiducial leaf $t_0 \in \mathcal{T}$ along the worldlines.

The motion of the dust is described by the mappings

$$T : \mathcal{M} \to \mathcal{T} \text{ by } x \mapsto t = T(x), \quad Z : \mathcal{M} \to \mathcal{S} \text{ by } x \mapsto z = Z(x). \qquad (3)$$

The four-velocity of the dust considered as a covector field on $\mathcal{M}$ is a combination of their differentials. In local coordinates,

$$U_\alpha = -T_{,\alpha} + W_k Z^k{}_{,\alpha}. \qquad (4)$$

The coefficients W_k are the spatial components of the covector U_α in the Lagrangian coordinates z^k. Globally, these again can be considered as mappings:

$$W : \mathcal{M} \to T^*\mathcal{S} \text{ by } x \mapsto W(x). \qquad (5)$$

Besides all this kinematic information, one also needs to know the distribution of dust in space. This is described by the rest mass density

$$M : \mathcal{M} \to R^+ \text{ by } x \mapsto M(x). \qquad (6)$$

One can use the mappings T, Z and M, W as state variables in the dust action

$$S^D[T, Z; M, W; \gamma] = \int_\mathcal{M} d^4x \ -\frac{1}{2}|\gamma|^{1/2} M(\gamma^{\alpha\beta} U_\alpha U_\beta + 1). \qquad (7)$$

Their variation yields the Euler equations of motion

$$-\frac{1}{2}|\gamma|^{1/2}(\gamma^{\alpha\beta} U_\alpha U_\beta + 1) = \delta S^D/\delta M = 0, \qquad (8)$$

$$-|\gamma|^{1/2} Z^k{}_{,\alpha} U^\alpha = M^{-1} \delta S^D/\delta W_k = 0, \qquad (9)$$

$$-(|\gamma|^{1/2} M U^\alpha)_{,\alpha} = \delta S^D/\delta T = 0, \qquad (10)$$

$$(|\gamma|^{1/2} M W_k U^\alpha)_{,\alpha} = \delta S^D/\delta Z^k = 0. \qquad (11)$$

The first two equations reassert the meaning of the frame and time variables: By Eq.(9), Z^k are the comoving coordinates. By Eq.(8), U is a unit timelike vector field. Taken together, Eqs.(8) and (9) imply that T is Lie propagated along U:

$$T_{,\alpha} U^\alpha = 0. \qquad (12)$$

The remaining two equations are continuity equations for the mass current $J^\alpha = M U^\alpha$ and the momentum current $J^\alpha_k = W_k J^\alpha$. They imply that the velocity projections W_k stay unchanged along the dust worldlines:

$$W_{k,\alpha} U^\alpha = 0. \qquad (13)$$

The variation of the action with respect to the metric yields the energy-momentum tensor $T^{\alpha\beta}$. Modulo the normalization equation (8),

$$T^{\alpha\beta} := 2|\gamma|^{-1/2} \delta S^D / \delta\gamma_{\alpha\beta} = MU^\alpha U^\beta. \tag{14}$$

Because S^D is invariant under spacetime diffeomorphisms, the energy-momentum tensor is covariantly conserved:

$$\nabla_\beta T^\beta_\alpha = 0. \tag{15}$$

Equations (10) and (15) imply that the dust particles move along geodesics:

$$U^\beta \nabla_\beta U^\alpha = 0. \tag{16}$$

To couple dust to geometry, one adds its action S^D to the Hilbert action

$$S^G[\gamma] = \int_{\mathcal{M}} d^4x \, |\gamma|^{1/2} R(x; \gamma). \tag{17}$$

The variation of the total action $S = S^G + S^D$ with respect to the metric gives then the Einstein law of gravitation

$$R^{\alpha\beta} - \frac{1}{2} R\gamma^{\alpha\beta} = \frac{1}{2} T^{\alpha\beta}. \tag{18}$$

7.   Canonical description of dust

Arnowitt, Deser and Misner established an algorithm for casting an arbitrary covariant action into Hamiltonian form. It is a tribute to the impact of their work that one no longer needs to explain its details. It is enough to state the results:

Foliate $\mathcal{M}$ by a one-parameter family of spacelike embeddings

$$X : \mathrm{R} \times \Sigma \to \mathcal{M} \quad \text{by} \quad (\tau, \sigma) \mapsto x = X(\tau, \sigma). \tag{19}$$

The transition from one leaf of the foliation (19) to another is described by the deformation vector $N^\alpha := \dot{X}^\alpha$. The decomposition of N^α into the normal n^α and tangential directions $X^\alpha_{,a}$ to the leaves yields the lapse function $N^\perp$ and the shift vector N^a:

$$N^\alpha := \dot{X}^\alpha = N^\perp n^\alpha + N^a X^\alpha_{,a}. \tag{20}$$

On each leaf, the spacetime metric $\gamma_{\alpha\beta}(x)$ induces an intrinsic metric $g_{ab}(\tau, \sigma)$. In the canonical formalism, the intrinsic metric becomes a canonical coordinate. The Legendre dual transformation introduces the momentum $p^{ab}(\tau, \sigma)$ conjugate to $g_{ab}(\tau, \sigma)$ and casts the gravitational action into canonical form:

$$S^G[g_{ab}, p^{ab}; N^\perp, N^a] = \int_{\mathrm{R}} d\tau \left(\int_\Sigma d^3\sigma \, p^{ab} \dot{g}_{ab} - H^G_\perp[N^\perp] - H^G[\vec{N}] \right). \tag{21}$$

The gravitational Hamiltonian

$$H^G_\perp[N^\perp] + H^G[\vec{N}] = \int_\Sigma d^3\sigma\, N^\perp(\sigma) H^G_\perp(\sigma) + \int_\sigma d^3\sigma\, N^a(\sigma) H^G_a(\sigma) \qquad (22)$$

is obtained by smearing the super-Hamiltonian

$$\begin{aligned}
H^G_\perp(\sigma; g, p] &= G_{ab\,cd}(\sigma; g) p^{ab}(\sigma) p^{cd}(\sigma) - |g|^{1/2} R(\sigma; g], \\
G_{ab\,cd} &= \frac{1}{2}|g|^{1/2}(g_{ac}g_{bd} + g_{ad}g_{bc} - g_{ab}g_{cd})
\end{aligned} \qquad (23)$$

by the lapse function $N^\perp$, and the supermomentum

$$H^G_a(\sigma; g, p] = -2\, D_b p^b_a(\sigma) \qquad (24)$$

by the shift vector N_a.

If the metric is non-derivatively coupled to a source, the canonical action on the extended phase space has the same form as in vacuum gravity, but the super-Hamiltonian and supermomentum are augmented by the energy density and momentum density of the source. I shall describe what these quantities turn out to be for incoherent dust.

The state variables M and W_k enter the dust action (7) as Lagrange multipliers. Their variation leads to the Euler equations (8)-(9) which guarantee the physical meaning of the complementary set of variables, T and Z^k. Conversely, the variation of the dynamical variables T and Z^k leads to the conservation laws, (10) and (11), for the Lagrange multipliers M and W_k. The duality of these two sets of variables carries over into the canonical formalism. The dynamical variables T and Z^k play the role of canonical coordinates, while the multipliers M and W_k neatly combine into their conjugate momenta, P and P_k.

As one would expect, the momentum P conjugate to the proper time T is the rest mass density on the hypersurfaces $X(\Sigma)$:

$$P = |\gamma|^{1/2} M \sqrt{1 + g^{ab} U_a U_b}\,, \qquad (25)$$

where the square root is just the relativistic 'gamma factor' from the Lagrangian observers with four-velocity $\mathbf{U}$ to the hypersurface observers with four-velocity $\mathbf{n}$. Similarly, the canonical momentum P_k conjugate to Z^k is (up to the sign) the momentum density, on $X(\Sigma)$, along the dust worldlines:

$$- P_k = P W_k\,. \qquad (26)$$

The canonical form of the dust action can again be obtained by the ADM algorithm:

$$S^D[T, Z^k, P, P_k; \gamma] = \int_R dt \left(\int_\Sigma d\sigma\, (P\dot{T} + P_k \dot{Z}^k) - H^D_\perp[N^\perp] - H^D[\vec{N}] \right). \qquad (27)$$

The structure of the Hamiltonian

$$H_\perp^D[N^\perp] + H^D[\vec{N}] = \int_\Sigma d^3\sigma\, N^\perp(\sigma)H_\perp^D(\sigma) + \int_\Sigma d^3\sigma\, N^a(\sigma)H_a^D(\sigma) \qquad (28)$$

is best understood by considering first a single relativistic particle of mass m following a geodesic on a Riemannian background $(\mathcal{M},\gamma)$. Such a motion is described by the mass-shell constraint

$$H := \frac{1}{2m}(\gamma^{\alpha\beta}\pi_\alpha\pi_\beta + m^2) = 0\,. \qquad (29)$$

By splitting the four-momentum π_α with respect to the foliation (19), one obtains the momenta

$$\pi_\tau = \dot{X}^\alpha\pi_\alpha \ \text{ and } \ \pi_a = X^\alpha{}_{,a}\pi_\alpha \qquad (30)$$

conjugate to the time τ and the position σ^a of the particle. The particle Hamiltonian is obtained by solving the mass-shell constraint for the energy $-\pi_\tau$: [32]

$$-\pi_\tau = -N^a(\tau,\sigma)\pi_a + N^\perp(\tau,\sigma)\sqrt{m^2 + g^{ab}(\tau,\sigma)\pi_a\pi_b}\,. \qquad (31)$$

Dust is a collection of geodesic motions (16). One gets the dust Hamiltonian from the particle Hamiltonian (31) by replacing the rest mass m by its density P, the momentum π_a by its density PU_a, and by summing the particle Hamiltonians over the hypersurface $X(\Sigma)$. This heuristic argument leads to the correct Hamiltonian (28). One identifies

$$H_\perp^D = \sqrt{P^2 + g^{ab}H_a^D H_b^D} \qquad (32)$$

with the energy density, and

$$-H_a^D = PU_a = -PT_{,a} - P_k Z^k{}_{,a} \qquad (33)$$

with the momentum density on $X(\Sigma)$.

The dynamics of geometry coupled to dust is described by the sum of the two actions, (21) and (27):

$$S[T, Z^k, g_{ab}, P, P_k, p^{ab}; N^\perp, N^a] = S^G + S^D\,. \qquad (34)$$

The variation of S with respect to the canonical variables gives the Hamilton equations. Its variation with respect to the lapse and the shift gives the constraints

$$H_\perp := H_\perp^G + H_\perp^D = 0\,, \qquad (35)$$

and

$$H_a := H_a^G + H_a^D = 0\,. \qquad (36)$$

8.   New constraints

Constraints can be written in many equivalent forms. The key idea behind the Schrödinger approach to canonical quantum gravity is that constraints must be solved with respect to the momentum conjugate to time prior to quantization.

The introduction of the dust time makes this task especially simple. The energy density (32) depends on the frame momentum P_k only through the momentum density H_a^D. This is determined by the supermomentum constraint (36). One can thus replace the super-Hamiltonian constraint (35) by a new constraint

$$H_\uparrow(\sigma) := P(\sigma) + h_\uparrow(\sigma; g, p] = 0 \,, \tag{37}$$

in which

$$h_\uparrow(\sigma) \;=\; -\sqrt{G(\sigma)}\,, \tag{38}$$
$$G(\sigma; g, p] \;:=\; (H^G(\sigma))^2 - g^{ab}(\sigma) H_a^G(\sigma) H_b^G(\sigma) \,. \tag{39}$$

As long as

$$H^G(\sigma) < 0 \quad \text{and} \quad G(\sigma) > 0 \,, \tag{40}$$

the constraints (35) and (37) are entirely equivalent. However, they generate different types of displacements. The generator $H_\perp[N^\perp]$ displaces embeddings by the proper time $N^\perp(\sigma)$ in the direction normal to the hypersurface. On the other hand,

$$H_\uparrow[N^\uparrow] := \int_\Sigma d^3x \, N^\uparrow(\sigma) H_\uparrow(\sigma) \tag{41}$$

displaces embeddings by the proper time $N^\uparrow(\sigma)$ along the worldlines of the dust particles:

$$\dot{T}(\sigma) \;:=\; \{T(\sigma), H_\uparrow[N^\uparrow]\} = N^\uparrow(\sigma) \,, \tag{42}$$
$$\dot{Z}^k(\sigma) \;:=\; \{Z^k(\sigma), H_\uparrow[N^\uparrow]\} = 0 \,.$$

This difference has an important consequence for the constraint algebra. Let two hypersurfaces pass through the same spacetime event $x \in \mathcal{M}$, but have different normals. The old generator $H_\perp[N^\perp]$ displaces them at x in two different directions. As a result, the generators $H_\perp[N^\perp]$ and $H_\perp[M^\perp]$ have a nonvanishing Poisson bracket. This bracket depends on the induced metric on Σ. Therefore, the old constraints (35) and (36) do not generate a true Lie algebra. On the other hand, the new generator displaces the two hypersurfaces at x by the same amount along the worldlines. As a result, the Poisson bracket of $H_\uparrow[N^\uparrow]$ and $H_\uparrow[M^\uparrow]$ vanishes, and the new constraints (36)-(37) do generate a Lie algebra:

$$\{H_\uparrow(\sigma), H_\uparrow(\sigma')\} = 0 \,. \tag{43}$$

(The remaining brackets are dictated by the facts that $H_a(\sigma)$ generate DiffΣ, and $H_\uparrow(\sigma)$ is a scalar density under DiffΣ).

Formally, Eq.(43) follows by a neat argument which circumvents tedious calculations [22]. Because the new constraints are equivalent to the old constraints, and the Poisson brackets of the old constraints weakly vanish, the Poisson brackets of the new constraints must also weakly vanish. However, because $P(\sigma)$ appears in the constraint (37) without any coefficient, the Poisson bracket (43) cannot depend on $P(\sigma)$. Thus, the new constraints cannot help in any way to make this bracket zero. If the bracket has to vanish, it must vanish strongly.

Similar conclusions hold for any choice of time variable. However, dust has a re-markable feature which I did not encounter in any other system, fundamental or phenomenological: The Hamiltonian density (38)-(39) depends only on the geomet-ric variables g, p, not on the dust variables T and Z^k. This remarkable property has an even more remarkable consequence: The Poisson bracket (43) does not have any cross terms between $P(\sigma)$ and $h_\uparrow(\sigma)$. Therefore, the $h_\uparrow(\sigma)$'s, and consequently the $G(\sigma)$'s, must have vanishing Poisson brackets among themselves:

$$\{G(\sigma; g,p], G(\sigma'; g,p]\} = 0. \tag{44}$$

This property is vital when one attempts to define the Hamilton operators $\hat{h}_\uparrow(\sigma)$ by spectral analysis.

I proved Eq.(44) by coupling geometry to dust. However, as it stands, Eq.(44) is an identity involving only the geometric variables g and p. As such, it must hold irrespective of whether geometry is coupled to dust, to any other matter or field system, or whether it is left alone in vacuum. Indeed, one can check directly that Eq.(44) is a consequence of the Dirac 'algebra' among the gravitational expressions $H_\perp^G(\sigma)$ and $H_a^G(\sigma)$.

In vacuum gravity, the imposition of the usual super-Hamiltonian and supermomen-tum constraints $H_\perp^G(\sigma) = 0 = H_a^G(\sigma)$ is globally equivalent to the imposition of a new set of constraints

$$G(\sigma) = 0 = H_a^G(\sigma). \tag{45}$$

Because $H_a^G(\sigma)$ generate DiffΣ, and $G(\sigma)$ transforms as a scalar density of weight 2, the new vacuum constraints (45) generate a true algebra. This is something which so far has been achieved only for systems in two-dimensional spacetimes.

In vacuum gravity, the new system (45) has a disadvantage that the Hamiltonian vector field generated by $G(\sigma)$ vanishes on the constraint surface and hence does not generate a motion. The inequalities (40) tell us that this difficulty does not exist in spacetimes filled with dust.

9.   Quantization

In the Dirac method of quantization, constraints are turned into operators and imposed as restrictions on states. The standard form $H_\perp$ of the super-Hamiltonian constraint yields the Wheeler-DeWitt equation, whose solution space carries no obvious Hilbert space structure [6]. The idea of matter clocks calls for the constraints to be resolved with respect to the time momentum before quantization. The imposition of the new constraint on quantum states then yields a functional Schrödinger equation in matter time.

For dust, this procedure can be carried out in detail. The resolved constraint takes the form (37). The requirement that it annihilate the physical states,

$$H_\uparrow(\sigma; \hat{P}, \hat{g}, \hat{p}] \, \Psi[T, Z, g] = 0 \,, \tag{46}$$

amounts to the Schrödinger equation

$$i \frac{\delta \Psi[T, Z, g]}{\delta T(\sigma)} = h_\uparrow(\sigma; \hat{g}, \hat{p}] \, \Psi[T, Z, g] \,. \tag{47}$$

The supermomentum constraint

$$(H_a^D(\sigma; \hat{T}, \hat{Z}^k, \hat{P}, \hat{P}_k] + H_a^G(\sigma; \hat{g}, \hat{p}]) \, \Psi[T, Z, g] = 0 \tag{48}$$

ensures that the state functional $\Psi[T, Z, g]$ remains unchanged by DiffΣ. Of course, the quantum constraints (46) and (48) are consistent only if the commutators of the operators $\hat{H}_\uparrow(\sigma)$ and $\hat{H}_a(\sigma)$ replicate the classical Poisson algebra (43), etc. I shall proceed under the assumption that there exists a factor ordering and regularization of the constraints that achieves this aim.

The remarkable properties of the generator $H_\uparrow(\sigma)$ enable one to do for dust what is difficult to do for any other system. Firstly, because the constraint operator $\hat{H}_\uparrow(\sigma)$ does not depend on the frame variables $\hat{Z}^k(\sigma)$ and $\hat{P}_k(\sigma)$ one can solve the Schrödinger equation alone, and then, by parametrizing this solution, satisfy both the constraints (46) and (48). Secondly, because $\hat{H}_\uparrow(\sigma)$ does not depend on the dust time $T(\sigma)$, one can solve the Schrödinger equation by separating time.

10.   Parametrization

As proved by Misner and Higgs, the supermomentum constraint (48) means that Ψ is invariant under spatial diffeomorphisms $\phi \in \text{Diff}\Sigma$:

$$\Psi[\phi_*T, \phi_*Z, \phi_*g] = \Psi[T, Z, g] \,. \tag{49}$$

Here, ϕ_* denotes the action of $\phi \in \text{Diff}\Sigma$ on the field variables T, Z and g. I adopt the global interpretation (3) of $Z(x)$ according to which $Z(\sigma)$ is a diffeomorphism that maps the fiducial space Σ to the dust space $\mathcal{S}$:

$$Z : \Sigma \to \mathcal{S} \quad \text{by} \quad \sigma \in \Sigma \mapsto z = Z(\sigma) \in \mathcal{S} \,. \tag{50}$$

This induces a mapping Z_* of tensor fields from Σ to $\mathcal{S}$. I shall denote all objects on Σ by light face symbols, and their counterparts on $\mathcal{S}$ by the corresponding bold face symbols. In particular,

$$\mathbf{T} = Z_* T \quad \text{and} \quad \mathbf{g} = Z_* g. \tag{51}$$

By restricting the state functional $\Psi[T, Z, g]$ to the identity $Z = I$, I obtain a functional $\Psi[\mathbf{T}, \mathbf{g}]$ of two arguments, $\mathbf{T}$ and $\mathbf{g}$:

$$\Psi[T, I, g] =: \Psi[\mathbf{T}, g] = \Psi[\mathbf{T}, \mathbf{g}]. \tag{52}$$

The restricted functional $\Psi[\mathbf{T}, \mathbf{g}]$ carries the same amount of information as $\Psi[T, Z, g]$. Indeed, one can reconstruct $\Psi[T, Z, g]$ from $\Psi[\mathbf{T}, \mathbf{g}]$ by putting $\phi = Z$ in Eq.(49):

$$\Psi[T, Z, g] = \Psi[Z_* T, Z_* g]. \tag{53}$$

Moreover, the functional (53) automatically satisfies Eq.(49). Hence, an arbitrary functional $\Psi[\mathbf{T}, \mathbf{g}]$ of $\mathbf{T}$ and $\mathbf{g}$ generates by Eq.(53) a diffeomorphism-invariant functional $\Psi[T, Z, g]$ of T, g and Z. This is the essence of the parametrization process.

Let me now take a $\Psi[T, Z, g]$ that satisfies not only the supermomentum constraint, but also the Schrödinger constraint (46). Because $\hat{H}_\uparrow(\sigma)$ does not operate on Z, Z merely labels different solutions $\Psi[T, Z, g]$ of Eq.(46). In particular, by choosing $Z = I$, one learns that $\Psi[\mathbf{T}, \mathbf{g}]$ of Eq.(52) solves the constraint

$$\mathbf{H}_\uparrow(z; \hat{\mathbf{P}}, \hat{\mathbf{g}}, \hat{\mathbf{p}}] \, \Psi[\mathbf{T}, \mathbf{g}] = 0, \tag{54}$$

i.e., satisfies the Schrödinger equation

$$i \frac{\delta \Psi[\mathbf{T}, \mathbf{g}]}{\delta \mathbf{T}(z)} = \mathbf{h}_\uparrow(z; \hat{\mathbf{g}}, \hat{\mathbf{p}}] \, \Psi[\mathbf{T}, \mathbf{g}]. \tag{55}$$

Conversely, from an arbitrary solution $\Psi[\mathbf{T}, \mathbf{g}]$ of the Schrödinger equation (55) I can construct by parametrization (53) a diffeomorphism-invariant solution $\Psi[T, Z, g]$ of the Schrödinger equation (47).

11. Separation of time

This reduces the task of solving the constraint system (46) and (48) to the task of finding the general solution of the Schrödinger equation (55). Because $\hat{\mathbf{H}}_\uparrow(z)$ is independent of $\mathbf{T}(z)$, I can separate the time variable. Because the operators $\hat{\mathbf{H}}_\uparrow(z)$ and $\mathbf{P}(z)$ all commute, the Schrödinger constraint (54) and the eigenvalue equation

$$\hat{\mathbf{P}}(z) \, \Psi = \mathbf{P}(z) \, \Psi \tag{56}$$

for the time momentum have common solutions

$$\Psi[\mathbf{P}; \mathbf{T}, \mathbf{g}] = \Phi[\mathbf{P}; \mathbf{g}] \exp\left(i \int_{\mathcal{S}} d^3 z \, \mathbf{P}(z) \mathbf{T}(z) \right), \tag{57}$$

whose coefficients $\Phi[\mathbf{P}; \mathbf{g}]$ satisfy the time-independent Schrödinger equation

$$\mathbf{h}_\uparrow(z; \mathbf{g}, \mathbf{p}) \, \Phi[\mathbf{P}; \mathbf{g}] = -\mathbf{P}(z) \, \Phi[\mathbf{P}; \mathbf{g}] \,. \tag{58}$$

The expressions (57) are functionals of $\mathbf{T}(z)$ and $\mathbf{g}(z)$ labeled by the eigenvalues $\mathbf{P}(z)$.

The operator $\mathbf{h}_\uparrow(z)$ must be defined by spectral analysis. Under the assumption that there exists a factor ordering of $\hat{\mathbf{g}}$ and $\hat{\mathbf{p}}$ and a regularization of $\hat{\mathbf{G}}(z)$ such that the classical equation (44) goes over into

$$[\mathbf{G}(z; \hat{\mathbf{g}}, \hat{\mathbf{p}}), \, \mathbf{G}(z'; \hat{\mathbf{g}}, \hat{\mathbf{p}})] = 0 \,, \tag{59}$$

the operators $\hat{\mathbf{G}}(z)$ have common eigenfunctions $\Phi[\mathbf{G}; \mathbf{g}]$:

$$\hat{\mathbf{G}}(z) \, \Phi[\mathbf{G}; \mathbf{g}] = \mathbf{G}(z) \, \Phi[\mathbf{G}; \mathbf{g}] \,. \tag{60}$$

The classical variables (39) are not positive definite, and hence the eigenvalues are not necessarily positive everywhere on $\mathcal{S}$. Let me span the Hilbert space $\mathcal{H}^+$ by the eigenfunctions (60) which do have everywhere positive eigenvalues $\mathbf{G}(z) > 0$. On $\mathcal{H}^+$, one can define the operator

$$\hat{\mathbf{h}}_\uparrow(z) = -\sqrt{\hat{\mathbf{G}}(z)} \tag{61}$$

by the requirement that it have the same eigenfunctions $\Phi[\mathbf{G}; \mathbf{g}]$, but with the eigenvalues

$$- \mathbf{P}(z) = -\sqrt{\mathbf{G}(z)} \,. \tag{62}$$

In other words, the eigenstates of Eq.(58) coincide with those of Eq.(60), but they are relabeled by Eq.(62).

On $\mathcal{H}^+$, the functional Schrödinger equation evolves an arbitrary initial state $\Psi[\mathbf{T_0}, \mathbf{g}]$ at $\mathbf{T_0}(z)$ into the final state $\Psi[\mathbf{T}, \mathbf{g}]$ at $\mathbf{T}(z)$,

$$\Psi[\mathbf{T}, \mathbf{g}] = \hat{\mathbf{U}}[\mathbf{T} - \mathbf{T_0}; \hat{\mathbf{g}}, \hat{\mathbf{p}}] \, \Psi[\mathbf{T_0}, \mathbf{g}] \,, \tag{63}$$

by the action of the many-fingered time evolution operator

$$\hat{\mathbf{U}}[\mathbf{T} - \mathbf{T_0}; \hat{\mathbf{g}}, \hat{\mathbf{p}}] := \exp\left(-i \int_{\mathcal{S}} d^3z \, (\mathbf{T}(z) - \mathbf{T_0}(z)) \, \hat{\mathbf{h}}_\uparrow(z)\right) \,. \tag{64}$$

Unlike the Wheeler-DeWitt equation, the Schrödinger equation (55) has a conserved positive-definite inner product on $\mathcal{H}^+$:

$$\langle \Psi_1 | \Psi_2 \rangle := \int \mathbf{Dg} \, \bar{\Psi}_1[\mathbf{T}, \mathbf{g}] \Psi_2[\mathbf{T}, \mathbf{g}] \,. \tag{65}$$

This was, after all, the main reason for introducing the matter time.

12.   Conclusions

The introduction of dust solves, at the phenomenological level, several outstanding problems of canonical quantum gravity:

- In vacuum gravity, it is not at all clear what combination of the phase-space variables g, p should play the role of internal time. After geometry is coupled to dust, one of the state variables of the coupled system, namely, $T(x)$, can be clearly identified as a time variable on $\mathcal{M}$. In globally hyperbolic spacetimes, $T(x)$ is defined on the whole spacetime. In canonical gravity, $T(\sigma)$ becomes one of the fundamental canonical coordinates.

- The dust time $T(x)$ is a spacetime scalar. The time coordinate of a spacetime event x therefore does not depend on the choice of a hypersurface through x. This removes the spacetime problem of the vacuum theory.

- The Hamiltonian constraint can be readily solved for the time momentum P. It is easy to state the conditions under which the old constraints are globally equivalent to the new constraints. The imposition of the resolved constraint as an operator restriction on states yields a functional Schrödinger equation.

- The ensuing Hamiltonian density $h_\uparrow(\sigma)$ has an extraordinary feature not encountered in other systems: it depends only on the geometric variables, not on the dust frame and time. This has three important consequences:

- Firstly, the functional Schrödinger equation can be solved by separating the dust time from the geometric variables.

- Secondly, the Hamiltonian densities strongly commute and can therefore be simultaneously defined by spectral analysis.

- Thirdly, the Schrödinger equation can be solved independently of the supermomentum constraint, which is then satisfied by parametrizing this solution.

- The Schrödinger equation has a conserved inner product. This sets the framework for statistical interpretation of quantum gravity.

Unfortunately, this solution to the problem of time is still incomplete. There are at least two issues that remain open: the problem of observables, and the problem of spacelike hypersurfaces.

1. *The problem of observables.* In classical theory, the dust frame components of the intrinsic metric and extrinsic curvature on a hypersurface specified by dust time become observable. One would hope that these classical observables can be turned into quantum observables whose statistical distributions are determined

by the Schrödinger inner product. Unfortunately, this is not as straightforward as it seems:

In quantum theory, a dynamical variable $\hat{\mathbf{F}} = \mathbf{F}[\hat{\mathbf{g}}, \hat{\mathbf{p}}]$ is an observable only if its action on a state Ψ from $\mathcal{H}^+$ again yields a state in $\mathcal{H}^+$. A simple example of a quantum observable is $\mathbf{G}(z; \hat{\mathbf{g}}, \hat{\mathbf{p}})$. Unfortunately, the fundamental geometric observables defined as the multiplication and differentiation operators

$$\hat{\mathbf{g}}(z) = \mathbf{g}(z)\times, \quad \hat{\mathbf{p}}(z) = -i\frac{\delta}{\delta\mathbf{g}(z)} \tag{66}$$

do not have this property. It is unclear what operators, if any, represent the classical observables $\mathbf{g}(z)$ and $\mathbf{p}(z)$.

2. *The problem of spacelike hypersurfaces.* The second problem of the functional Schrödinger approach is how to maintain the spacelike character of the hypersurfaces $\mathbf{T}(z)$. This problem already exists at the classical level: The Lie propagation of an initial spacelike hypersurface does not in general produce a spacelike foliation $\mathcal{T}$. Thus, e.g., in a dust-driven Friedmann universe, any spacelike hypersurface which is not everywhere orthogonal to the dust world-lines ultimately becomes somewhere timelike. The only hypersurface which is guaranteed to remain spacelike is one which is orthogonal to the worldlines. Unfortunately, if the dust rotates, such a hypersurface does not exist.

If we were required to stay only on the leaves of $\mathcal{T}$, the time propagation would drive us into hypersurfaces which are not spacelike. In the functional time formalism, we can steer our course to avoid them: We choose $N^\dagger(\tau, z)$ such that the hypersurface continues to be spacelike. This, of course, requires monitoring the signature of $\mathbf{g}(z)$ as we sail through spacetime. By choosing different $N^\dagger(\tau, z)$'s, we can reach all spacelike hypersurfaces.

In quantum theory, this is not that easy. The state functional describes an ensemble of metrics $\mathbf{g}(z)$, and a hypersurface $\mathbf{T}(z)$ which is spacelike with respect to one of them does not need to be spacelike with respect to another. One can imagine starting with a functional $\Psi[\mathbf{T_0}, \mathbf{g}]$ which has support only on Riemannian metrics, and admitting only hypersurfaces $\mathbf{T}(z)$ such that the evolved functional $\Psi[\mathbf{T}, \mathbf{g}]$ has again support only on Riemannian metrics. However, it is quite possible that most initial functionals immediately start leaking into timelike metrics, and that we get stuck on the initial hypersurface. If so, we should either abandon the Schrödinger approach, or learn how to live with hypersurfaces that are not necessarily spacelike.

13. Acknowledgments

The work on this paper has been partially supported by the NSF grant PHY-9207225 to the University of Utah. I am grateful to Julian Barbour for his comments on the final draft.

References

[1] Dirac P A M 1964 *Lectures on Quantum Mechanics* (New York: Yeshiva University); and references therein.

[2] Arnowitt R, Deser S and Misner C W 1962 The dynamics of general relativity *Gravitation: An Introduction to Current Research* ed L Witten (New York: Wiley); and references therein.

[3] See, however, Dirac P A M 1959 *Phys. Rev.* **114** 924

[4] Wheeler J A 1968 Superspace and the nature of quantum geometrodynamics *Batelle Rencontres: 1967 Lectures in Mathematics and Physics* ed C DeWitt and J A Wheeler (New York: Benjamin)

[5] DeWitt B S 1967 *Phys. Rev.* **160** 1113

[6] Kuchař K V 1992 Time and interpretations of quantum gravity *Proceedings of the 4th Canadian Conference on General Relativity and Relativistic Astrophysics* ed G Kunstatter *et al.* (Singapore: World Scientific)

[7] Kuchař K V 1972 *J. Math. Phys.* **13** 768

[8] Kuchař K V 1970 *J. Math. Phys.* **11** 3322

[9] Misner C W 1972 Minisuperspace *Magic Without Magic: John Archibald Wheeler, A Collection of Essays in Honor of his 60th Birthday* ed J Klauder (San Francisco: Freeman)
Ryan M 1972 *Hamiltonian Cosmology* (Berlin: Springer)
MacCallum M A H 1975 Quantum Cosmological Models *Quantum Gravity* ed C J Isham, R Penrose and D W Sciama (Oxford: Clarendon)

[10] Kuchař K V 1971 *Phys. Rev.* **D4** 955

[11] Kuchař K V 1989 *Phys. Rev.* **D39** 1579, 2263

[12] Torre C G 1991 *Class. Quantum Grav.* **8** 1895

[13] Hájíček P 1989 *J. Math. Phys.* **30** 2488
Hájíček P 1990 *Class. Quantum Grav.* **7** 871
Schön M and Hájíček P 1990 *Class. Quantum Grav.* **7** 861
Hájíček P 1990 *Lecture Notes on Quantum Cosmology* (Bern: University of Bern) unpublished

[14] Torre C G 1992 *Phys. Rev.* **D46** R3231

[15] Kuchař K V 1991 The Problem of Time in Canonical Quantization of Relativistic Systems *Conceptual Problems in Quantum Gravity* ed A Ashtekar and J Stachel (Boston: Birkhäuser)

[16] Isham C J 1992 Conceptual and geometrical problems in canonical quantum gravity *Recent Aspects of Quantum Fields* ed H Mitter and H Gausterer (Berlin: Springer)

[17] Einstein A 1920 *Über die spezielle und die allgemeine Relativitätstheorie* (Braunschweig: Vieweg) 67
Einstein A 1961 *Relativity: The Special and the General Theory* transl R W Lawson (New York: Crown)

[18] Hilbert D 1917 Grundlagen der Physik *2 Mitt., Nachr. Ges. Wiss. Göttingen* **53**

[19] Isham C J and Kuchař V K 1985 *Ann. Phys., NY* **164** 288, 316

[20] Hartle J B 1989 Time and Prediction in Quantum Cosmology *Proceedings of the Fifth Marcel Grossmann Meeting* ed D Blair and M Buckingham (Singapore: World Scientific)

[21] Halliwell J and Hartle J B 1991 *Phys. Rev.* **D43** 1170

[22] Kuchař K V and Torre C G 1991 *Phys. Rev.* **D43** 419

[23] Kuchař K V and Torre C G 1991 *Phys. Rev.* **D44** 3116
Stone C L and Kuchař K V 1991 *Class. Quantum Grav.* **9** 757

[24] Kuchař K V and Torre C G 1989 *J. Math. Phys.* **30** 1769

[25] Torre C G 1989 *Phys. Rev.* **D40** 2558

[26] Kuchař K V 1992 *Phys. Rev.* **D44** 43

[27] Kuchař K V 1991 *Phys. Rev.* **D43** 3332

[28] DeWitt B S 1962 The Quantization of Geometry *Gravitation: An Introduction to Current Research* ed L Witten (New York: Wiley)

[29] Rovelli C 1991 *Class. Quantum Grav.* **8** 297, 317

[30] Brown J D 1992 Action Functionals for Relativistic Perfect Fluids *Preprint* (Raleigh: North Carolina State University)

[31] Brown J D and Kuchař K V 1993 Dust as A Standard of Space and Time in Canonical Quantum Gravity *Paper in preparation*

[32] Kuchař K V 1981 Canonical methods of quantization *Quantum Gravity 2: A Second Oxford Symposium* ed C J Isham et al. (Oxford: Clarendon)

[33] Misner C W 1957 *Rev. Mod. Phys.* **29** 497

[34] Higgs P W 1958 *Phys. Rev. Lett.* **1** 373

The Isotropy and Homogeneity of the Universe

Richard A. Matzner *

Abstract

I discuss current observational limits on the inhomogeneity and the isotropy of the universe. Isotropy *observations* come from the COBE differential microwave radiometer. COBE results are consistent with prior estimates based on cosmic nucleosynthesis. The COBE results on the present structure can be used to limit the range of background density, in particular the closure of the described universe.

Examples from the literature are given whereby a 30 eV massive neutrino simultaneously fits both the observed structure on small scales, and the level of observed quadrupole anisotropy. Further simulations are needed to verify these theoretical fits to the observations.

This paper is dedicated to Charles Misner on his sixtieth birthday.

1. Introduction

In 1966, prompted by the apparent anisostropic distribution of the three or four then known QSOs, Charles Misner began investigating the behavior of anisotropic universes. These had been studied before, by Kasner [1], Zel'dovich [2], Thorne [3], Taub [4], but Misner's development was a tour de force combining differential geometry, classical mechanics, and astrophysics. One track of his research led to the Mixmaster universe [5], a closed 3-spherical universe in which the ratios of the principal circumferences oscillate as the universe expands and recollapses. This oscillation can lead to very large horizon lengths in particular directions, and gave the hope of explaining the horizon problem. The Mixmaster results led directly to Hamiltonian cosmology and the Quantum Cosmology research effort. The more astrophysical branch of this research [6] developed into studies of dissipation in anisotropic Bianchi type I cosmologies. In these cosmologies the expansion rate (the Hubble constant) is anisotropic. Collisionless matter in the cosmology can be differentially blue shifted so much that the back reaction can reverse the anisotropy, in a way similar to the Mixmaster oscillations. At the same time large differences in energy of collisionless

---

*Center for Relativity, University of Texas at Austin, Austin, TX, 78712

particles streaming in different directions set the stage for dramatically dissipative interactions, and Misner introduced the idea of neutrino 'viscosity' which reduces residual anisotropy, until a redshift of approximately 10^{10}, when the neutrino mean free path became so large that effective neutrino decoupling occurred. This epoch is also at the beginning of the nucleosynthesis era. Work by Rothman and Matzner [7] (for instance) finds that the anisotropy of the expansion, measured in the anisotropy of the Hubble constant and of the neutrino distribution, is bounded to be less than 1.5×10^{-2} during nucleosynthesis ($z = 3 \times 10^8, T = 10^9 K$). Larger values would have overproduced ^{4}He for the observed value of deuterium. Misner's [6] predicted present amplitude of quadrupole anisotropy, $\Delta T/T \lesssim 3 \times 10^{-4}$, based on the assumption that Hubble anisotropy was less than unity at $z = 5 \times 10^9$ is also consistent with other estimates limiting anisotropy by its effect on nuclear abundances, if one assumes evolution with low baryon density ($\leq 10^{31} gm/cm^3$ now) and massless neutrinos. I shall return to this particular result but anticipate somewhat by noting that mapping anisotropy limits at nucleosynthesis to those today requires an understanding of the physics of the universe in the interim.

In a recent preprint the COBE differential microwave radiometer (DMR) results are given and analyzed [8]. The first year's DMR results give, after extensive processing to remove galaxy and other known sources, $\Delta T/T \sim (1 \pm 0.2) \times 10^{-5}$ on 10° FWHM gaussian smoothing, and give $\Delta T/T \sim (1 \pm 0.3) \times 6 \times 10^{-6}$ for the quadrupole result. This is consistant [8] with a Harrison-Zel'dovich ($n = 1$) spectrum, which predicts $\sigma(10°) \sim 2 \times$ Quadrupole.

The real value of the COBE results lies in the fact that they are <u>observations</u> rather than just upper limits. One now has definite values against which to test various theories of the microwave background structure. Here I display the result of such a test, which shows that aspects of a universe containing a 30eV neutrino (a hot dark matter model) are consistent with the COBE observations.

2. Small Scale Microwave Temperature Anisotropy

We shall begin with the small scale structure. Recently, Anninos, Matzner, Tuluie, and Centrella [9] published the results of a hot dark matter structure simulation. In this simulation, we close the universe with 30eV neutrinos. The neutrinos are initially perturbed by a Harrison-Zel'dovich spectrum with a short wavelength cutoff to form structure giving "cluster–cluster" correlation with slope -1.77 at present. At present the density contrast has an rms value of order 2, and a peak value ~ 200. Thus this model reproduces some of the features seen in the large scale structure of the universe. Because of limitation on the computational size, the simulations model only structure smaller than $\sim 165 Mpc$ (current size). Structures formed are typically $\sim 50 Mpc$.

Although the universe is closed in neutrinos, a small fraction of baryon matter is also assumed, and the ionization fraction of this matter is tracked. After determining and recording the history of the structure, Anninos et al perform a photon calculation, propogating photons through the forming structure. At the beginning of a photon run, a number τ_0 is selected from an exponential distribution for each photon. The photons will travel through this optical depth before scattering. As the photons evolve through the model universe, $\Delta\tau$ is computed for each photon using the local (different photons see different densities) electron density $n_e = x_H \rho_\nu \Omega_{0bar}/m_p$. Here x_H is the ionization fraction. Any photon for which the resulting $\Delta\tau$ exceeds τ_0 is considered to have instantaneously scattered and is assigned a new value for τ_0.

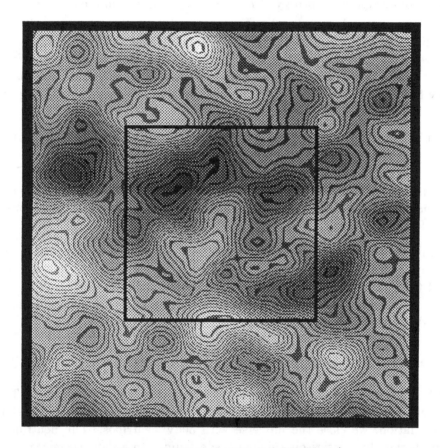

Figure 1: Plot of plate 7 (cf Table 1.)
The image is displayed with a resolution of 380×380 pixels by interpolating the 60×60 smoothed grid data and superposing a dark gray contour to highlight structure by delineating the 32 level scale. Each contour corresponds to $\Delta T/T = 2.2 \times 10^{-6}$. Darker is redshifted; lighter is blue-shifted temperature. This image is 4×4 with a 2×2 delineated inner square. It was constructed a "random" direction in an $\Omega_0 = 1$, $\Omega_{baryon} = 0.02$ structure model, imaged with 10^6 photons.

The photon is considered to have scattered again when further changes in $\Delta\tau$ exceed the new τ_0. The scattering process is performed in the rest frame of the scattering electron whose velocity is given by the local weighted average of the velocities of matter overlapping the zone in which the photon is scattered. Photon orbits naturally given in comoving coordinates are Lorentz transformed into the rest frame of the scattering electron.

The scattering process is accomplished with Monte Carlo probabilistic techniques. The scattering angles (Θ, Φ) are defined as spherical coordinates relative to the incident ray in the scattering electron's rest frame. We choose Θ from a distribution in $cos\Theta$ that varies as $\sim (1 + cos^2\Theta)$ (that is, we randomly choose $cos\Theta$ in the range $-1 \le cos\Theta \le 1$ and use the method of rejection to enforce the correct distribution). The azimuthal angle Φ is chosen from a uniform distribution over $0 \le \Phi \le 2\pi$. We have verified that such a construction results in the correct distribution over a sphere. After scattering the photon trajectories are transformed back to the comoving frame. In this way we couple to electron velocities as well as to electron density enhancements.

The physical effects correctly modelled here are: the slow, diffuse decoupling surface, (rather than a sharp epoch of decoupling); the interaction of photons with the gravitational field and with the nonlinear growing structure in the matter content, including the (small) differential Doppler shifts that arise from scattering on moving electrons. One finds that the microwave fluctuations can be small, and substantially below the density inhomogeneity.

The simulations found a typical rms temperature fluctuation of 1×10^{-5}, on the scales relevant to the simulation, i.e., $\sim 0°.5$ to $\sim 2°$, which is exactly consistent with the COBE results at $10°$, assuming a Harrison-Zel'dovich $n = 1$ spectrum (which predicts approximately flat $(\Delta T/T)_{rms}$ as a function of angular scale). Figure 1 shows a typical derived map from this simulation. Table 1 gives the rms $\Delta T/T$ values for a number of simulations, defined by averaging the squares of the differences from the "plate" mean over 3600 pixels. The data was first smoothed the 3×3 pixel areas, so the effective resolution here is 20×20 ($\sim 0°.1 \times 0°.1$). Both from the table and from the image it can be seen that the fluctuations are smooth and well defined on this scale, providing in fact a visualization of the early structure formation in the model.

3. Quadrupole Temperature Anisotropy

Is this suggestion of a universe closed by 30 eV neutrinos otherwise consistent with COBE observations? We can consider the quadrupole observations, which COBE determines at $Q_{rms} \sim 6 \times 10^{-6}$. The point of our calculation here will be the following: suppose the universe were as anisotropic as possible while still consistent with nucleosynthesis results. This gives [7], [10] an anistropy at $T = 10^9 K$ of less than 1.5×10^{-2}.

Plate	rms	Minimum	Maximum
1	+1.0	−2.9	+2.6
2	+1.0	−3.1	+2.1
3	+1.0	−3.1	+2.9
4	+1.0	−3.3	+2.3
5	+0.8	−3.2	+1.9
6	+0.9	−3.1	+1.9
7	+0.8	−2.5	+2.5
8	+0.9	−1.7	+3.2

TABLE 1

TEMPERATURE FLUCTUATIONS $\Delta T/T \times 10^5$

From Anninos et al [9]. Each 'Plate' is a computer simulation. Sizes of plates are either $4° \times 4°$ or $2° \times 2°$. The statistics are similar because the fundamental cube of our calculation, $\lesssim 165 Mpc$ now, subtends of order $1°$ at a redshift of 10^3. Hence larger plates sample an (almost) repetitive image of the sky. The plates sample several directions and several realizations of the matter structure. The results are consistent with those found at $\sim 10°$ scale by COBE [8].

The component of this anisotropy corresponding to the largest scales (the quadrupole anisotropy) evolves in a radiation filled universe in a decaying oscillating manner. If collisionless particles (neutrinos, photons) are too small a constituent of the universe then the oscillations are quickly damped away. There is also some range of possible behavior after the nucleosynthesis limit, effectively depending on the phase of the oscillating anisotropy.

We now want to consider the anisotropy evolution from nucleosynthesis to the present. Misner showed that for the small anisotropy case as we now consider, there are two distinct behaviors, depending on whether the gross expansion of the Universe is dominated by radiation, or by pressureless matter. For the radiation-dominated case Misner [6] defined the following quantity,

$$\frac{8\pi G}{3}\rho_\beta = \frac{1}{6}\dot{\beta}_{ij}^2 + \frac{8\pi G}{3}\rho_\nu \left(\frac{4}{15}\beta_{ij}^2\right) \tilde{\propto} R^{-5} \tag{1}$$

where this form holds for small β_{ij} and small $|\dot{\beta}|/H$. (Here $R \equiv e^\alpha$ is the scale factor.) The two conditions imply that the universe is only slightly anisotropic, and the Hubble constant is dominated by an isotropic matter component. $\dot{\beta}_{ij}/H_0$ is related to the differential Hubble rate via $(\Delta H)^2 \equiv (H_1 - H_2)^2 + (H_2 - H_3)^2 + (H_3 - H_1)^2$ and $\dot{\beta}_{ij}^2/H^2 = (\Delta H/H)^2$, and when suitably normalized β_{ij} is related to the microwave temperature quadrupole anisotropy. The R^{-5} dependence is not obvious, but holds in the radiation dominated epoch of a universe with two or three neutrino flavors; this was derived by [6]. The form just above can be simplified by dividing by the square of the Hubble parameter:

$$H^2 = \frac{8\pi G}{3}\rho \tag{2}$$

where ρ is the isotropic density driving the expansion. Then, assuming $\rho \propto R^{-4}$ (radiation domination)

$$\frac{\rho_\beta}{\rho} = \frac{1}{6}\frac{\dot{\beta}_{ij}\dot{\beta}_{ij}}{H^2} + \frac{\rho_\nu}{\rho}\frac{4}{15}\beta_{ij}\beta_{ij} \,\tilde{\propto}\, R^{-1} \tag{3}$$

The ratio ρ_ν/ρ is a constant, determined by the number of neutrino species and the ratio of the neutrino and photon temperatures. Also ρ_ν decays as R^{-4} under the conditions considered here so we have

$$\frac{\rho_\beta}{\rho} = \frac{1}{6}\left(\frac{\Delta H}{H}\right)^2_{rms} + \frac{a_\nu}{a}\cdot\frac{4}{15}\beta_{ij}\beta_{ij} \,\tilde{\propto}\, R^{-1} \tag{4}$$

while the universe is radiation dominated. (a_ν and a are "radiation constants")

In the matter-dominated case, the evolution is of the form [6]

$$(\beta - \beta_1)_{ij} = \frac{2}{3}\left[\frac{\dot{\beta}_{ij}}{H}\right]_1 (1 - e^{-(3/2)(\alpha-\alpha_1)}), \tag{5}$$

where the subscript 1 denotes the "instant" from which we begin to measure β. For late times this obviously becomes $(\beta - \beta_1)_{ij} \to \frac{2}{3}\left[\dot{\beta}_{ij}/H\right]_1$.

Finally, the connection between β and the quadrupole temperature anisotropy follows from the small anisotropy limit

$$T_\gamma(n) \approx T_{c\gamma}(1 - \beta_{ij}n_in_j) \tag{6}$$

where now β is the anisotropy measured since the *photons* became collisionless [6]. This is the normalization referred to above. Hence this β is a direct measure of the photon-temperature anisotropy. $T_{c\gamma}$ is a constant, the average on the sky of the temperature $\sim 2.7°K$. (Note that β is measured from the photon-decoupling epoch, $z \sim 1000$. We suppose that any subsequent reheating provides so little optical path that the photons are effectively collisionless after $z = 1000$.) If all neutrinos are massless, then the baryon density (10^{-31} gm/cm now) begins to dominate at a red-shift $z \sim$ 200–300. If, on the other hand, one neutrino has a mass $m_\nu \sim 30$ eV, it will become nonrelativistic at $z \sim 10^5$. Then this constituent of matter is providing closure now, and acting like a non-radiation source; the universe is thus closed by matter domination at present with neutrinos of mass $m_\nu = 30eV \left(H_0/\left(55km/sec/Mpc\right)\right)^2$ giving $\rho_H = 4.7 \times 10^{-30} \left(H_0/\left(55km/sec/Mpc\right)\right)^2$, where ρ_H is the density of the 'heavy' constituent. Comparing the contribution of this constituent to the (photon plus two neutrinos) radiation contribution, we find that the transition to matter domination occured at $z \simeq 7500 \left(H_0/55\right)^2$. (The 30eV neutrino became nonrelativistic well before this epoch, so the picture is so far consistent.)

The initial upper limit (from [7], [10]) is $\rho_\beta/\rho = 0.015$ at $z = 3 \times 10^8$. Thus

$$\left(\frac{\Delta H}{H}\right)^2_{z=7500(H_0/55)^2} \leq 6\left(\frac{\rho_\beta}{\rho}\right)_{z=3\times10^8} \times \left(\frac{7500\left(H_0/55\right)^2}{3 \times 10^8}\right)$$
$$= 2.3 \times 10^{-6}\left(H_0/55\right)^2 \tag{7}$$

After matter domination Eq. (5) above gives $\dot\beta_{ij}/H \propto e^{-(3/2)\alpha}$, or $(\Delta H/H)^2 \propto R^{-3}$:

$$\left(\frac{\Delta H}{H}\right)^2_{z=10^3} \leq 2.3 \times 10^{-6}\left(H_0/55\right)^2 \cdot \left(\frac{1000}{7500\left(H_0/55\right)^2}\right)^3$$
$$= 5.3 \times 10^{-9}\left(55/H_0\right)^4 \tag{8}$$

At this point the photons become collisionless, and we use Eq. (5) to measure their temperature evolution.

Equations (5) and (6) above, together with the definitions for Q_{rms} used with the COBE data analysis [8] lead to

$$Q_{rms}^2 \leq \frac{8}{135} \left(\frac{\Delta H}{H} \right)_1^2$$

With Eq. (8) above, this is:

$$Q_{rms} \leq 1.8 \times 10^{-5} \left(\frac{55}{H_0} \right)^2$$

This is a result derived assuming maximal consistent anisotropy at nucleosynthesis, and is then an upper limit on the allowed quadrupole contribution to the anisotropy. This form makes clear how the result depends on the Hubble constant under the assumption $\Omega = 1$. With the choice 30eV the neutrinos close the universe, and as we have seen, such a model is consistent with structure formation and with observed COBE microwave structure, in particular with the quadrupole anisotropy. However, if the Hubble constant were significantly greater than $75 km/sec/Mpc$, it would be very difficult to argue that there is sufficient nonrelativistic mass content to provide $\Omega = 1$, while maintaining quadrupole consistency with the COBE results, at least if the nonrelativistic domination has held since $z \sim 7500$. In fact this limit can be read as bounding the mass of a dark matter neutrino to less than 90 eV, which would demand $H_0 \sim 95 km/sec/Mpc$. Notice that the quadrupole anisotropy that raises the nucleosynthesis ^{4}He production is local to small regions of a few light-minutes scale (horizon size) during nucleosynthesis. Such anisotropy is the sum of all contributions of wavelength exceeding the horizon size at that time. Although there are special low baryon density inhomogeneous situations that lead to slight Helium suppresion, the general behavior is for every deviation from homogeneity and isotropy to increase the net Helium production. We certainly expect an incoherent addition of modes, so that the global quadrupole is bounded from above by the anisotropy during nucleosynthesis, and bounds based on nucleosynthesis for the quadrupole amplitude should be valid.

It is also clear that it will be very interesting to consider larger scale structure and microwave simulations, since those described by Anninos et al [9] do not cover the range accessible to COBE, and to carry out a more detailed model of the evolution of the quadrupole anisotropy, to refine the implications of the nucleosynthesis limit.

4.   Conclusion

The COBE results finally present a definite observation against which to compare theory. I have presented here two aspects of a hot dark matter structure model which fit well (consistently) with the COBE results. Contrary to popular first impression, the COBE results are understandable from the viewpoint of structure formation in a fairly standard theory.

Misner's contribution in bringing to light the connection between physics at $z = 10^{10}$, at $z = 10^3$, and at $z = 0$ has played a very important part in driving the field. I am pleased in this paper to have shown some of the developments from Misner's pioneering work.

5.   Acknowledgements

Professor Dennis Sciama made some very valuable suggestions on an earlier version of this paper. I also thank R. Tuluie and P. Anninos for very helpful comments. I am very grateful for the hospitality of Prof. R. Penrose and the Mathematical Institute, Oxford, where some of this work was carried out. This work was supported by NSF grant PHY 8806567, by Texas Advanced Research Grant ARP–085, and by a Cray University Research Grant to Richard Matzner.

References

[1] Kasner, E., *Geometrical Theorems on Einstein's Cosmological Equations*. Am. Jour. Math. **43** (1921) p. 217

[2] Zel'dovich, Ya B., *Magnetical Model of the Universe*. Sov. Phys.–JETP **21** (1965) p. 656

[3] Thorne, K., Ph.D. thesis, Princeton University. (University Microfilms, Ann Arbor; 1966)

[4] Taub A.H., *Empty Space-Times Admitting a Three Parameter Group of Motions*. Ann. Math. **53** (1951) p. 472

[5] Misner C.W., *Mixmaster Universe*. Phys. Rev. Lett. **22** (1969) p. 1071

[6] Misner C.W., *The Isotropy of the Universe*. Astrophys. J. **151** (1968) p. 431

[7] Rothman, T., and Matzner, R.A., *Nucleosynthesis in Anisotropic Cosmologies Revisited*. Phys. Rev. **D 30** (1984) p. 1649.

[8] Smoot G.F, Bennett C.L., Kogut A., Wright E.L., Aymon J., Boggess N.W., Cheng E.S., DeAmici G., Gulkis S., Hauser M.G., Hinshaw G., Lineweaver C., Loewenstein K., Jackson P.D., Janssen M., Kaita E., Kelsall T., Keegstra P., Lubin P., Mather J.C., Meyer S.S., Moseley S.H., Murdock T.L., Rokke L., Silverberg R.F., Tenorio L., Weiss R., Wilkinson D.T., *Structure in the COBE DMR First Year Maps*. Astrophys. Jour. Lett. (1992), submitted.

[9] Anninos, P., Matzner, R., Tuluie, R., and Centrella, J., *Anisotropies of the Cosmic Background Radiation in a "Hot" Dark Matter Universe*. Astrophys. Jour. **382**, (1991) p. 71

[10] Kurki-Suonio, H. and Matzner, R., *Anisotropy and Cosmic Nucleosynthesis of Light Isotopes including ^7Li*. Phys. Rev. **D31** (1985) p. 1811.

Recent Advances in ADM Reduction

V. Moncrief [*]

1. Introduction

One of Charles Misner's most influential contributions to the mathematical development of general relativity was his proposal, along with R. Arnowitt and S. Deser, of the ADM program of (Hamiltonian) reduction of Einstein's equations [1,2]. By ADM reduction we mean not simply the elegant but essentially elementary reexpression of Hilbert's action in terms of canonical variables but rather the complete reduction of Einstein's equations to an unconstrained Hamiltonian system defined over an appropriate phase space of "true degrees of freedom." This program, though developed over the years by many people, beginning of course with Arnowitt, Deser and Misner themselves, remains largely unfinished even today. Nevertheless many model problems have been worked out, to a large extent by Misner and his former students and other collaborators, and the reduction program has proven itself of value, within these models, for the study of both classical and quantum dynamics. In this article we shall describe some recent developments of the ADM program and discuss some of the problems which remain to be solved.

Some of the work described below in Sect. II is based in part on a Yale senior physics project by Juan Lin, whereas much of that described in Sect. III is based upon the Yale Mathematics Ph.D. thesis of John Cameron (1991). The work described in Sect. IV represents research in progress in collaboration with Arthur Fischer (University of California, Santa Cruz). Section V mentions some recent joint work with Yvonne Choquet-Bruhat (Université Paris VI) and James Isenberg (University of Oregon).

2. ADM Reduction in 2 + 1 Dimensions

As a model for subsequent developments, let us first consider ADM reduction of the (vacuum) Einstein equations in 2 + 1 dimensions. Einstein's equations in dimension 3 imply that spacetime is flat but the solutions need not be (globally) stationary and indeed can exhibit complicated dynamics. To be specific let us consider only spacetimes which are foliated by compact, orientable hypersurfaces Σ of constant mean (extrinsic) curvature τ.

[*]Department of Physics, Yale University, 217 Prospect Street, New Haven, Connecticut 06511

If the genus of Σ is zero (i.e., $\Sigma \approx \mathbf{S^2}$), it is easy to show, upon integrating the Hamiltonian constraint over Σ, using the Gauss-Bonnet theorem and the fact that transverse-traceless symmetric tensor fields vanish identically on $\mathbf{S^2}$ (since the corresponding Teichmüller space, for which such fields would represent the tangent space, is zero dimensional), that there are *no solutions*. One would have to introduce matter sources or a cosmological constant to get a non-vacuous problem on $\mathbf{S^2}$. Therefore turn to $\mathbf{T^2}$. In this case some solutions admit *maximal* Cauchy hypersurfaces (with $\tau = 0$). But these solutions are always stationary and therefore admit global foliations by such surfaces. Roughly speaking these spacetimes merely correspond to imposing periodic spatial identifications on $2 + 1$ dimensional Minkowski space and are thus essentially trivial.

More interesting behavior is found by looking for solutions on $\mathbf{T^2} \times \mathbf{R}$ which have $\tau = $ constant $\neq 0$ Cauchy surfaces. ADM reduction leads in this case to a reduced configuration space which is topologically $\mathbf{R^2}$ but which is most naturally identified with $\mathbf{H^2}$—hyperbolic 2-space (having constant negative curvature). The reason for this is that the reduced Hamiltonian (obtained upon fixing a time coordinate condition in which $\tau = \exp[t]$) is simply the square root of the usual Hamiltonian for geodesic motion on $\mathbf{H^2}$. This Hamiltonian generates motion equivalent to geodesic motion (for non-degenerate geodesics whose velocities are non-zero) but relative to a non-conventional parameterization (which depends upon the energy of the geodesic). $\mathbf{H^2}$ may, as is well known, be identified with the Teichmüller space $\mathcal{T}(\mathbf{T^2})$ for $\mathbf{T^2}$ and, in this case, the reduced Einstein equations generate simply geodesic motion on the Teichmüller space [3,4]. The reduced phase space is just $T^*\mathcal{T}(\mathbf{T^2})$ and the reduced Hamiltonian (as described above) is simply a time-dependent multiple of the area function of the $\tau = $ constant hypersurfaces.

Now consider the case of higher genus surfaces Σ . It is easy to prove, by the same method outlined above for the $\mathbf{S^2}$ case, that *no solutions exist admitting a maximal hypersurface*. Every solution must correspond to either an expanding or a contracting universe. We can carry out ADM reduction relative to the time coordinate condition $t = \tau$. One finds after a lengthy but straightforward analysis which uses the classical *uniformization theorem*, the Lichnerowicz-Choquet-Bruhat-York *conformal method* for solving the constraints [5], the Fischer-Tromba characterization of the Teichmüller space $\mathcal{T}(\Sigma)$ in terms of the conformal structures on Σ [6] and the Eells, Earle and Sampson proof of the existence of a *global cross section* of the "principle bundle" of constant negative curvature Riemannian metrics on Σ (with group defined by the smooth diffeomorphisms of Σ isotopic to the identity) [7], that the reduced configuration space is again the associated Teichmüller space $\mathcal{T}(\Sigma)$ and the reduced phase space its cotangent bundle $T^*\mathcal{T}(\Sigma)$. With the choice $t = \tau$ the reduced Hamiltonian becomes simply the area function of the $\tau = $ constant hypersurfaces.

For a surface Σ of genus $g > 1, T^*\mathcal{T}(\Sigma)$ is a manifold diffeomorphic to $\mathbf{R^{12g-12}}$. The (time dependent) Hamiltonian flow generated on this space appears to be quite complicated in general though the solutions of Hamilton's equations can apparently be determined indirectly by evaluating a complete set of (gauge invariant) traces of holonomies and expressing them (as is always possible in principle) in terms of coordinates on $T^*\mathcal{T}(\Sigma)$ and τ [8,9,10]. Setting the 12 g - 12 independent functions equal to the constant values which define a specific spacetime, one could in principle solve for the τ-dependence of the 12 g - 12 canonical variables without the need for explicitly writing out Hamilton's equations.

Since the procedure outlined above has only been carried out explicitly for the case of $\Sigma \approx \mathbf{T^2}$, there are many open questions concerning the behavior of the higher genus solutions (which do not seem to have been resolved by other methods either). Does every expanding solution evolve from a "big bang" (of infinite mean curvature) to a state of infinite area (as the mean curvature tends to zero) or can "topology-changing" singularities intrude whereby a genus g surface "pinches off" to yield either a surface of lower genus or perhaps a disconnected union of surfaces of lower genus? This would correspond to solution curves in $T^*\mathcal{T}(\Sigma)$ running to the boundary of Teichmüller space at some finite, non-zero value of the mean curvature τ .

It is not difficult to derive, from the integrated Hamiltonian constraint, a (τ - dependent) bound on the magnitude of the velocity of the solution curve projected down onto Teichmüller space (equipped with the Weil-Peterson metric) but, unfortunately, the Weil-Peterson metric is known to be incomplete so this bound alone does not suffice to prevent solution curves from running to the boundary. Some more detailed features of the Einstein evolution would seemingly need to be taken into account. Perhaps a more detailed study of the traces of holonomies (computed for example around minimal geodesics of the conformal metric of constant negative curvature), utilizing the fact that they are constants of the motion (a key feature in Witten's approach to 2 + 1 gravity [9]), could be used to rule out such topology change, at least at the classical level. Such topology changing behavior, if possible, would not be limited to 2 + 1 dimensions since one can take the product of any solution with a circle and get a genuine, spatially compact (and still flat) Lorentzian solution of Einstein's equations in 3 + 1 dimensions.

The nature of "big bang" singularities in these flat spacetimes is also somewhat mysterious. They cannot of course be curvature singularities but they also seem not to be (except in special cases) compact Cauchy horizons of the Taub-NUT type either. Indeed, except for the case of genus g = 1, any candidate surface for the compact Cauchy horizon would not support the smooth, nowhere vanishing direction field induced upon this surface by the (future directed) null geodesic generators of this hypothetical Cauchy horizon. In addition there are some rather general arguments by Isenberg and the author which strongly suggest (though the complete proof is still lacking) that any vacuum spacetime with a compact Cauchy horizon must in fact admit a non-vanishing Killing field and the horizon must in fact be a Killing horizon (with null generators tangent to the Killing field) [11,12,13]. But vacuum spacetimes containing higher genus Cauchy surfaces (of constant mean extrinsic curvature) cannot admit (global) Killing fields since it is straightforward to show (using uniqueness of the conformal projection in the higher genus cases) that the first fundamental form of the Cauchy surface cannot admit a Killing symmetry unless the conformal metric does also and one can easily exclude that possibility upon taking the divergence of Killing's equations and appealing to the constancy of the (Ricci) curvature of the conformal metric. This result, together with the fact that a higher genus surface cannot admit a flat metric, eliminates the possibility that it can support non-trivial "Cauchy data" for a Killing field.

If one considers the special solutions obtained by taking quotients of $2+1$ dimensional Minkowski space by suitable discrete subgroups of the Lorentz group (chosen so as to compactify each spacelike hyperboloid $t^2 - x^2 - y^2 = c^2 > 0$ to a genus $g > 1$ surface) then it is straightforward to reduce the problem of motion for null geodesics (i.e., test light rays) in these spacetimes to that of ordinary geodesic motion of a fixed compact surface of genus g and constant negative sectional curvature. In the course of transforming one problem to the other one finds that an infinite lapse of affine parameter in the latter problem has been compressed into a finite lapse (measured from the big bang) in the former problem. This, combined with the well known results implying chaotic behavior for the ordinary geodesics on a compact, negatively curved surface, leads to the surprising result (also true in higher dimensional analogues) that a generic test photon *visits any region of the expanding hyperboloidal surface infinitely many times in an arbitrarily short interval of proper time as measured from the big bang*. In $2 + 1$ dimensions (but not in $3 + 1$) this mixing property even survives the introduction of dust matter (but not radiation itself). Thus, while not apparently of physical interest as a possible solution to the horizon problem, this mixing property of photon trajectories in these particular solutions renders the study of their singularities by the tracking of light rays highly chaotic.

It is interesting to speculate whether the relaxation of conventional energy conditions (perhaps because of quantum effects in matter near the big bang) could allow such mixing behavior in non-vacuum $3 + 1$ dimensional spacetimes. In a sense such mixing represents the complete antithesis of the usual horizon behavior– every region of the universe communicates with every other region (at least virtually) infinitely many times during an arbitrarily short interval of proper time as measured from the big bang.

3.   Reduction in $3 + 1$ Dimensional Vacuum Spacetimes having a Spacelike U(1)-Isometry Group

Before turning to the general case of reduction in $3 + 1$ dimensions it is of interest to look at a special case: reduction for spacetime manifolds which are diffeomorphic to S^1 - bundles over bases of the type $\Sigma \times R$ where Σ (as above) is a compact, orientable surface and where the spacetime metric is assumed to be invariant with respect to the $U(1)$ action of translation along the fibers of the bundle. The simplest cases of (arbitrary) bundles over $S^2 \times R$ and (trivial) bundles over $T^2 \times R$ were considered by the author [14,15] whereas the general case of arbitrary bundles over arbitrary surfaces has been treated in depth by Cameron [16]. Cameron's results are the subject of a recent, extensive review [17] so we shall not describe them in detail here. We shall simply observe that the field equations can always be reduced (in Kaluza-Klein fashion) to a system defined on the base whose specific structure depends of course upon the bundle from which they were derived but which consists essentially of a set of *harmonic map* fields (with hyperbolic 2-space H^2 as their target manifold) coupled to the Teichmüller parameters (and their conjugate momenta) defined over $T^*T(\Sigma)$. The full phase space is of course infinite dimensional in this case since the *harmonic map* fields contribute (roughly) two degrees of freedom per spatial point. These are in fact just the genuine two degrees of freedom of $3 + 1$ gravity disguised by the procedure of Kaluza-Klein reduction to appear as matter fields in a system of $2 + 1$ dimensional, non-vacuum Einstein equations, which of course contribute the Teichmüller parameters, as above, upon ADM reduction.

In the special case of bundles over $S^2 \times R$ the Teichmüller parameters are absent and the reduced field equations take the form of pure *harmonic map* equations (with however a domain space Lorentzian metric which is not given a priori but is instead a known functional of the *harmonic map* fields under consideration). One of the constants of motion, inherited from the 3 dimensional isometry group of hyperbolic 2-space, must be fixed to be integral in value in order to determine a solution on the S^1 - bundle of specified Chern class.

At first sight this elegant form for the reduced field equations in terms of *harmonic maps* (representing the true gravitational degrees of freedom) seems to provide an attractive pattern for potential generalization to the fully general (non-symmetric) case. However the harmonic map description is intimately tied to the assumed existence of a Killing field. Indeed one of these harmonic map scalars is just the norm of the Killing field and the other is its "twist potential." Thus, elegant or not, this is not the structure we should expect to find in the general case. It is not the harmonic maps we should seek to generalize. *It is the Teichmüller parameters.*

4. ADM Reduction in 3 + 1 Dimensions

We saw above that ADM reduction led to a conceptually (though not analytically) simple picture for 3-dimensional manifolds of the form $\Sigma \times R$ where Σ is a compact, orientable surface of genus $g > 1$. One got for each of these manifolds a reduced Hamiltonian system defined on the associated cotangent bundle $T^*T(\Sigma)$ where $T(\Sigma)$ is the Teichmüller space of the surface Σ. For $\Sigma \approx S^2$ the problem degenerated in that there were no solutions. For $\Sigma \approx T^2$ there were solutions of two types—stationary solutions (globally foliated by maximal hypersurfaces) and evolving solutions determined by a well defined Hamiltonian flow on $T^*\mathcal{T}(\mathbf{T^2}) \approx T^*\mathbf{H^2} \approx \mathbf{R^4}$. But is there a phase manifold which incorporates all of these solutions together? The answer is that the full space of solutions for $\mathbf{T^2} \times \mathbf{R}$ is actually the product of a manifold and a cone—the singular points of which correspond to the special solutions with (timelike) Killing symmetry.

In fact it is known quite generally (in $3 + 1$ dimensions) that conical singularities in the space of solutions of the Einstein constraint equations occur at precisely those sets of Cauchy data (on a compact, orientable 3-manifold) for which the associated Einstein spacetime has a (global) Killing symmetry [18,19,20]. It is not so much the symmetry of the solutions that produces the singularities (all solutions of fixed isometry type do form a manifold) but rather the fact that some nearby solutions (in $3 + 1$ dimensions, at least) always have lower symmetry. The conical singularities arise (in the spatially compact case considered here) whenever solutions with a particular isometry group occur at the boundary of regions of lower symmetry. In $2 + 1$ dimensions the rigidity of Einstein's equations forces all of the (non-stationary) vacuum solutions on $\mathbf{T^2} \times \mathbf{R}$ to share the same $\mathbf{U(1)} \times \mathbf{U(1)}$ isometry group. Since this symmetry cannot be broken it does not give rise to conical singularities (which thus arise only at the stationary solutions) and indeed the reduced phase space for the non-stationary solutions is globally a manifold as we have seen. For higher genus surfaces no global, continuous isometries are possible at all and thus the reduced phase space is everywhere a smooth manifold.

Thus to generalize the simple picture described above (for the higher genus surfaces) one should first look at those 3-manifolds for which *every solution of Einstein's equations is devoid of global Killing symmetries*. For these alone can we expect the corresponding spaces of solutions to be globally non-singular and perhaps identifiable as the cotangent bundles of some suitable (infinite dimensional) spaces which would then play the role of Teichmüller spaces.

If we consider only vacuum spacetimes which admit Cauchy hypersurfaces $\{M, g, \pi\}$ of constant mean curvature τ, then we know that the solutions of the constraint equations which generate vacuum spacetimes with (globally defined) Killing symmetries all fall into two classes:

(i) $\pi = 0$ and g is flat,

(ii) $\mathcal{L}_{\mathbf{X}} g = \mathcal{L}_{\mathbf{X}} \pi = 0$ for some vector field $\mathbf{X}$ defined on M.

Here g is the induced Riemannian metric and π the ADM gravitational momentum defined on M whereas $\mathcal{L}_{\mathbf{X}}$ signifies the Lie derivative.

There are only 6 compact, orientable 3-manifolds which admit flat metrics (the simplest being, of course, the 3-torus). If we exclude them we eliminate case (i) which corresponds to stationary spacetimes.

If $\{M, g, \pi\}$ admits a (non-zero) vector field X satisfying (ii) above then the corresponding vacuum spacetime can be shown to admit a spacelike Killing field, $^{(4)}X$, which induces X on the initial Cauchy surface. In fact there must be a compact Lie group of isometries of $\{M, g\}$ of which at least some continuous subgroup (containing X as a generator) also leaves π invariant [21]. Every such Lie group contains an $SO(2)$ subgroup so it certainly suffices to discard those (compact, orientable) 3-manifolds M which can support effective $SO(2)$ actions in order to exclude all manifolds of the form $M \times R$ which can support globally hyperbolic, vacuum spacetimes with Killing symmetries (and at least one Cauchy hypersurface of constant mean curvature). We believe that this exclusion can also be shown to be necessary (i.e., that every M which admits an effective $SO(2)$ action admits vacuum Cauchy data invariant under this action) but we have not yet completed a proof of this conjecture.

Orlik and Raymond have classified all compact 3-manifolds which can support an effective $SO(2)$ action [22]. If we discard these together with the 6 manifolds of *flat type* (5 of which already occur on Orlik and Raymond's list) then we have excluded every compact, orientable 3-manifold which can support Cauchy data for any vacuum spacetime having Killing symmetries (and at least one $\tau =$ constant hypersurface). Indeed this is true even if the *Poincaré conjecture* should prove to be false since neither the characterization of flat 3-manifolds (due to Calabi) nor that of 3-manifolds supporting $SO(2)$ actions required the validity of this conjecture. As we shall see below, even though we have thereby discarded an infinite collection of 3-manifolds, *most such manifolds still remain at our disposal* and one knows from the general arguments alluded to above that the space of solutions of the constraint equations (and hence of the full vacuum Einstein equations) for any remaining 3-manifold will then necessarily be itself an (infinite dimensional) manifold and one could hope to identify (after taking an appropriate quotient by the diffeomorphism group of M) the reduced ADM configuration and phase spaces (the analogues of $\mathcal{T}(\Sigma)$ and $T^*\mathcal{T}(\Sigma)$ in the $2 + 1$ dimensional case) for the chosen 3-manifold M.

To carry out reduction using the conformal method of Lichnerowicz, Choquet-Bruhat and York and be thus led to reduced configuration manifolds analogous to the Teichmüller spaces obtained for the higher genus surfaces seems to require however another restriction on the choice of M. All of the higher genus surfaces shared the property that *every Riemannian metric on each of these surfaces was globally uniquely conformal to one of constant negative scalar curvature*. This fact allowed one to uniquely decompose an arbitrary metric g into a conformal factor and another, "conformal," metric h with the property that the scalar curvature of h satisfies $R(h) = -1$. Furthermore, one could show, using the implicit function theorem, that the space of metrics having scalar curvature equal to negative unity forms a smooth submanifold of the space of all Riemannian metrics on the surface [6]. Finally this manifold was shown to be a $\mathcal{D}_0$-bundle over Teichmüller space (where $\mathcal{D}_0$ is the group of smooth diffeomorphisms of the surface which are isotopic to the identity).

Thanks to work by Yamabe, Trudinger, Aubin and Schoen [23] it is now known, even in higher dimensions, that every Riemannian metric on a compact, connected, orientable manifold is globally conformal to one of constant scalar curvature. However, the decomposition of a metric g into conformal metric h and a conformal factor may fail to be unique unless the constant value of $\mathbf{R}(\mathbf{h})$ is negative. Fortunately, the decomposition is always unique when the constant is negative. The Yamabe class of any given metric (i.e., the sign of the scalar curvature of an associated "conformal" metric) is always well-defined (and conformally invariant) but, on some manifolds, may vary from one conformal class to another. Work by Gromov and Lawson [24] and by Kazdan and Warner [25] has shown however that many (in fact most) compact 3-manifolds admit only metrics of the negative Yamabe class. For these manifolds the conformal decomposition of arbitrary metrics works just as in the case of higher genus surfaces and one can show that the subset of all metrics satisfying the "normalization" condition $\mathbf{R}(\mathbf{h}) = -1$ is a smooth submanifold of the space of all Riemannian metrics on any one of the chosen 3-manifolds.

What we want is the intersection of the set of 3-manifolds admitting no Killing-symmetric vacuum Cauchy data and the set admitting metrics of only the strictly negative Yamabe class. This intersection turns out to be quite large. Consider 3-manifolds expressible in terms of their "prime-decomposition"

$$\mathbf{M} \approx \mathbf{S}^3/\Gamma_1\#\ldots\#\,\mathbf{S}^3/\Gamma_q\#(\mathbf{S}^3\times\mathbf{S}^1)_1\#\ldots\#(\mathbf{S}^3\times\mathbf{S}^1)_r\#\,\mathbf{K}_1\#\ldots\#\mathbf{K}_s$$

where each Γ_k is a finite group acting freely on $\mathbf{S}^3$, where each $\mathbf{K}_j$ is a $\mathbf{K}(\pi,1)$ - manifold (i.e., one for which all the homotopy groups above the fundamental group are trivial) and where $\#$ signifies *connected sum*. According to Milnor, if the Poincaré conjecture is true, every compact, connected, orientable 3-manifold has a unique such decomposition (excluding the 3-sphere itself which behaves as an identity element with respect to connected summation).

The set of manifolds we want (assumming that each of the finite groups occurring above is a subgroup of $\mathbf{SO(4)}$) are obtained by requiring that:

(i) there is at least one $\mathbf{K}(\pi,1)$ - manifold in the connected sum decomposition of M and

(ii) if M is a "stand-alone" $\mathbf{K}(\pi,1)$ - manifold then it is not of flat type and its fundamental group does not have infinite cyclic center.

Condition (i) assures (according to the results of Gromov and Lawson) that every metric on M lies in the negative Yamabe class whereas condition (ii) guarantees (according to the results of Orlik and Raymond) that M cannot support Cauchy data for a vacuum spacetime with Killing-symmetries. In particular, (ii) excludes both the manifolds of flat type and those other pure $K(\pi, 1)$-manifolds (such as S^1-bundles over the torus and surfaces of higher genus) which admit $SO(2)$ actions. There is a remarkable partial overlap of the sets of manifolds defined by Gromov and Lawson on the one hand and by Orlik and Raymond on the other such that condition (i), which is needed to exclude metrics of the positive and zero Yamabe classes, also eliminates all the remaining manifolds on the list of those admitting $SO(2)$-actions. If counterexamples to the Poincaré conjecture should exist and admit metrics only in the negative Yamabe class then these hypothetical manifolds should also be added to our list of admissible manifolds.

The list of manifolds which we have defined seems to be the most natural analogue, from the standpoint of ADM reduction at least, of the set of higher genus surfaces for the 2 + 1 dimensional case. One has sound reasons for believing that, for each of these *admissible* manifolds, the reduced phase space for Einstein's equations is devoid of (conical-) singularities and can be realized as the cotangent bundle of an associated "Teichmüller space" defined as the quotient of the "manifold of conformal metrics" (having scalar curvature equal to -1) $\mathcal{M}_{-1}(M)$ by the group $\mathcal{D}_0(M)$ of smooth diffeomorphisms of M which are isotopic to the identity. Thus we are led to define, for each such M, a reduced configuration space $T(M) = \mathcal{M}_{-1}(M)/\mathcal{D}_0(M)$ and a reduced phase space $T^*T(M)$ and to formulate the reduced Einstein equations as a (time-dependent) Hamiltonian system on $T^*T(M)$.

Most of the analysis needed to carry out this reduction goes through as in the 2 + 1 dimensional case. In the final step however (i.e., taking the quotient of $\mathcal{M}_{-1}(M)$ by $\mathcal{D}_0(M)$) there is the chance that artificial, "orbifold-type" singularities may be introduced by the possibility that some metrics lying in $\mathcal{M}_{-1}(M)$ may have discrete isometry groups isotopic to the identity. This could be ruled out in the 2 + 1 dimensional problem and was the main reason for taking quotients with respect to $\mathcal{D}_0(\Sigma)$ instead of the full diffeomorphism group $\mathcal{D}(\Sigma)$ since the latter would have given rise to Riemann moduli space (with its orbifold singularities) in place of Teichmüller space. A theorem due to Fraenkel [26] shows that metrics on M cannot support discrete isometries lying in $\mathcal{D}_0(M)$ if their Ricci tensors and sectional curvatures are negative definite but of course, in 3 and higher dimensions, this is not implied by the negativity of the scalar curvature. Thus one is left with an open question of whether $T(M) = \mathcal{M}_{-1}(M)/\mathcal{D}_0(M)$ is globally a manifold or has perhaps orbifold singularities. In the latter eventuality one would be tempted to take the full $\mathcal{D}(M)$-quotient and pass to the analogue of *Riemann moduli space* and its "cotangent bundle" as reduced configuration and phase spaces. In our present state of limited knowledge however we can afford to be optimistic that $T(M)$ will prove to be a manifold.

Even if $T(M)$ proves to be everywhere smooth there is still the wide open problem of characterizing it topologically and of realizing it (at least locally if not globally) as a cross section of the "principal-bundle" $\mathcal{M}_{-1}(M) \to \mathcal{T}(M)$. In 2 + 1 dimensions $\mathcal{T}(\Sigma)$ is topologically trivial and can be realized as a global cross section of the trivial principal bundle $\mathcal{M}_{-1}(\Sigma) \to \mathcal{T}(\Sigma)$. Such a realization of the abstractly defined quotient $\mathcal{T}(M)$ is of direct interest in the study of the ADM reduced Einstein equations since one is ultimately interested in reconstructing Lorentzian geometries on $M \times R$ rather than merely tracking their projections onto "Teichmüller space". A local chart for $\mathcal{T}(M)$ near any $\mathcal{D}_0(M)$-class [h] can always be expressed in terms of the space of *transverse-traceless*, symmetric tensors relative to h, an important feature of t-t tensors already implicit in the original work of Arnowitt, Deser and Misner.

5.   Outlook

Of course one cannot afford to discard some of the most familiar 3-manifolds just because they have the unpleasant feature of admitting flat metrics, metrics with positive scalar curvature or $SO(2)$-actions. The actual universe could after all have a simple topology (even though the odds, a priori, seem stacked against it). The next simplest collection of compact 3-manifolds (from the point of view considered here) would seem to be those which, though admitting neither flat metrics nor $SO(2)$-actions, nevertheless admit metrics of the positive Yamabe class. For these one knows that the space of solutions of the Einstein constraint equations is still everywhere a manifold (since Killing symmetries continue to be excluded) but the sharp parallel with reduction in 2 + 1 dimensions is now broken since one must of course broaden the definition of "Teichmüller space" to include classes of conformal metrics with positive scalar curvature. Can the reduced phase space still be realized as the cotangent bundle of some naturally defined configuration manifold or is it a symplectic manifold of more general type?

How can one treat the (singular) spaces of solutions associated with manifolds admitting flat metrics or continuous group actions from the ADM point of view? Evidently there is a heirarchy of manifolds beginning with those which admit $SO(2)$-, but no higher dimensional, group actions and no flat metrics and ending with the lowly three-sphere which admits group actions of the highest allowed dimension (though of course no flat metrics). There is also the (finite) collection of manifolds of flat type which admit group actions ranging from zero to three dimensions (with the torus being of course the most symmetric). Can one define reduced configuration and phase spaces for these 3-manifolds and, if so, how do the conical singularities of the spaces of solutions of the constraint equations re-emerge from this picture? These questions seem to be mixed up with problems from pure differential geometry which concern the lack of uniqueness of the "Yamabe projection" (to a metric of constant scalar curvature) in the case of positive or zero Yamabe class. Is there a true relationship between these various problems or is the apparent difficulty of carrying out reduction for the remaining Yamabe classes an artifact of the particular technique being used?

What would be the implications for the reduction of Einstein's equations of the existence of counterexamples to the Poincaré conjecture? Since such hypothetical manifolds could not support either flat metrics or continuous group actions they could not support Cauchy data for vacuum spacetimes with Killing symmetries (and at least one τ = constant Cauchy surface) and hence would determine spaces of solutions of the Einstein constraints devoid of conical singularities. If they supported only metrics of the negative Yamabe class they would fall into the category of our *admissible* 3-manifolds for which reduction could be carried through much as for the higher genus surfaces in 2 + 1 dimensions. If they supported metrics of positive scalar curvature however, they would apparently require a "generalized Teichmüller space" (incorporating all three Yamabe classes) as configuration manifold. As mentioned above there are known 3-manifolds for which such a generalization also seems to be necessary. Can one learn something about the Poincaré conjecture (and its relation to the Yamabe classification) by studying the geometrical properties of these spaces?

In our discussions of reduction we have repeatedly made use of the assumption that the spacetimes of interest had Cauchy hypersurfaces of constant mean extrinsic curvature. One of the principal reasons for making this assumption is that it leads to a decoupling of the constraint equations which can then be solved by the conventional conformal method. This is an artificial reason however and one would much prefer to generalize the conformal method to the case of non-constant mean curvature. Some (non- exhaustive) results in this direction have recently been obtained by Yvonne Choquet-Bruhat, James Isenberg and the author [27]. One can use several independent techniques (Leray-Schauder degree theory, the contraction mapping principle, the implicit function theorem, for example) to prove existence, uniqueness and smoothness of solutions of the constraints (parameterized, as usual, by a conformal metric and a t-t tensor field) provided suitable bounds are placed upon the magnitude and gradient of the mean curvature. These conditions are sufficient but certainly not necessary (for example, they exclude a mean curvature function which changes sign over the Cauchy surface). Can one remove these restrictions and prove a more comprehensive existence theorem? In either case, can one carry out ADM reduction relative to the more general slicings allowed by these results and, if so, identify the reduced configuration and phase spaces as above? How does non-constancy of the mean curvature interact with the $\mathcal{D}_0(\mathrm{M})$-quotient that one wishes to take in the last step of reduction?

For each of our *admissible* 3-manifolds we have proposed, as reduced configuration space, an analogue of Teichmüller space. There are, however, many open questions associated with the structure of these "Teichmüller spaces." Are they topologically trivial (e.g., open balls in some Banach space, by analogy with finite dimensional Teichmüller spaces) or not? Does the (weak, $\mathbf{L^2}$) analogue of the Weil-Peterson metric on "Teichmüller space," which emerges naturally from a term in the integrated

Hamiltonian constraint, play a useful role in the analysis of the reduced Einstein equations? What is the Riemannian geometry of "Teichmüller space" defined by this metric and does it have properties analogous to the finite dimensional case?

These are some of the questions which might be illuminated by a deeper study of the problem of the ADM reduction of Einstein's equations.

ACKNOWLEDGMENT

This work was supported in part by NSF grant PHY- 8903939 to Yale University.

REFERENCES

(1) R. Arnowitt, S. Deser and C. Misner, "The dynamics of general relativity" in Gravitation: An Introduction to Current Research ed L. Witten (Wiley, New York, 1962), 227-265.

(2) C. Misner, K. Thorne and J. Wheeler, Gravitation (Freeman, San Francisco, 1973), chap. 21.

(3) V. Moncrief, "Reduction of the Einstein equations in 2 + 1 Dimensions to a Hamiltonian System over Teichmüller Space," J. Math. Phys. **30**, 2907-2914 (1989).

(4) V. Moncrief, How solvable is (2 + 1)-dimensional Einstein gravity?, J. Math. Phys. **31**, 2978-2982 (1990).

(5) Y. Choquet-Bruhat and J. York, "The Cauchy Problem" in General Relativity and Gravitation ed A. Held (Plenum, New York,1980).

(6) A. Fischer and A. Tromba, Math. Ann. **267**, 311-345 (1984).

(7) C. Earle and J. Eells, J. Diff. Geom. **3**, 19-43 (1969); see also Ref. [6] for a discussion of Sampson's contribution to the proof of the smoothness of the global cross-section constructed herein.

(8) This procedure is discussed and carried out explicitly in Ref.[4].

(9) E. Witten, Nucl. Phys. B **311**, 46 (1988).

(10) A Hosoya and K. Nakao, Class. Quantum Grav. **7**, 163-176 (1990).

(11) V. Moncrief and J. Isenberg, Commun. Math. Phys. **89**, 387- 413 (1983).

(12) J. Isenberg and V. Moncrief, J. Math. Phys. **26**, 1024-7 (1985).

(13) J. Isenberg and V. Moncrief, "On spacetimes containing Killing vector fields with non-closed orbits," Class. Quantum Grav. **9**, 1683-91 (1992).

(14) V. Moncrief, Ann. Phys. **167**, 118-142 (1986).

(15) V. Moncrief, Class. Quantum Grav. **7**, 329-352 (1990).

(16) J. Cameron, "The Reduction of Einstein's Vacuum Equations on Spacetimes with Spacelike U(1)-Isometry Groups", Doctoral Thesis, Yale University (1991).

(17) J. Cameron and V. Moncrief, "The Reduction of Einstein's Vacuum Equations on Spacetimes with Spacelike U(1)- Isometry Groups," to appear in the proceedings of the AMS conference on Classical Field Theory, Seattle, July 1991.

(18) J. Arms, J. Marsden and V. Moncrief, Ann. Phys. **144**, 81-106 (1982).

(19) J. Arms, J. Marsden and V. Moncrief, Commun. Math. Phys. **78**, 445-478 (1981).

(20) A. Fischer, J. Marsden and V. Moncrief, Ann. Inst. Henri Poincare **33**, 147-194 (1980); see also the original references cited herein.

(21) For a recent discussion of this see Ref. [13].

(22) P. Orlik and F. Raymond, "Actions of SO(2) on 3-Manifolds," in the Proceedings of the Conference on Transformation Groups, New Orleans (Springer-Verlag, New York, 1967).

(23) For a recent treatment, which includes references to the original articles, see J. Lee and T. Parker, "The Yamabe Problem," Bull. Amer. Math. Soc. **17**, 37-81 (1987).

(24) M. Gromov and B. Lawson, Jr., Ann. of Math. **111**, 209-230 and 423-434.

(25) For a complete discussion of these results see J. Kazdan, "Prescribing the Curvature of a Riemannian Manifold," Conference Board of the Mathematical Sciences, Regional Conference Series in Mathematics, # 57, Amer. Math. Soc., Providence, RI (1985).

(26) T. Frankel, Jour. Math. and Mechanics **15**, 373-377 (1966).

(27) Y. Choquet-Bruhat, J. Isenberg and V. Moncrief, "Solutions of Constraints for Einstein's Equations," to appear in Comptes Rendus (1992).

Some Progress in Classical Canonical Gravity

JAMES M. NESTER

Department of Physics, National Central University, Chung-Li, Taiwan 32054

1. INTRODUCTION

In China, I have learned, there is a valuable tradition: to appreciate and show great respect for one's teachers. Undoubtedly, the most successful method of teaching is by example. Certainly this was true in my case. Charles W. Misner was a pioneer in applying modern differential geometry (Misner 1964) especially differential forms (Misner and Wheeler 1957) to gravitational theory, in investigating the canonical Hamiltonian formulation of gravity, and in formulating suitable expressions for conserved quantities, in particular mass-energy (Arnowitt, Deser and Misner (ADM) 1962). I have, as this work indicates, at least to some extent, followed my teacher's directions.

The outline of this work is as follows. First, differential form methods are used to obtain a covariant Hamiltonian formulation for any gravitational theory. The Hamiltonian includes a covariant expression for the conserved quantities of an asymptotically flat or constant curvature space. Next the positive total energy test, a promising and appropriate theoretical test for alternate gravitational theories, is described. The final topic concerns application to Einstein gravity of new rotational gauge conditions and their associated special orthonormal frames. These frames provide a good localization of energy and parameterization of solutions.

2. COVARIANT HAMILTONIAN FORMALISM

There are significant benefits in the canonical Hamiltonian formulation of a theory, in particular the identification of constraints, gauges, degrees of freedom and conserved quantities (Isenberg and Nester 1980), however the usual approach (e.g., in Misner, Thorne and Wheeler (MTW) 1973 and ADM 1962) exacts a heavy price: the loss of manifest 4-covariance. Differential form techniques can be used to largely avoid this cost (Nester 1984). Here we outline our method which leads to the 4-covariant Hamiltonian 3-form for general theories of dynamic geometry. The Hamiltonian 3-form includes a boundary term which yields the conserved quantities, energy-momentum and angular momentum, for solutions with an asymptotic region (i.e., flat or constant negative curvature). The formalism can be applied to quite general geometries including non-metric compatible theories, non-symmetric metric theories or indeed theories in which the metric does not appear as a fundamental variable.

Differential forms have several advantages which make them a suitable tool for our purpose here. First they provide a convenient representation of the geometry in terms of a general linear co-frame θ^α, the metric tensor $g = g_{\mu\nu}\theta^\mu \otimes \theta^\nu$ and the connection one-form $\omega^\alpha{}_\beta$. These potentials determine the geometric field strengths: the *torsion* 2-form $D\theta^\alpha := d\theta^\alpha + \omega^\alpha{}_\beta \wedge \theta^\beta$, the *metricity* one form $Dg_{\mu\nu} := dg_{\mu\nu} - \omega^a{}_\mu g_{\alpha\nu} - \omega^a{}_\nu g_{\mu\alpha}$ and the *curvature* 2-form $\Omega^\alpha{}_\beta := d\omega^\alpha{}_\beta + \omega^\alpha{}_\gamma \wedge \omega^\gamma{}_\beta$.

Since differential forms are designed for integration they are very suitable for variational principles. The geometrodynamic variational principle can be based on a Lagrangian 4-form $\mathcal{L}(g_{\mu\nu}, \theta^\alpha, \omega^\mu{}_\nu, \psi; Dg_{\mu\nu}, \Theta^\alpha, \Omega^\mu{}_\nu, D\psi)$. At this stage the source field ψ and the covariant second rank tensor $g_{\mu\nu}$ are treated on the same footing; we do not yet assign to the latter any geometric role. Succinctly, the relation

$$
\begin{aligned}
\delta\mathcal{L} &= \delta(Dg_{\mu\nu}) \wedge \pi^{\mu\nu} + \delta\Theta^\alpha \wedge \tau_\alpha + \delta\Omega^\mu{}_\nu \wedge \rho_\mu{}^\nu + \delta(D\psi) \wedge p \\
&\quad + \delta g_{\mu\nu}\frac{\partial\mathcal{L}}{\partial g_{\mu\nu}} + +\delta\theta^\alpha \wedge \frac{\partial\mathcal{L}}{\partial\theta^\alpha} + \delta\omega^\mu{}_\nu \wedge \frac{\partial\mathcal{L}}{\partial\omega^\mu{}_\nu} + \delta\psi \wedge \frac{\partial\mathcal{L}}{\partial\psi} \\
&= d(\delta g_{\mu\nu}\pi^{\mu\nu} + \delta\theta^\alpha \wedge \tau_\alpha + \delta\omega^\mu{}_\nu \wedge \rho_\mu{}^\nu + \delta\psi \wedge p) \\
&\quad + \delta g_{\mu\nu}\frac{\delta\mathcal{L}}{\delta g_{\mu\nu}} + \delta\theta^\alpha \wedge \frac{\delta\mathcal{L}}{\delta\theta^\alpha} + \delta\omega^\mu{}_\nu \wedge \frac{\delta\mathcal{L}}{\delta\omega^\mu{}_\nu} + \delta\psi \wedge \frac{\delta\mathcal{L}}{\delta\psi},
\end{aligned}
\tag{1}
$$

implicitly defines the conjugate momenta (which are forms of appropriate rank) and the variational derivatives (which will serve as field equations). It is straightforward to work out the field equations explicitly as well as the Noether identities associated with the local general linear frame freedom and local diffeomorphism freedom. We shall spare the reader the details here but a few features are noteworthy.

In general it is necessary to allow for an explicit dependance on the frame θ^α in order to assure the necessary invariance of the Lagrangian 4-form. Of course θ^α is an arbitrary reference frame not a physical field hence one wonders about the "field equation" status of the variational derivative $\delta\mathcal{L}/\delta\theta^\alpha$. Holding the frame fixed and varying all the other quantities leads to all of the necessary physical field equations. But a Noether identity gives $\delta\mathcal{L}/\delta\theta^\alpha$ as a homogeneous linear combination of the other variational derivatives and hence it must also vanish. In particular, for *non-symmetric* g, if the other field equations are satisfied, $\delta\mathcal{L}/\delta\theta^\alpha = 0 \iff \delta\mathcal{L}/\delta g_{\mu\nu} = 0$. So in this case θ^α and $g_{\mu\nu}$ are *equivalent* variables. If g is symmetric (e.g., the metric) then the 10 symmetric components of $\delta\mathcal{L}/\delta\theta^\alpha = 0$ are equivalent (if the other field equations are satisfied) to $\delta\mathcal{L}/\delta g_{\mu\nu} = 0$ and the 6 antisymmetric components are related, as usual, to local Lorentz invariance and angular momentum conservation. In this latter case it is thus justifiable to keep the metric coefficients constant, e.g., Minkowski, and vary the frame which is now (by definition) orthonormal. From the gauge theory viewpoint this is an attractive scheme for then the geometric potentials are all one-forms. It is worth noting that one can use the orthonormal frame as a basic variable *without* imposing the metriciy condition $Dg_{\mu\nu} = -\omega_{\nu\mu} - \omega_{\mu\nu} = 0$. For *metric compatible theories* an attractive compact formulation is to use orthonormal frames and anti-symmetric connection coefficients (Nester 1991b). Here we consider the general case with no restrictions on the frame, metric or connection; it is easily restricted to special cases by dropping the appropriate terms.

The Lagrangian formalism just outlined is not convenient for obtaining a general Hamiltonian formulation. To this end (taking a hint from Kuchař 1976) we can replace the Lagrangian 4-form (above) by a first order formalism which is linear in the derivatives of the fields, e.g., $\mathcal{L} = dA \wedge *F + \frac{1}{2}F \wedge *F$ for electromagnetism. Similarly, dynamical equations for geometry can be obtained from the first order Lagrangian 4-form

$$\mathcal{L} := D\psi \wedge p + Dg_{\mu\nu} \wedge \pi^{\mu\nu} + D\theta^\alpha \wedge \tau_\alpha + \Omega^\mu{}_\nu \wedge \rho_\mu{}^\nu - \Lambda(\psi, p, g, \pi, \theta, \tau, \rho). \quad (2)$$

The least action principle yields the vacuum field equations:

$$\frac{\delta\mathcal{L}}{\delta\pi^{\mu\nu}} := -Dg_{\mu\nu} - \frac{\partial\Lambda}{\partial\pi^{\mu\nu}} = 0 = \frac{\delta\mathcal{L}}{\delta g_{\mu\nu}} := D\pi^{\mu\nu} - \frac{\partial\Lambda}{\partial g_{\mu\nu}},$$

$$\frac{\delta\mathcal{L}}{\delta\tau_\alpha} := D\theta^\alpha - \frac{\partial\Lambda}{\partial\tau_\alpha} = 0 = \frac{\delta\mathcal{L}}{\delta\theta^\alpha} := D\tau_\alpha - \frac{\partial\Lambda}{\partial\theta^\alpha}, \qquad (3)$$

$$\frac{\delta\mathcal{L}}{\delta\rho_\alpha{}^\beta} := \Omega^\alpha{}_\beta - \frac{\partial\Lambda}{\partial\rho_\alpha{}^\beta} = 0 = \frac{\delta\mathcal{L}}{\delta\omega^\alpha{}_\beta} := D\rho_\alpha{}^\beta + \theta^\alpha \wedge \tau_\beta - g_{\alpha\nu}\pi^{\beta\nu} - g_{\mu\alpha}\pi^{\mu\beta}.$$

Such a first order formulation has certain advantages. One is that *a priori* restrictions such as metric compatibility, vanishing torsion (symmetric connection) or vanishing curvature (teleparallel connection) are easily imposed by just selecting the potential Λ to be independent of the corresponding field momentum. Another is outlined here. Our method uses this first order Lagrangian formulation for dynamic geometry to construct a covariant Hamiltonian formulation. (For another Hamiltonian approach using differential forms see Wallner 1989.)

The canonical Hamiltonian analysis separates the 4-covariant dynamical equations into evolution and constraint equations. The essential thing is to split the space and time derivatives. Our principal reason for using differential forms is that they facilitate this splitting. For differential form equations such as (3) this separation can easily be done by the following method. First foliate the spacetime by constant

t spacelike hypersurfaces and choose *any* timelike vector field N such that $i_N dt = 1$. (The vector field N will play the role of $\partial/\partial t$.) Each differential form σ is split into a "time component" $\hat{\sigma} = i_N \sigma$ and a "spatial component" $\underline{\sigma} := \sigma - dt \wedge \hat{\sigma}$, the latter determines the pullback (restriction) of the form to the spatial hypersurfaces. Note that the splitting used here, in contrast to other approaches, does *not* depend on the metric or any other dynamic variable. It applies equally well to theories which do not have a metric.

One advantage of this approach is that there is a direct correspondence between the 4-covariant equations and their splitting into Hamiltonian equations. The Hamiltonian constraint and evolution equations are just the space and time components of the first order 4-covariant equations (3). The spatial components contain only *spatial* derivatives and hence are (primary) *initial value constraints*. The time components are the *evolution* equations; they are linear in the *time* derivatives of the fields. (The appropriate time derivative operator here is the *Lie derivative* along the timelike vector field N. On differential forms: $\mathcal{L}_N = i_N d + d i_N$.) Thus, for example the time component of Eq. (3d) contains $i_N D\theta^\alpha$ which contains the term $i_N d\theta^\alpha = \mathcal{L}_N \theta^\alpha - d i_N \theta^\alpha$; all of the time derivatives in Eq. (3) arise in this way. The Hamiltonian constraint and evolution equations obtained by splitting the first order covariant equations can also be obtained by varying a Hamiltonian derived via a spacetime splitting of the action.

Just as in the classical mechanics relation $L dt = (p\dot{q} - H)dt$ the first order Lagrangian 4-form for any k-form field $\mathcal{L} = d\phi \wedge p - \Lambda(\phi, p)$ can be rearranged in a simple way which implicitly defines the Hamiltonian 3-form:

$$\mathcal{L} \equiv dt \wedge i_N \mathcal{L} = dt \wedge (i_N d\phi \wedge p - (-1)^k d\phi \wedge i_N p - i_N \Lambda) = dt \wedge (\mathcal{L}_N \phi \wedge p - \mathcal{H}). \quad (4)$$

The first order dynamic geometry action integral splits in a similar straightforward fashion:

$$S = \int \mathcal{L} = \int dt \wedge i_N \mathcal{L} = \int dt \int i_N \mathcal{L}$$

$$= \int dt \int i_N Dg_{\mu\nu} \wedge \pi^{\mu\nu} + i_N D\theta^\alpha \wedge \tau_\alpha + i_N \Omega^\mu{}_\nu \wedge \rho_\mu{}^\nu$$
$$- Dg_{\mu\nu} \wedge i_N \pi^{\mu\nu} + D\theta^\alpha \wedge i_N \tau_\alpha + \Omega^\mu{}_\nu \wedge i_N \rho_\mu{}^\nu - i_N \Lambda$$

$$= \int dt \int \Big((\mathcal{L}_N g_{\mu\nu} - \hat{\omega}^\alpha{}_\mu g_{\alpha\nu} - \hat{\omega}^\alpha{}_\nu g_{\mu\alpha}) \wedge \pi^{\mu\nu} + (\mathcal{L}_N \theta^\alpha + \hat{\omega}^\alpha{}_\beta \theta^\beta - Di_N\theta^\alpha) \wedge \tau_\alpha$$
$$+ (\mathcal{L}_N \omega^\mu{}_\nu - D\hat{\omega}^\mu{}_\nu) \wedge \rho_\nu{}^\mu - Dg_{\mu\nu} \wedge i_N \pi^{\mu\nu} + D\theta^\alpha \wedge i_N \tau_\alpha + \Omega^\mu{}_\nu \wedge i_N \rho_\mu{}^\nu - i_N \Lambda \Big)$$

$$= \int dt \int \mathcal{L}_N g_{\mu\nu} \wedge \pi^{\mu\nu} + \mathcal{L}_N \theta^\alpha \wedge \tau_\alpha + \mathcal{L}_N \omega^\mu{}_\nu \wedge \rho_\mu{}^\nu - \mathcal{H}(N), \quad (5)$$

and thereby constructs the Hamiltonian. (A technical detail here is that $\hat{\omega}^\mu{}_\nu$ has been treated as a tensor field. This parameter controls the general linear frame gauge freedom.) The 3-form

$$
\begin{aligned}
\mathcal{H}(N) &:= i_N \Lambda + Dg_{\mu\nu} \wedge i_N \pi^{\mu\nu} - D\theta^\alpha \wedge i_N \tau_\alpha - i_N \theta^\alpha \wedge D\tau_\alpha - \Omega^\mu{}_\nu \wedge i_N \rho_\mu{}^\nu \\
&\quad - \hat{\omega}^\alpha{}_\beta (D\rho_\alpha{}^\beta - g_{\alpha\nu}\pi^{\beta\nu} - g_{\mu\alpha}\pi^{\mu\beta} + \theta^\beta \wedge \tau_\alpha) + d(i_N \theta^\alpha \wedge \tau_\alpha + \hat{\omega}^\alpha{}_\beta \wedge \rho_\alpha{}^\beta),
\end{aligned} \tag{6}
$$

is a 4-covariant expression for the geometrodynamic Hamiltonian. (Manifest covariance is ultimately lost when a specific N and spatial surfaces are chosen and $\mathcal{H}$ is *restricted* to the spatial surface.) Varying expressions (5,6) wrt $\underline{\theta}$, $\underline{\omega}$, $\underline{\tau}$, $\underline{\rho}$, $\underline{g}\,(=g)$, $\underline{\pi}$ gives the time evolution equations; varying wrt $\hat{\theta}$, $\hat{\omega}$, $\hat{\tau}$, $\hat{\rho}$, $\hat{\pi}$ gives the primary constraint equations.

The Hamiltonian 3-form (6) can be obtained in a different way. The same quantity (essentially the generalization to differential forms of the canonical energy-momentum tensor) results from a Noether translation invariance argument applied to the Lagrangian 4-form (2). As is well known, the Noether expressions are not unique; they can be adjusted by adding an exact differential. At first sight the Hamiltonian reflects this ambiguity but the formalism contains a further principle which removes this freedom.

The Hamiltonian 3-form contains a term of the form dB which *does not* affect the equations of motion. One role of this important exact differential term is revealed by the Noether identity

$$
\mathcal{H}(N) \equiv -i_N \pi \wedge \frac{\delta \mathcal{L}}{\delta \pi} - i_N \theta \wedge \frac{\delta \mathcal{L}}{\delta \theta} - i_N \tau \wedge \frac{\delta \mathcal{L}}{\delta \tau} - i_N \omega \wedge \frac{\delta \mathcal{L}}{\delta \omega} - i_N \rho \wedge \frac{\delta \mathcal{L}}{\delta \rho} + dB. \tag{7}
$$

(This identity can be verified by direct calculation; the key is that $i_N \Lambda = i_N \pi \wedge \frac{\partial \Lambda}{\partial \pi} + i_N \theta \wedge \frac{\partial \Lambda}{\partial \theta} + i_N \tau \wedge \frac{\partial \Lambda}{\partial \tau} + i_N \rho \wedge \frac{\partial \Lambda}{\partial \rho}$ since $\Lambda(g, \pi, \theta, \tau, \rho)$ is a scalar 4-form and the interior product i_N is a derivation.) Hence, aside from the dB term, the Hamiltonian 3-form *vanishes* on a solution to the field equations. The *value* of the Hamiltonian on a solution, i.e., the integral of the Hamiltonian 3-form $\mathcal{H}$ over a spacelike hypersurface, consequently depends *only* on the exact differential term. Stokes theorem then gives the value of the Hamiltonian as the integral of the 2-form B over the 2-surface bounding the spatial hypersurface.

The exact differential term plays another important role which fixes its form. The variational derivatives of the Hamiltonian are well defined only if the boundary term in the variation vanishes (Regge and Teitelboim 1974). The exact differential which arises in the variation of the Hamiltonian 3-form (6) is

$$
d(\delta g_{\mu\nu} i_N \pi^{\mu\nu} - \delta \theta^\alpha \wedge i_N \tau_\alpha - \delta \omega^\alpha{}_\beta \wedge i_N \rho_\alpha{}^\beta + \delta(i_N \theta^\alpha)\tau_\alpha + \delta \hat{\omega}^\alpha{}_\beta \rho_\alpha{}^\beta). \tag{8}
$$

This expression leads to a boundary term which is generally non-vanishing. In particular it does not vanish in Einstein's theory nor in the general Poincaré gauge theory

(PGT, see e.g., Hehl 1980, Hehl, Nitsch and von der Heyde 1980, Hayashi and Shira-fuji 1980, 1981) for solutions which have asymptotically zero or constant curvature. To correct for this the exact differential term must be adjusted. We can replace the exact differential term in (6) by dB with

$$B := i_N\theta^\alpha \Delta\tau_\alpha + \Delta\omega^\alpha{}_\beta \wedge i_N\rho_\alpha{}^\beta + \hat{\omega}^\alpha{}_\beta\Delta\rho_\alpha{}^\beta, \tag{9}$$

where $\Delta\omega^\alpha{}_\beta := \omega^\alpha{}_\beta - \omega^\alpha{}_\beta(\text{asymptotic})$ is an asymptotically well defined tensor valued form; $\Delta\rho_\alpha{}^\beta$ and $\Delta\tau_\alpha$ are defined similarly. With this adjustment the exact differential term in the variation of $\mathcal{H}$ takes the form

$$d(\delta g_{\mu\nu}i_N\pi^{\mu\nu} - \delta\theta^\alpha \wedge i_N\tau_\alpha + \delta(i_N\theta^\alpha)\Delta\tau_\alpha + \Delta\omega^\alpha{}_\beta \wedge \delta(i_N\rho_\alpha{}^\beta) + \delta\hat{\omega}^\alpha{}_\beta\Delta\rho_\alpha{}^\beta). \tag{10}$$

This expression leads to a boundary integral which should vanish at spatial infinity. (It vanishes in particular for asymptotically flat or constant curvature PGT solutions.) The corrected covariant Hamiltonian 3-form is thus

$$\begin{aligned}\mathcal{H}(N) =& i_N\Lambda + Dg_{\mu\nu} \wedge i_N\pi^{\mu\nu} - D\theta^\alpha \wedge i_N\tau_\alpha - \Omega^\alpha{}_\beta \wedge i_N\rho_\alpha{}^\beta - i_N\theta^\alpha D\tau_\alpha \\ &- \hat{\omega}^\alpha{}_\beta(D\rho_\alpha{}^\beta - g_{\alpha\nu}\pi^{\beta\nu} - g_{\mu\alpha}\pi^{\mu\beta} + \theta^\beta \wedge \tau_\alpha) \\ &+ d(i_N\theta^\alpha \Delta\tau_\alpha + \Delta\omega^\alpha{}_\beta \wedge i_N\rho_\alpha{}^\beta + \hat{\omega}^\alpha{}_\beta\Delta\rho_\alpha{}^\beta).\end{aligned} \tag{11}$$

As noted above, the value of the Hamiltonian on a solution depends only on the integral of the 2-form B (9) over the 2-surface bounding the spacelike hypersurface. At spatial infinity, with N asymptotic to a Killing vector, this surface integral gives values for the conserved quantities: energy, momentum and angular momentum.

Suitable expressions for the conserved quantities were first known only for asymp-totically flat spaces (ADM 1962, Regge and Teitelboim 1974 for Einstein's theory, Hayashi and Shirafuji 1985, Blagojević and Vasilić 1988 for the PGT). Later inves-tigators showed how to evaluate the conserved quantities in Einstein's theory for solutions with asymptotically constant (negative) curvature (Abbot and Deser 1982, Henneaux and Teitelboim 1985). Certain expressions have been proposed for evalu-ating the conserved quantities of an asymptotically constant curvature PGT solution but they did not prove to be completely successful (Baekler, Hecht, Hehl and Shirafuji 1987, Mielke and Wallner 1988).

R. Hecht and I have been using REDUCE and EXCALC (see Schruefer, Hehl and McCrea 1987) to investigate expression (9) and find that it gives precisely the desired values for the known Kerr anti-de Sitter PGT exact solution (Baekler, Gurses, Hehl and McCrea 1988) and reduces to the aforementioned Einstein constant curvature expression and the flat PGT expressions in the appropriate limits. It can be evaluated in any frame, in particular in a coordinate basis or an orthonormal frame as long as

$$\mathcal{L}_N\theta^\alpha \equiv i_N\Theta^\alpha + DN^\alpha - \hat{\omega}^\alpha{}_\beta\theta^\beta = 0, \tag{12}$$

asymptotically. This latter restriction means, in particular, that expression (9) *may not* give the correct value for the angular momentum in rectangular frames. From the point of view of generating the correct Hamiltonian equations it is not possible to tamper with the boundary term any further. However, for the purposes of calculating its value we can adjust it by using the condition (12) to replace the non-covariant looking factor $\hat{\omega}^{\alpha}{}_{\beta}$ in the boundary expression to give

$$B := i_N\theta^\alpha \, \Delta\tau_\alpha + \Delta\omega^\alpha{}_\beta \wedge i_N\rho_\alpha{}^\beta + (D_\beta N^\alpha + \Theta^\alpha(N, e_\beta))\Delta\rho_\alpha{}^\beta, \qquad (13)$$

which is a manifestly 4-covariant geometric expression for the conserved quantities. (Note that it does not depend on the metric explicitly.) This expression can be evaluated on any 2-surface and hence also at future null infinity. I have found that (9) or (13) also gives the correct Bondi mass, linear and angular momentum (the latter had long been a problem see e.g., Winicour 1980, Goldberg 1980, 1990) at least for the the simplest cases.

The exact differential term is, in general, only defined asymptotically. For asymptotically flat solutions, however, the fields fall off fast enough to permit a globally covariant Hamiltonian (at least for N asymptotically constant and timelike). To achieve this we must consider a special case, the Einstein theory, for which the Lagrangian density (the scalar curvature) has an exceptional asymptotic fall off. For this special case a spinor parameterization of the evolution vector $N^\mu = \overline{\psi}\gamma^\mu\psi$ allows the boundary term to be written in the globally covariant form

$$B := D(\overline{\psi}\gamma_5\gamma)\psi - \overline{\psi}D(\gamma_5\gamma\psi), \qquad (14)$$

(here $\gamma = \gamma_\mu\theta^\mu$) which permits a globally covariant expression for the Einstein Hamiltonian 3-form (Nester 1984):

$$\mathcal{H} = D(\overline{\psi}\gamma_5\gamma) \wedge D\psi - D\overline{\psi} \wedge D(\gamma_5\gamma\psi) - \frac{1}{2}\hat{\omega}^\alpha{}_\beta D * (g_{\alpha\mu}\theta^\mu \wedge \theta^\beta). \qquad (15)$$

Any scalar curvature $a_0 R * 1$ contribution to the Lagrangian (2) can be extracted and dealt with in this fashion in order to achieve a globally covariant Hamiltonian. We need simply reparameterize the field momentum 2-form conjugate to the connection by extracting a constant term: $\rho'_\alpha{}^\beta = \rho_\alpha{}^\beta - a_0 g_{\alpha\mu} * (\theta^\mu \wedge \theta^\beta)$. The boundary term (9) is then replaced by the fully covariant expression

$$B := i_N\theta^\alpha\tau_\alpha + a_0\{D(\overline{\psi}\gamma_5\gamma)\psi - \overline{\psi}D(\gamma_5\gamma\psi)\}. \qquad (16)$$

(In this case $\Delta\tau = \tau$ and the $\hat{\omega}$ term vanishes asymptotically.) The dB term in the Hamiltonian 3-form (11) can then be expanded to give a globally 4-covariant

Hamiltonian 3-form for dynamic geometry:

$$\mathcal{H}(N) = i_N \Lambda + Dg_{\mu\nu} \wedge i_N \pi^{\mu\nu} - D\theta^\alpha \wedge i_N \tau_\alpha - \Omega^\alpha{}_\beta \wedge i_N \rho'^{\ \beta}_\alpha + D(i_N \theta^\alpha) \wedge \tau_\alpha$$
$$- \hat{\omega}^\alpha{}_\beta (D\rho_\alpha{}^\beta - g_{\alpha\nu}\pi^{\beta\nu} - g_{\mu\alpha}\pi^{\mu\beta} + \theta^\beta \wedge \tau_\alpha) \tag{17}$$
$$+ a_0 \{ D(\overline{\psi}\gamma_5\gamma) \wedge D\psi - D\overline{\psi}\gamma_5\gamma \wedge D\psi) \}.$$

The procedure outlined here should yield the correct covariant Hamiltonian for all dynamic geometry theories including special cases like the PGT (*a priori* metric compatible) and its degenerate special cases: the Einstein theory, Einstein-Cartan-Sciama-Kibble theory, and teleparallel theory. Applications of this formalism includes local energy investigations and the positive total energy test.

3. POSITIVE ENERGY TEST

Einstein's theory of gravity has proved to be very successful, yet alternative gravitational theories have been proposed (in particular the PGT) which not only pass all the current observational tests but also agree with the Einstein theory to post-Newtonian order (Schwitzer and Strauman 1979, Schweitzer, Strauman and Wipf 1980). Consequently no foreseeable observational test can distinguish them from the standard theory; they are equally viable. Experimental tests are just not capable of selecting a unique theory. Theoretical tests are also needed. The usual theoretical tests, such as no faster than light signals and no negative energy waves (Sezgin and van Nieuwenhuizen 1980, Hayashi and Shirafuji 1980d) did not seem to be sufficiently restrictive so the positive energy test was proposed (Chern and Nester 1986) as an effective theoretical test for gravitational theories.

For gravitating systems which are asymptotically Newtonian there is a conserved total energy. This energy is given (with $c = 1$) by $E = M$ where M is the apparent active gravitational mass of the asymptotically Newtonian gravitational field. Hence positive total energy simply means that gravity is purely attractive asymptotically, negative total energy is antigravity — repulsive.

Fundamental theoretical reasons (in particular stability and thermodynamics) require of any good gravitational theory that all of its asymptotically Newtonian solutions (with positive mass density sources) have positive total energy with zero energy being empty Minkowski space. It can be very difficult to prove that *all* of the asymptotically Newtonian solutions for a given theory have positive total energy (Schoen and Yau 1979, Witten 1981, Choquet-Bruhat 1984). It is much easier to use this requirement as a test that rejects any theory which permits a negative total energy solution or even a non-trivial zero energy solution. In this way only one "bad" solution need be found.

The application of the positive total energy test to a particular theory usually begins with the Hamiltonian formulation. This formulation gives ready access to an expression for the total energy (in terms of an integral over the 2-sphere at spatial infinity) and to all of the solutions by way of the solutions to the initial value constraints. The initial value constraints generally form a coupled non-linear elliptic system. It is not necessary to undertake the formidable task of trying to find a solution to the constraints with negative total energy. A non-trivial zero energy solution is sufficient. (Such solutions are usually simpler mathematically and are also more transparent physically; moreover a "nearby" negative total energy solution can be constructed by perturbation.) It is generally easier to select the geometry so that the total energy is zero and then use the constraints to algebraically determine the source. If the source thus obtained is physically permissible, then the theory permits an unacceptable solution and hence must be rejected.

A seriously considered alternative gravitational theory is the 10-parameter Poincaré Gauge Theory (PGT) developed by Hehl (1980) and by Hayashi and Shirafuji (1980). As mentioned above, for a wide range of the parameters this theory is experimentally viable.

We have shown the effectiveness of the positive total energy test when applied to the Poincaré gauge theory. We found that: (i) the special subclass the *teleparallel* theories are not viable for generic values of the parameters (Chen, Chern and Nester 1987, Cheng and Nester 1987), (ii) the PGT is not viable in the generic asymptotically flat or generic asymptotically constant curvature case (Chern, Nester and Yo 1992), (iii) the PGT is not viable in the asymptotically flat case with extra primary but no secondary "if-constraints" (Yo, 1991).

Hence only degenerate special cases of the PGT might be viable. Such a degeneracy leads to additional "if" constraints (Blagojević and Nikolić 1983, Nikolić 1984) which may prevent the construction of the "bad" solutions we have found. There are many possible degenerate special cases; each must be considered separately. In the asymptotically flat case we know that a viable PGT must have some secondary "if constraints".

These results hardly reflect the full scope of the test. Only simple ansatz were used to find "bad" solutions. The positive total energy test is quite capable of yielding more restrictions but the choices are more complicated and "bad solutions" leading to further restrictions are not obtained quite so easily.

A comparison of the full positive total energy test on the PGT with the results of other tests, in particular with the linearized theory based no negative-energy or faster-than-light waves tests cited above and especially with their more recent stricter form (Kuhfuss and Nitsch 1986) is still in progress. So far the results are compatible and, to some extent, complimentary. Some qualitative comparative remarks are in order here.

A noteworthy feature of the positive energy test is that it considers the full nonlinear theory and is thus not subject to such doubts as may linger regarding a result of a test derived from the linearized theory. The test, however, can be applied only if the total energy is well defined. Hence it applies only to the asymptotically flat or constant negative curvature configurations. For these configurations, in principal, the positive total energy test actually includes all the valid results of the linearized theory "no negative energy waves" test. For if a linearized wave solution carrying negative local energy is an approximation to an exact wave solution then the exact wave is a negative total energy solution.

Among other theoretical tests which may be applied to gravitational theories the positive total energy test has certain special distinctions: (1) it applies only to gravititational theories, (2) it relies on the distinctive essential fundamental properties of the gravitational interaction: gravity is universal and purely attractive (my students remember that it is like love) — not a trace of repulsion is acceptable. No matter what other desirable properties an alternative theory may possess if it does not pass this test it is just not attractive enough to be pursued.

4. SPECIAL ORTHONORMAL FRAMES

A parallelizable Riemannian geometry may be described by using orthonormal frames e_a and their associated dual basis one-forms θ^a. The connection coefficients $\Gamma^a{}_{bc} := \theta^a(\nabla_c e_b)$ are given in terms of the commutator coefficients $C^a{}_{bc} := -d\theta^a(e_b, e_c)$ by $\Gamma_{abc} = \omega_{ab}(e_c) = \frac{1}{2}(C_{cab} - C_{bca} - C_{abc})$.

The choice of orthonormal frame has rotational gauge freedom. The *rotational state* of the frame can be characterized by a one form $\tilde{q} := e_a \rfloor d\theta^a$ and a 3-form $\hat{q} := \frac{1}{2}\theta_a \wedge d\theta^a$. The components of these forms

$$q_c = C^a{}_{ca} = -\Gamma^a{}_{ca}, \qquad q_{abc} = -\frac{3}{2}C_{[abc]} = 3\Gamma_{[abc]}, \tag{18}$$

appear explicitly in the Dirac operator:

$$\gamma^c \nabla_c \psi = \gamma^c(\psi_{,c} + \tfrac{1}{4}\Gamma^{ab}{}_c \gamma_{ab}\psi) = \gamma^c\psi_{,c} - \tfrac{1}{2}q_b\gamma^b\psi + \tfrac{1}{12}q_{abc}\gamma^{abc}\psi. \tag{19}$$

A natural gauge condition (Nester 1989a, 1992),

$$\delta\hat{q} = 0 = d\tilde{q} \tag{20a, b}$$

(here $\delta = \pm * d*$ is the co-differential) selects a *special orthonormal frame*. The combination

$$\delta \hat{q} + d \tilde{q} = 0 \qquad (21)$$

is a nonlinear elliptic (or hyperbolic) equation for a rotation $R^{a'}{}_{b}$, taking any given frame θ^b to a special orthonormal frame $\theta^{a'} = R^{a'}{}_{b}\theta^b$.

The frame is determined only up to a constant rotation. Hence, the conditions determine a preferred global parallelism. In 2-dimensions special orthornormal frames are just those frames which make the metric conformally flat.

For the 3-dimensional case Dimakis and Müller-Hoissen (1989) related the non-linear gauge conditions (20) to the 3-dimensional Dirac equation (which is linear) and established that unique solutions exist in that case. (An exception is at any points where $\psi = 0$; 3-metrics which permit such points are believed to be quite exceptional.) Generalizing their idea we can show that unique solutions also exist in the 4-dimensional case.

Under a conformal transformation $\theta^a \rightarrow \varphi \theta^a$ the rotational quantities transform as $\tilde{q} \rightarrow \tilde{q} - (n-1)d\ln\varphi$ and $\hat{q} \rightarrow \varphi^2 \hat{q}$. Consequently the gauge condition (20b) is *conformally invariant*. The gauge condition (20a) also is conformally invariant in 2-dimensions and in the gravitationally important cases: (i) 4-dimensions, and (ii) asymptotically flat 3-dimensional spaces (in this case $\hat{q}$ is the dual of a constant q which must vanish asymptotically).

For any special orthonormal frame the rotational state quantity $\tilde{q}$ is closed and hence locally exact. Thus a special orthonormal frame (at least locally) determines a *special function* (modulo a constant). In 2-dimensions the function is the conformal factor. For the 3-dimensional case the choice $q_k = 4\partial_k \ln \Phi$ is convenient for gravitational applications. The special function is related to the spinor field of equation (19) by $\Phi = |\psi|^{-1/2}$.

5. APPLICATIONS OF SOF TO GRAVITY

The Einstein Hamiltonian (with vanishing *shift vector* N^k) has the form (see ADM 1962, MTW 1973, Isenberg and Nester 1980):

$$H(N) = \int d^3x \, N\mathcal{H} + \oint N \, \delta^{kc}_{am} \, g^{mb} \Gamma^a{}_{bc} dS_k, \qquad (22)$$

with *lapse* N, and *superhamiltonian*

$$\mathcal{H} = g^{-\frac{1}{2}}(\pi^{mn}\pi_{mn} - \tfrac{1}{2}\pi^2) - g^{\frac{1}{2}}R, \qquad (23)$$

for asymptotically flat spaces the boundary term at spatial infinity (given here for an asymptotically cartesian frame) determines the total energy.

In terms of orthonormal frames the scalar curvature term is

$$-g^{\frac{1}{2}}R = -2\partial_m(g^{\frac{1}{2}}q^m) + g^{\frac{1}{2}}(q^{ab}q_{ab} - \tfrac{1}{2}q^m q_m - \tfrac{1}{2}q^2), \tag{24}$$

where $q_{ac} = \tfrac{1}{2}\epsilon_{mn(a}\Gamma^{mn}{}_{c)}$, and the boundary integrand becomes $2Ng^{\frac{1}{2}}q^m$. In terms of *special* orthonormal frames with $q_k = 4\partial_k \ln \Phi$ these quantites take the form

$$-g^{\frac{1}{2}}R = 8\Phi^{-1}\partial_m(g^{\frac{1}{2}}g^{mn}\partial_n\Phi) + g^{\frac{1}{2}}(q^{ab}q_{ab} - \tfrac{1}{2}q^2), \tag{25}$$

and $8Ng^{\frac{1}{2}}g^{mk}\partial_k \ln \Phi$ respectively.

Using the divergence theorem to eliminate the boundary term then leads to the Hamiltonian (i.e., energy) density in the form (Nester 1989abc, 1991a)

$$\begin{aligned}\mathcal{H}(N) =\,&8Ng^{\frac{1}{2}}g^{nm}\partial_n \ln(N\Phi^{-1})\partial_m \ln \Phi \\ &+ N\{g^{-\frac{1}{2}}(\pi^{ab}\pi_{ab} - \tfrac{1}{2}\pi^2) + g^{\frac{1}{2}}(q^{ab}q_{ab} - \tfrac{1}{2}q^2)\}.\end{aligned} \tag{26}$$

This expression is good for both compact spatial surfaces (in which case q is constant by the gauge condition) and for asymptotically flat spatial surfaces (in which case q vanishes).

A good choice for the lapse is $N = \Phi$ which makes the *gravitational energy density*

$$\mathcal{H}(\Phi) = \Phi\{g^{-\frac{1}{2}}(\pi^{ab}\pi_{ab} - \tfrac{1}{2}\pi^2) + g^{\frac{1}{2}}(q^{ab}q_{ab} - \tfrac{1}{2}q^2)\}, \tag{27}$$

locally non-negative on maximal ($\pi = 0$) asymptotically flat spatial hypersurfaces thereby affording (in addition to a positive energy proof) a *quasi-localization* (the Hamiltonian density is a *local* function on the *space of special orthonormal frames*) of gravitational energy (Penrose 1982, 1986, Christodoulou and Yau 1988) which clearly displays the physical degrees of freedom of the gravitational field. The total gravitational energy within a volume V with surface S is given by

$$16\pi G E = \int_V \mathcal{H}(\Phi)\,d^3x = 8\oint_S g^{\frac{1}{2}}g^{km}\partial_m\Phi dS_k. \tag{28}$$

For the Schwarzschild black hole, this localizes all of the energy inside.

The special orthonormal frame expression for the Hamiltonian initial value constraint

$$8g^{\frac{1}{2}}\nabla^2\Phi = \mathcal{H}(\Phi) + 16\pi G g^{\frac{1}{2}}\Phi\rho, \tag{29}$$

generalizes the Poisson equation of Newtonian gravity, reveals the special function as a generalization of the Newtonian potential, and justifies our identification of the gravitational energy density (27).

Because the gauge conditions are conformally invariant, special orthonormal frames mesh with the standard analysis of the initial value problem (see e.g., Choquet-Bruhat and York 1980) and consequently are good variables for describing the space of solutions. Indeed, equation (29) is just the usual scale equation for a conformal factor such that $g = \phi^4 \bar{g}$, except that our $\phi = \Phi^{-1}$ and $\bar{g}$ have an *absolute geometric* meaning: the corresponding base orthonormal frame $\bar{\theta}^a = \Phi^2 \theta^a$ has vanishing $\bar{q}$. In terms of this geometrically selected base metric, the left hand side of equation (29) has just the usual form $-8\bar{g}^{\frac{1}{2}} \bar{\nabla}^2 \phi$ and the energy flux integrand is $-8\bar{g}^{\frac{1}{2}} \bar{g}^{mk} \partial_m \phi$.

We are now investigating the applications of special orthonormal 4-frames to Einstein gravity. Many standard frames for familiar solutions are special orthonormal frames, e.g., the cartesian frames for the static, spherically symmetric metric and the homogeneous cosmologies. The special function associated with condition (20b) now satisfies a wave equation (the 4-dimensional version of Eq. (25)) related to the scalar curvature and hence to the trace of the matter energy-momentum tensor. Gauge condition (20a) requires that $d * \hat{q} = 0$, hence (locally) $*\hat{q} = d\beta$ where β is an $n - 4$ form. Thus in 4-dimensions (20a) determines a second special function; i.e. $q^{\alpha\beta\gamma} = \epsilon^{\alpha\beta\gamma\delta} \nabla_\delta \beta$. Hence each 4-metric determines *two* special functions. Aside from the fact that it vanishes for diagonal metrics, little is yet known concerning the special function β. We wonder: what is the physical interpretation of this second special function?

ACKNOWLEDGEMENT

It was very fortunate for me to have had Charlie Misner as my advisor. His treatment seemed optimal to encourage me to bring out my best. With gratitude, this is dedicated to C.W. Misner in honor of his 60th birthday. This work was supported by the National Science Council of the R.O.C. under Contract No. NSC 81-0208-M-008-03.

REFERENCES

Abbott L F and Deser S (1982) *Nuc. Phys.* B **195** 76

Arnowitt R, Deser S and Misner C W (1962) in *Gravitation: An Introduction to Current Research* ed L Witten (Wiley, New York) pp 227-265

Baekler P, Gurses M, Hehl F W and McCrea J D (1988) *Phys. Lett.* A **128** 245

Baekler P, Hecht R, Hehl F W and Shirafuji T (1987) *Prog. Theor. Phys.* **78** 16

Blagojević M and Nikolić I A (1983) *Phys. Rev.* **D 28** 2455

Blagojević M and Vasilić M (1988) *Class. Quant. Grav.* **5** 1241

Chen H H, Chern D C and Nester J M (1987) *Chinese J. Phys.* **25** 481

Cheng W H and Nester J M (1987) *Chinese J. Phys.* **25** 601

Chern D C and Nester J M (1986) in *Gravitational Collapse and Relativity* ed H Sato and T Nakamura (World Scientific, Singapore) p 106

Chern D C, Nester J M and Yo H J (1992) *Int. Jour. Mod. Phys.* **A 7** 1993

Choquet-Bruhat Y (1984) in *Relativity, Groups and Topology* II ed B de Witt and R Stora (North Holland, Amsterdam) p 739

Choquet-Bruhat Y and York J W Jr. (1980) in *General Relativity and Gravitation: One Hundred Years After the Birth of Albert Einstein* ed A Held (Plenum, New York) Vol 1 p 99

Christodoulou D and Yau S T (1988) in *Mathematics and General Relativity* ed J Isenberg (American Math Society, Providence) p 9

Dimakis A and Müller-Hoissen F (1989) *Phys. Lett.* **A 142** 73

Goldberg J (1980) in *General Relativity and Gravitation: One Hundred Years After the Birth of Albert Einstein* ed A Held (Plenum, New York) Vol 1 p 459

Goldberg J (1990) *Phys. Rev.* **D 41** 410

Hayashi K and Shirafuji T (1980) *Prog. Theor. Phys.* **64** 866, 883, 1435, 2222

Hayashi K and Shirafuji T (1981) *Prog. Theor. Phys.* **65** 525, 318; **66** 2258

Hayashi K and Shirafuji T (1985) *Prog. Theor. Phys.* **73** 54

Hehl F W (1980) in *Cosmology and Gravitation: Spin, Torsion, Rotation and Supergravity* ed P G Bergmann and V de Sabbata (Plenum, New York) p 5

Hehl F W, Nitsch J and von der Heyde P (1980) in *General Relativity and Gravitation: One Hundred Years After the Birth of Albert Einstein* ed A Held (Plenum, New York) Vol 1 p 329

Henneaux M and Teitelboim C (1985) *Comm. Math. Phys.* **98** 391

Isenberg J and Nester J M (1980) in *General Relativity and Gravitation: One Hundred Years After the Birth of Albert Einstein* ed A Held (Plenum, New York) Vol 1 p 23

Kuchař K (1976) *J. Math. Phys.* **17** 77, 792, 801

Kuhfuss R and Nitsch J (1986) *Gen. Rel. Grav.* **18** 947

Mielke E W and Wallner R P (1988) *Nuovo Cimento* **101B** 607

Misner C W (1964) in *Relativity, Groups and Topology* ed C DeWitt and B S DeWitt (Gordon and Breech, New York) pp 881-929

Misner C W, Thorne K and Wheeler J A (1973) *Gravitation* (Freeman, San Francisco) Ch 21

Misner C W and Wheeler J A (1957) *Ann. Phys. (U.S.A.)* **2** 525-603

Nester J M (1984) in *Asymptotic Behavior of Mass and Space-time Geometry* (Lecture Notes in Physics **202**) ed F Flaherty (Springer, Heidelberg) p 155

Nester J M (1989a) *J. Math Phys.* **30** 624

Nester J M (1989b) *Int. J. Mod Phys. A* **4** 1755

Nester J M (1989c) *Phys. Lett. A* **139** 112

Nester J M (1991a) *Class. Quant. Grav.* **8** L19

Nester J M (1991b) *Mod. Phys. Lett. A* **6** 2655

Nester J M (1992) *J. Math Phys.* **33** 910

Nikolić I A (1984) *Phys. Rev. D* **30** 2508

Penrose R (1982) *Proc. Roy. Soc. Lond. A* **381** 52

Penrose R (1986) in *Gravitational Collapse and Relativity* ed H Sato and T Nakamura (World Scientific, Singapore) p 43

Regge T and Teitelboim C (1974) *Ann. Phys. (N. Y.)* **88** 286

Schoen R and Yau S T (1979) *Comm. Math. Phys.* **65** 45

Schruefer E, Hehl F W and McCrea J D (1987) *Gen. Rel. Grav.* **19** 197

Schweitzer M and Strauman N (1979) *Phys. Lett. A* **71** 493

Schweitzer M, Strauman N and Wipf A (1980) *Gen. Rel. Grav.* **12** 951

Sezgin E and van Nieuwenhuizen P (1980) *Phys. Rev. D* **21** 3269

Wallner R P (1989) "Forms in Field Theory, the canonical formalism" Univ. of Cologne preprint

Winicour J (1980) in *General Relativity and Gravitation: One Hundred Years After the Birth of Albert Einstein* ed A Held (Plenum, New York) Vol 2 p 71

Witten E (1981) *Comm. Math. Phys.* **80** 381

Yo H J (1991) "Positive Energy Test Applied to Poincaré Gauge Theory" MSc. Thesis, National Central University

Harmonic Map Formulation of Colliding Electrovac Plane Waves

Y. Nutku *

The formulation of the Einstein field equations admitting two Killing vectors in terms of harmonic mappings of Riemannian manifolds is a subject in which Charlie Misner has played a pioneering role. We shall consider the hyperbolic case of the Einstein-Maxwell equations admitting two hypersurface orthogonal Killing vectors which physically describes the interaction of two electrovac plane waves. Following Penrose's discussion of the Cauchy problem we shall present the initial data appropriate to this collision problem. We shall also present three different ways in which the Einstein-Maxwell equations for colliding plane wave spacetimes can be recognized as a harmonic map. The goal is to cast the Einstein-Maxwell equations into a form adopted to the initial data for colliding impulsive gravitational and electromagnetic shock waves in such a way that a simple harmonic map will directly yield the metric and the Maxwell potential 1-form of physical interest.

for Charlie W. Misner on his 60th birthday

1. Introduction

Charlie Misner was the first to recognize that the subject of harmonic mappings of Riemannian manifolds finds an important application in general relativity. In a pioneering paper with Richard Matzner [1] he found that stationary, axially symmetric Einstein field equations can be formulated as a harmonic map. Eells and Sampson's theory of harmonic mappings of Riemannian manifolds [2] provides a geometrical framework for thinking of a set of *pde*'s, in the same spirit as "mini-superspace" that Charlie was to introduce [3] for *ode* Einstein equations a little later. The subsequent developement of the subject of space-times admitting two Killing vectors, that eventually led to its recognition as a completely integrable system [4] - [7] has employed another formulation of the stationary, axi-symmetric field equations due to Ernst [8] which is equivalent to that of Matzner and Misner.

Charlie's later work on harmonic maps [9] encompasses a scope much broader than this specific problem and its power and elegance is bound to make a major impact

*Department of Mathematics, Bilkent University, 06533 Ankara, Turkey

on theoretical physics.

I was privileged to be in contact with Charlie's ideas at that time and worked on the two Killing vector problem [10], [11]. It was the hyperbolic version of gravitational fields admitting two Killing vectors that attracted my attention. This is the problem of colliding impulsive plane gravitational waves for which Khan and Penrose had presented a famous solution [12]. My work finally led to the exact solution for colliding impulsive plane gravitational waves with non-collinear polarizations [13] which is physically the most general solution of this type. It turned out that the Matzner-Misner formalism was the one more readily amenable to the hyperbolic problem, even though it was originally intended for the elliptic case, whereas the Ernst formulation fitted the elliptic problem, ie the exterior field of rotating stars, best. The relationship between these two formalisms is given by a Neugebauer-Kramer involution [14].

A few years after the solution [13] appeared, there was a remarkable avalanche of papers on colliding plane gravitational waves. There were important papers examining the singularity structure of spacetimes resulting from the collision of gravitational waves [18] - [20]. However, there was also a mass of new exact solutions which are all essentially devoid of any physical interest because their authors had chosen not to solve the Cauchy problem with the initial data appropriate to generic plane waves, but rather they started with a "solution" and derived (!) the initial data. This type of derived initial data for the collision problem describes some very peculiar plane waves indeed. An inordinately large number of such references can be found in [15].

Nevertheless, physically interesting colliding wave problems are still open and waiting for an exact solution! Remarkably enough, the interaction of plane impulsive gravitational and electromagnetic shock waves is in this category. We have the Khan-Penrose and Bell-Szekeres [16] solutions describing the interaction of either two impulsive gravitational, or two electromagnetic shock waves alone and also the solution of Griffiths [17] for the interaction of an impulsive gravitational wave with an electromagnetic shock wave. But the generic case where we must consider the collision of both type of waves is missing even in the case of collinear polarization. The important open problem here is the construction of an exact solution of the Einstein-Maxwell equations that reduces to all, the Khan-Penrose, Bell-Szekeres and Griffiths solutions. There are various unsatisfactory treatments of this problem in the literature [21], [22]. I shall give its harmonic map formulation.

2.   Initial Data

The problem of colliding plane gravitational waves was proposed and in essence solved by Penrose [23] in 1965 even though most people writing on this subject do not seem to be familiar with it. We shall use Penrose's formulation of the Cauchy problem [24], [25] to discuss the interaction of two plane waves, every one of which will consist of a superposition of an impulsive gravitational wave and an electromagnetic shock wave. The interaction will be determined by an integration of the Einstein-Maxwell equations with initial data defined on a pair of intersecting null characteristics. The initial values of the fields will be those appropriate to a plane wave which is given by the Brinkman metric [26], [27]

$$ds^2 = 2\,du'\,dv' - dx'^2 - dy'^2 + 2\,H(v',x',y')\,dv'^2 \tag{1}$$

and the superposition of an impulsive gravitational wave and an electromagnetic shock wave, with amplitudes proportional to a, b respectively, is obtained for

$$H = \frac{a}{2}\left(y'^2 - x'^2\right)\delta(v') - \frac{b^2}{2}\left(x'^2 + y'^2\right)\theta(v') \tag{2}$$

where δ is the Dirac delta-function and θ is the Heaviside unit step-function.

The Brinkman coordinate system employed in eq.(1) is useful because the superposition of waves travelling in the same direction is obtained simply by addition. However, the Brinkman coordinates are not suitable for the collision problem because of the explicit dependence of the metric on x', y'. For this purpose we must transform to the Rosen form where the metric coefficients will depend on v alone. This is accomplished by the Khan-Penrose transformation

$$\begin{aligned}
v' &= v\,, \\
u' &= u + \tfrac{1}{2}x^2\,F\,F_v + \tfrac{1}{2}y^2\,G\,G_v\,, \\
x' &= x\,F\,, \\
y' &= y\,G\,,
\end{aligned} \tag{3}$$

which results in

$$ds^2 = 2\,du\,dv - F^2\,dx^2 - G^2\,dy^2\,, \tag{4}$$

provided

$$\begin{aligned}
F_{vv} &= [\,-a\,\delta(v) - b\,\theta(v)\,]\,F\,, \\
G_{vv} &= [\,-a\,\delta(v) + b\,\theta(v)\,]\,G\,.
\end{aligned} \tag{5}$$

These are linear, distribution-valued ordinary differential equations which can be solved using the Laplace transform

$$\mathcal{F}(s) = \int_0^\infty e^{sv}\,F(v)\,dv \tag{6}$$

and from eq.(5) we find

$$\mathcal{F}(s) = \frac{1}{s^2 + b^2}\,[(s - a)F(0) + F_v(0)] \tag{7}$$

where of $F(0)$, $F_v(0)$ are initial values. They are obtained from the continuity of the metric and its first derivatives across $v = 0$ which requires $F(0) = 1$, $F_v(0) = 0$. In this case inverting the Laplace transform we get

$$F = \cos(bv\theta(v)) - \frac{a}{b}\,\sin(bv\theta(v)) \tag{8}$$

and the result for G is obtained by letting $a \to -a$ in eq.(8) as indicated by eqs.(5).

In Rosen coordinates the general form of the metric that admits two hypersurface orthogonal Killing vectors is given by

$$ds^2 = 2\, e^{-M}\, du\, dv\, -\, e^{-U}\left(\, e^V\, dx^2\, +\, e^{-V}\, dy^2\,\right) \qquad (9)$$

where U, V, M depend on only u, v and comparison with eqs.(4) and (8) shows that for the initial value problem the data is given by

$$
\begin{aligned}
e^{-U} &= \cos^2(bv\theta(v)) - \frac{a^2}{b^2}\sin^2(bv\theta(v)) \\
e^{-V} &= \frac{b + a\,\tan\left(bv\theta(v)\right)}{b - a\,\tan\left(bv\theta(v)\right)} \\
e^{-M} &= 1\,.
\end{aligned}
\qquad (10)
$$

The limiting values of this result are familiar. If we have just an impulsive gravitational wave, we must pass to the limit $b \to 0$ which yields

$$
\begin{aligned}
e^{-U} &= 1 - a\,v^2\,\theta(v) \\
e^{-V} &= \frac{1 + av\theta(v)}{1 - av\theta(v)} \\
e^{-M} &= 1
\end{aligned}
\qquad (11)
$$

as in the case of Khan and Penrose. Furthermore, in the limit $a \to 0$ we have only an electromagnetic shock wave

$$
\begin{aligned}
e^{-U} &= 1 - \sin^2(bv\theta(v)) \\
e^{-V} &= 1 \\
e^{-M} &= 1
\end{aligned}
\qquad (12)
$$

which is the result for the Bell-Szekeres case.

Spacetimes describing colliding plane waves are divided into four regions:

Region I : $u < 0, v < 0$ empty space before the collision
Region II: $u > 0, v < 0$ a plane wave Region III: $u < 0, v > 0$ another wave travelling in the opposite direction
Region IV: $u > 0, v > 0$ the interaction region

The initial values given above are on $v = 0$, the boundary between Regions III, IV and similar results hold on $u = 0$, the boundary between Regions II, IV, determining the u-dependence. In the latter case the amplitudes of these waves will be given by different constants, say $a \to p$ and $b \to q$, cf eq.(28) in sequel. The case of Griffiths' solution is a mixture where we have eqs.(11) between Regions III, IV and eqs.(12) with u replacing v on the boundary between Regions II, IV.

3.   Einstein-Maxwell Equations

The Einstein-Maxwell field equations governing the interaction of two plane waves is well-known [28]. Starting with the metric (9) and the Maxwell potential 1-form $\mathcal{A}$

$$\mathcal{A} = A dx \tag{13}$$

where A depends only on u, v, we find a set of Einstein field equations which can be grouped into two categories. First we have the initial value equations

$$\begin{aligned} 2U_{vv} - U_v^2 + 2M_v U_v - V_v^2 &= 2\kappa e^{U-V} A_v^2 \\ 2U_{uu} - U_u^2 + 2M_u U_u - V_u^2 &= 2\kappa e^{U-V} A_u^2 \end{aligned} \tag{14}$$

and their integrability conditions

$$\begin{aligned} U_{uv} - U_v U_u &= 0 \\ 2A_{uv} - V_v A_u - V_u A_v &= 0 \\ 2V_{uv} - U_v V_u - U_u V_v + 2\kappa e^{U-V} A_u A_v &= 0 \\ 2M_{uv} + U_v U_u - V_v V_u + 2\kappa e^{U-V} A_u A_v &= 0 \end{aligned} \tag{15}$$

where κ is Newton's constant in geometrical units.

In eqs.(15) we have the wave equation for e^{-U} and its solution is immediate from the initial values. The following two equations are the main equations and the last equations is irrelevant as M can be obtained from quadratures once the main equations are solved.

The problem consists of finding a solution to eqs.(15) satisfying the initial data (10). Finally, we shall remark that eqs.(11), or (12) can be regarded as the solution of an initial value problem themselves, namely one between Region I and either a gravitational impulsive wave, or an electromagnetic shock wave across the null plane $v = 0$ in Region III. In this mini-problem the first one of eqs.(14) serves as the field equation.

4.   Harmonic Maps

We refer to [2] and [29] for a review and survey of the principal results on harmonic mappings of Riemannian manifolds. Here we shall briefly recall the most basic definitions in order to fix the notation. We shall consider two Riemannian manifolds endowed with metrics

$$\begin{aligned} ds^2 &= g_{ik} \, dx^i \, dx^k \,, & i &= 1, ..., n \\ ds'^2 &= g'_{\alpha\beta} \, dy^\alpha \, dy^\beta \,, & \alpha &= 1, ..., n' \end{aligned} \tag{16}$$

and a map

$$f : \mathcal{M} \to \mathcal{M}' \tag{17}$$

between them. This map is called harmonic if it extremizes the energy fuctional of Eells and Sampson, $\delta \mathcal{I} = 0$,

$$\mathcal{I}(f) = \int g'_{\alpha\beta} \frac{\partial f^{\alpha}}{\partial x^i} \frac{\partial f^{\beta}}{\partial x^k} g^{ik} \sqrt{|g|}\, d^n x \tag{18}$$

where the Lagrangian consists of the trace with respect to the metric g of the induced metric $f * g$ on $\mathcal{M}$. When the target space $\mathcal{M}'$ is 1-dimensional, harmonic maps satisfy Laplace's equation on the background of $\mathcal{M}$ and on the other hand if $\mathcal{M}$ is 1-dimensional, then harmonic maps coincide with the geodesics on $\mathcal{M}'$. The nonlinear sigma model corresponds to the harmonic map $f : R^2 \to S^2$.

There are at least three different ways in which the Einstein-Maxwell equations (15) can be formulated as a harmonic map.

1. We can take a 4-dimensional target space $\mathcal{M}'$ with the metric

$$ds'^2 = e^{-U}\left(2\,dM\,dU + dU^2 - dV^2\right) - 2\kappa\, e^{-V} dA^2 \tag{19}$$

which has sections of constant curvature and the flat metric

$$ds^2 = 2\,du\,dv \tag{20}$$

on $\mathcal{M}$. This is the electrovac analogue of the formulation given in [13]. It is not the most economical approach because M, which can be obtained by quadratures, appears explicitly in the metric (19).

2. We can get rid of M and consider a 2-dimensional target space with constant curvature at the expense of regarding U as a given function on $\mathcal{M}$ which does not enter into the variational problem as one of the local components of the harmonic map. Thus assuming that e^{-U} satisfies the wave equation, as in eqs.(15), we can take the metric on the target space as

$$ds'^2 = e^{-U} \frac{d\mathcal{E}\, d\bar{\mathcal{E}}}{|Re\mathcal{E}|^2} \tag{21}$$

where

$$\mathcal{E} = e^{(V-U)/2} + i\sqrt{\frac{\kappa}{2}}\, A \tag{22}$$

is an Ernst potential type of complex coordinate. The metric (21) is that of a space of constant negative curvature and $\mathcal{E}$ is the complex coordinate for the Poincaré upper half plane. There exists another representation, namely Klein's unit disk for the space of constant negative curvature. This is obtained by the transformation

$$\mathcal{E} = \frac{\xi + 1}{\xi - 1} \tag{23}$$

and

$$ds'^2 = e^{-U} \frac{d\xi \, d\bar{\xi}}{(1 - \xi\bar{\xi})^2} \tag{24}$$

is the resulting form of the metric. Frequently this is the most convenient representation. The original Matzner-Misner [1] as well as the Neugebauer-Kramer [14] formulations are of this type.

The metric on $\mathcal{M}$ is the same as the one in eq.(20).

This approach is by far the most common procedure followed in the literature.

3. It is possible to reformulate the reduced problem by avoiding the *ad hoc* introduction of U on $\mathcal{M}'$ at the expense of adding a new Killing direction to $\mathcal{M}$ where the magnitude of the Killing vector is e^{-U}. In this case we have

$$ds'^2 = \frac{d\xi \, d\bar{\xi}}{(1 - \xi\bar{\xi})^2} \tag{25}$$

where the definition of ξ is the same as before and

$$ds^2 = 2 \, du \, dv - e^{-2U} \, dz^2 \tag{26}$$

where we consider the mapping to be independent of z and once again U is specified *a priori*. It appears that this possibility has not been discussed before in the literature.

The first formulation where it is not necessary to introduce information from outside the variational principle is attractive. However, the close resemblance of the latter formulations to non-linear σ-models is an advantage.

5. Solutions

The metric (20) on M is not in a form most suitable for the construction of harmonic maps. We need to rewite it using new coordinates in such a way that some information about the initial data (10) is already incorporated into the system with the result that we can look for simple harmonic maps automatically satisfying the initial data. For this purpose we start with the wave equation for e^{-U} and following Szekeres [28] write its solution as

$$e^{-U} = f(u) + g(v) \tag{27}$$

where from eq.(8) we know that

$$\begin{aligned} f &= \tfrac{1}{2} - \left(1 + \tfrac{a^2}{b^2}\right) \sin^2(bv\theta(v)), \\ g &= \tfrac{1}{2} - \left(1 + \tfrac{p^2}{q^2}\right) \sin^2(qv\theta(v)). \end{aligned} \tag{28}$$

We can now consider a trivial coordinate transformation

$$u \to f(u) \qquad v \to g(v)$$

which amouts to a replacement of u, v by f, g in the Einstein-Maxwell equations (15). We shall further introduce the following definitions which are formally the same as those given by Khan and Penrose

$$
\begin{aligned}
p &= \sqrt{\tfrac{1}{2} - f} & q &= \sqrt{\tfrac{1}{2} - g} \\
r &= \sqrt{\tfrac{1}{2} + f} & w &= \sqrt{\tfrac{1}{2} + g}
\end{aligned}
\tag{29}
$$

in terms of which it will be convenient to introduce new coordinates on M.

In the vacuum case with non-collinear polarization we had found that

$$
\sigma = pw - qr \qquad \tau = pw + qr
\tag{30}
$$

were useful new coordinates because of two reasons. First of all, σ and τ can be recognized as prolate spheroidal coordinates [30] and we have the Kerr-Tomimatsu-Sato solutions of the main equations. Furthermore, the simplest and the most familiar solution of this type

$$
\xi = cos(\alpha)\tau + isin(\alpha)\sigma
\tag{31}
$$

satisfies the initial data.

This situation changes drastically for the electrovac case. If we were to follow blindly the approach that was successful for vacuum, we get a "solution" that does not satisfy the initial data. So we must look for something new. It turns out that it will be useful to consider

$$
\sigma = \frac{p - q}{r + w} \qquad \tau = \frac{p + q}{r + w}
\tag{32}
$$

as new coordinates on M. In terms of these σ and τ the metric on M is given by

$$
ds^2 = \frac{d\tau^2}{(1 + \tau^2)^2} - \frac{d\sigma^2}{(1 + \sigma^2)^2}
\tag{33}
$$

and from eqs.(27) and (28) we have

$$
\begin{aligned}
e^{-U} &= \frac{(1 - \sigma^2)(1 - \tau^2)}{(1 + \sigma^2)(1 + \tau^2)} \\
&= 1 - p^2 - q^2.
\end{aligned}
\tag{34}
$$

Then the main equations become

$$
\begin{aligned}
&-\frac{(1+\sigma^2)^2}{(1-\sigma^2)} \frac{\partial}{\partial\sigma}\left((1-\sigma^2)\frac{\partial\xi}{\partial\sigma}\right) + \frac{(1+\tau^2)^2}{(1-\tau^2)} \frac{\partial}{\partial\tau}\left((1-\tau^2)\frac{\partial\xi}{\partial\tau}\right) \\
&= \frac{2\bar{\xi}}{\xi\bar{\xi} - 1}\left[-(1+\sigma^2)^2\left(\frac{\partial\xi}{\partial\sigma}\right)^2 + (1+\tau^2)^2\left(\frac{\partial\xi}{\partial\tau}\right)^2\right]
\end{aligned}
\tag{35}
$$

which is similar to the prolate spheroidal case but differs from it in some important respects. Its advantage lies in the fact that

$$\xi = \epsilon\sigma, \qquad \epsilon^2 = \pm 1 \tag{36}$$

is a solution of eq.(35) that leads to the Bell-Szekeres solution. Furthermore it can be readily verified that $\xi = \epsilon\tau$ is also a solution as in eq.(31) which is again the Bell-Szekeres solution. In terms of these coordinates the Bell-Szekeres solution is given by

$$ds^2 = \frac{d\tau^2}{(1+\tau^2)^2} - \frac{d\sigma^2}{(1+\sigma^2)^2} - \left(\frac{1-\tau^2}{1+\tau^2}\right)^2 dx^2 - \left(\frac{1-\sigma^2}{1+\sigma^2}\right)^2 dy^2 \tag{37}$$

which may help to clarify the meaning of σ and τ.

So the remaining problem is to a find a one-parameter complex solution of eq.(35) that reduces to eq.(36). Such a solution will be of physical interest as it will automatically satisfy the proper initial data.

6. Conclusion

The only proper conclusion of a paper such as this, namely the exact solution of eqs.(35) satisfying the initial data in eqs.(10) is missing. In this paper I have given a list of the essential properties that we must require from a physically acceptable solution describing the interaction of plane impulsive gravitational and electromagnetic shock waves and presented some preparatory material towards such a solution. In my case this solution has been missing since 1978 and that is why I felt it inappropriate to publish the work reported here earlier. However, I now feel that the abundance of so many irrelevant "solutions" of this problem in the literature has made the presentation of the real problem imperative.

7. Acknowledgement

This work was in part supported by The Turkish Scientific Research Council TÜBITAK under tbag-cg-1.

References

[1] R. A. Matzner and C. W. Misner, Phys. Rev. **154**, 1229 (1967)

[2] J. Eells, Jr. and R. H. Sampson, Am. J. Math. **86**, 109 (1964)

[3] C. W. Misner, in *Magic without Magic: John Archibald Wheeler; A Collection of Essays in Honor of His 60th Birthday*, edited by J. R. Klauder (Freeman, San Francisco, 1972)

[4] D. Maison, Phys. Rev. Lett. **41**, 521 (1978)

[5] V. A. Belinskii and V. E. Zakharov, Zh. Eksp. Teor. Fiz. **75** (1978) 1952; Sov. Phys. JETP **48** (1978) 984

[6] G. Neugebauer, J. Phys. A: Math. and Gen. **12**, 167 (1979)

[7] G. Neugebauer and D. Kramer, J. Phys. A: Math. and Gen. **16**, 1927 (1983)

[8] F. J. Ernst, Phys. Rev. **167**, 1175 (1968); **168**, 1415 (1968)

[9] C. W. Misner, Phys. Rev. **D 18**, 4510 (1978)

[10] Y. Nutku, Ann. Inst. H. Poincaré, **A 21**, 175 (1974)

[11] A. Eris and Y. Nutku, J. Math. Phys. **16**, 1431 (1976)

[12] K. Khan and R. Penrose, Nature (London)**229**, 185 (1971)

[13] Y. Nutku and M. Halil, Phys Rev. Lett. **39**, 1379 (1977)

[14] G. Neugebauer and D. Kramer, Annalen der Physik **24**, 62 (1969)

[15] J. B. Griffiths, *Colliding Waves in General Relativity*, Oxford University Press 1990

[16] P. Bell and P. Szekeres, Gen. Rel. and Grav. **5**, 275 (1974)

[17] J. B. Griffiths, Ann. Phys. (N.Y.) **102**, 388 (1976)

[18] F. J. Tipler, Phys. Rev. **D 22**, 2929 (1980)

[19] R. A. Matzner and F. J. Tipler, Phys. Rev. **D 29**, 1575 (1984)

[20] U. Yurtsever, Phys. Rev. **D 36**, 1662 (1987); **D 38**, 1706 (1988); **D 40**, 329 (1989)

[21] S. Chandrasekhar and B. Xanthopoulos, Proc. Roy. Soc. (London), **A 398**, 223 (1985); **A 410**, 331 (1987)

[22] J. B. Griffiths, J. Phys. A: Math. Gen. **16**, 1175 (1983)

[23] R. Penrose, Rev. Mod. Phys. **37**, 215 (1965)

[24] R. Penrose, "Null hypersurface initial data for classical fields of arbitrary spin and for general relativity", in *Quantization of generally covariant fields* A. R. L. Technical Documents, Report no. 63-56, ed. P. G. Bergman (1963)

[25] R. Penrose, in "General Relativity, Papers in Honour of J. L. Synge" ed. L. O'Raifertaigh. Clarendon Press Oxford (1972)

[26] H. W. Brinkman, Proc. Natl. Acad. Sci. (USA) **9**, 1 (1923)

[27] J. Ehlers and W. Kundt, in *Gravitation, An Introduction to Current Research*, edited by L. Witten (Wiley, New York, 1962)

[28] P. Szekeres, J. Math. Phys. **13**, 286 (1972)

[29] J. Eells and L. Lemaire, Bull. London Math. Soc. **10**, 1 (1968)

[30] D. Zipoy, J. Math. Phys. **7**, 1137 (1966)

Geometry, the Renormalization Group and Gravity

Denjoe O'Connor *   *C. R. Stephens* †

Abstract

We discuss the relationship between geometry, the renormalization group (RG) and gravity. We begin by reviewing our recent work on crossover problems in field theory. By crossover we mean the interpolation between different representations of the conformal group by the action of relevant operators. At the level of the RG this crossover is manifest in the flow between different fixed points induced by these operators. The description of such flows requires a RG which is capable of interpolating between qualitatively different degrees of freedom. Using the conceptual notion of course graining we construct some simple examples of such a group introducing the concept of a "floating" fixed point around which one constructs a perturbation theory. Our consideration of crossovers indicates that one should consider classes of field theories, described by a set of parameters, rather than focus on a particular one. The space of parameters has a natural metric structure. We examine the geometry of this space in some simple models and draw some analogies between this space, superspace and minisuperspace.

1.   Introduction

The cosmopolitan nature of Charlie Misner's work is one of its chief features. It is with this in mind that we dedicate this article on the occasion of his 60th birthday. There are several recurring leitmotifs throughout theoretical physics; prominent amongst these would be geometry, symmetry, and fluctuations. Geometry clarifies and systematizes the relations between the quantities entering into a theory, e.g. Riemannian geometry in the theory of gravity and symplectic geometry in the case of classical mechanics. Symmetry performs a similar role, and in the case of continuous symmetries is often intimately tied to geometrical notions. For instance in the above examples Riemannian geometry and symplectic geometry are intimately related to

*D. I. A. S., 10 Burlington Road, Dublin 4, Ireland.
†Instituut voor Theor. Fysica, Universiteit Utrecht, 3508TA Utrecht, Netherlands.

272

the diffeomorphism and canonical groups respectively. Our third leitmotif, fluctuations, enters ubiquitously through the quantum principle, or classically in statistical physics. The key underlying idea here is that because of the fluctuations physics must be described in a probabilistic manner.

Having stated our prejudices let us be a little less ambitious than to consider all of theoretical physics and restrict our attention to field theory. We make no pretension to mathematical rigour taking the point of view that a field theory on a manifold $\mathcal{M}$ can be defined via a functional integral with a probability measure which is a functional of a set of possibly position dependent parameters $\{g^i\}$, e.g. coupling constants, masses, background fields etc. Physical quantities can be expressed as combinations of moments which in turn can be written as functions of the $\{g^i\}$. If we think of these parameters as coordinates on a parameter space $\mathcal{G}$ it is clear that physics should be invariant under changes in these coordinates. A particular type of coordinate change is engendered by a renormalization, e.g. between bare and renormalized g^i's. Other possible symmetry group transformations such as coordinate transformations on $\mathcal{M}$ or gauge transformations act as diffeomorphisms on $\mathcal{G}$. Here we are concerned exclusively with the behaviour under RG transformations, and hence under scale transformations. We investigate some geometrical structures on $\mathcal{G}$, in particular defining a metric and associated connection. We look at the change in the geometry under renormalization, thereby introducing all three of our leitmotifs. The geometry is a result of the fluctuations in the system, i.e the probabilistic description. Without fluctuations the metric is identically zero. The RG induces a flow on $\mathcal{G}$ the fixed points of which are of particular interest as they represent conformally invariant systems. This flow with respect to a given parameter can be either centrifugal or centripetal for a particular fixed point. If the former the parameter is said to be relevant, and irrelevant for the latter. The relevance or irrelevance can change according to the fixed point.

RG flows between different fixed points, i.e different conformal field theories, are especially interesting. The reason for this is the following: one of the most important tasks confronting a theory is to identify correctly the degrees of freedom (DOF) of a physical system. It is a fact of life that all physically relevant theories have qualitatively different effective DOF at different scales. For instance, in QCD the high energy DOF are quark, gluon DOF, whilst at low energy they are hadron, meson ... DOF. In gravity at low energy, gravitons are the low energy DOF, whereas at high energies, who knows...topological foam, strings .... The only thing that is reasonably certain is that it won't be gravitons. A closer to earth example would be liquid helium in a 3 dimensional (3D) slab geometry. For correlation lengths much less than the slab thickness helium atoms are the relevant DOF whereas in the opposite limit it is vortices. An example we will treat here is that of a $\lambda\phi^4$ theory on a manifold $S^1 \times R^{d-1}$ of size L. Suitably altered this model model can describe, amongst others, the Higgs model at finite temperature, the Casimir effect for an interacting quantum field theory or the critical behaviour of an Ising ferromagnet in a slab geometry. Here there is a change in DOF as the variable $x = mL$ changes, where m is the "mass" (inverse correlation length) in the physical system. As $x \to \infty$ the DOF are effectively d dimensional and as $x \to 0$, $d - 1$ dimensional. We will also briefly discuss similar considerations in more realistic "cosmologies".

One of the first questions one must confront with a crossover problem is: how should one renormalize? If one accepts the fairly common point of view that renormalization means the consistent removal of ultraviolet (UV) divergences one generically finds a resultant RG which is independent of the parameter inducing the crossover, e.g. L in the above example. The β functions and anomalous dimensions of the problem are all then L independent. One also finds that the theory gives perturbative nonsense as $x \to 0$. The reason for this is relatively simple. Let us take a more physical picture of renormalization, as a "course graining" such as decimation/block spinning[1] [1]. Here we imagine integrating out DOF between one scale and another. For the finite system at scales $\ll L$ one would integrate out d dimensional DOF. However, as one course grains further one is eventually integrating out DOF with scales $\sim L$. In the finite direction there are no DOF with scales $> L$, therefore one cannot integrate them out. The only DOF left are $d - 1$ dimensional and these are the physically relevant ones. So, a physically intuitive renormalization procedure takes into account the qualitatively changing nature of the effective DOF. It should be clear then why a L independent RG is badly behaved. Such a group is equivalent to integrating out only d dimensional DOF **for all scales**. The moral is that one should try to develop a RG that is capable of interpolating between qualitatively different DOF. In this paper we will show how this can be achieved in a wide class of crossover problems.

The outline of this article will be as follows: in section 2 we will give a short, intuitive exposition of renormalization and the RG with a view to the treatment of crossovers. In section 3 we will develop the concept of a RG that can interpolate between qualitatively different DOF, introducing the concept of a "floating" fixed point and illustrating our ideas with $\lambda\phi^4$ on $S^1 \times R^{d-1}$. In section 4 we will describe the beginnings of a geometrical framework for field theory wherein a much more general theory of crossovers may be built illustrating the concepts using a Gaussian model. Finally in section 5 we take an opportunity to make some speculative remarks and draw some conclusions.

2.   Renormalization and the Renormalization Group

In this section as well as setting notation we would like to give an extremely brief and hopefully intuitive account of renormalization, hoping that the unconventional viewpoint will not prove unintelligible. As a concrete example consider a self interacting scalar field theory described by a partition function (generating functional) on

---

[1]Strictly speaking such renormalizations form a semigroup not a group.

$\mathcal{M} = R^d$

$$Z[m_B, \lambda_B, \Lambda] = \int [D\phi_B]_\Lambda e^{-\int_\Lambda d^d x L(\phi_B, m_B, \lambda_B)} \qquad (1)$$

where

$$L = \frac{1}{2}(\nabla\phi)^2 + \frac{1}{2}m_B^2\phi_B^2 + \frac{\lambda_B}{4!}\Lambda^{4-d}\phi_B^4 \qquad (2)$$

For the sake of making sense out of the theory we will assume there is always an UV cutoff Λ. In (1) we have a probability measure which is a function of two parameters and a cutoff. The parameters m_B and λ_B are good descriptors of the physics at scales $\sim \Lambda$. What this means is the following; if one could calculate the 2 and 4 point vertex functions exactly one would find them to be very complicated functions of m_B, λ_B and Λ. At scales $\sim \Lambda$, however, one would find that for $\lambda_B << 1$ $\Gamma^{(2)} \sim m_B$ and $\Gamma^{(4)} \sim \lambda_B$. On the other hand at scales $\kappa << \Lambda$ the parameters m_B and λ_B are in no way a good description of the associated correlation functions. Obviously as $\frac{\Lambda}{\kappa} \to \infty$ they get worse and worse. The deep underlying reason behind this is of course the existence of fluctuations. It is the dressing due to quantum or thermal fluctuations that changes the correlation functions as one changes scale. We emphasize though that if one can calculate in the theory exactly the bare parameters are as good as any others. What one would like is to describe the correlation functions using a more suitable set of parameters, in particular if we are considering physics at a scale $\sim \kappa$ it would seem to make good sense to describe the physics using new parameters m and λ which are a more natural description of the physics at this scale. An obvious natural choice would be to describe the physics at the scale κ in terms of the 2 and 4 point vertex functions at a scale κ' where $\kappa \sim \kappa'$. Thus one would require

$$\Gamma^{(2)}(k = 0, m, \lambda, \kappa') = m^2 \qquad \Gamma^{(4)}(k = 0, m, \lambda, \kappa') = \bar{\lambda} = \lambda\kappa'^{4-d} \qquad (3)$$

The physics at the scale κ, i.e the correlation functions at that scale, would now be described in terms of the correlation functions at a nearby fiducial scale, κ'.

In the above we have loosely outlined the renormalization program for this model. Why renormalize? There are two answers to this, one perturbative, and one not. Perturbatively as $\frac{\Lambda}{\kappa}$ grows perturbation theory in terms of the bare coupling becomes worse and worse. This is the well known problem of "UV divergences". In terms of fluctuations the bare parameters are being perturbatively dressed by fluctuations between the scales Λ and κ. The recipe for getting round this problem is as outlined above; to perturb with a "small" coupling rather than a large one, i.e. the renormalized coupling. Thus one uses the value of $\Gamma^{(4)}$ at some scale κ' as one's small parameter. This perturbation theory is then reasonably well defined as long as κ is not too different from κ' as. In 4D, for example, the correction terms are proportional to powers of $\ln\frac{\kappa}{\kappa'}$. Thus it is perturbatively better to dress the correlation functions a small amount. The dressing between Λ and κ is large and therefore difficult to compute perturbatively whilst the dressing between κ and κ' is smaller. The optimum approach is to consider an infinitesimal dressing and to integrate the resulting differential equation. So, if one wishes to implement perturbation theory renormalization is essential. The non-perturbative reason is somewhat subtler and depends ultimately on whether one believes there is a fundamental cutoff or not. One puts it in to make mathematical sense of the theory and then asks if it can be sensibly removed again. It seems to be the case that this is only possible for special values of the bare parameters — their fixed point values. To understand this we must consider the RG.

One can think of renormalization as a mapping of correlation functions between two different "scales". These mappings have an abelian group structure and this group is known as the RG. The group action on $\mathcal{G}$ generates a flow. Of particular interest are the fixed points of this flow as they imply a system possesses scale invariance. The fundamental relation between bare and renormalized vertex functions is

$$\Gamma^{(N)}(k, m, \lambda, \kappa) = Z_\phi^{\frac{N}{2}} \Gamma_B^{(N)}(k, m_B, \lambda_B, \Lambda) \tag{4}$$

where the renormalized parameters are defined at some arbitrary scale κ, and Z_ϕ denotes a wavefunction renormalization factor. The bare theory's independence from κ leads to the RG equation

$$\left(\kappa \frac{\partial}{\partial \kappa} + \beta \frac{\partial}{\partial \lambda} + \gamma_{\phi^2} m^2 \frac{\partial}{\partial m^2} - \frac{N}{2} \gamma_\phi \right) \Gamma^{(N)}(k_i, m^2, \lambda, \kappa) = 0 \tag{5}$$

where $\gamma_{\phi^2} = -\frac{\partial \ln Z_{\phi^2}}{\partial \ln \kappa}$, Z_{ϕ^2} being the renormalization constant associated with the operator ϕ^2 and $\gamma_\phi = \frac{\partial \ln Z_\phi}{\partial \ln \kappa}$ are the anomalous dimensions of the operators ϕ^2 and ϕ respectively. It is important to note that (5) results from an **exact** symmetry even though it expresses an apparent triviality, the reparameterization invariance of the correlation functions. Equation (5) can be solved by the method of characteristics and together with dimensional analysis yields

$$\Gamma^{(N)}(k_i, \lambda, m, \kappa) = (\kappa\rho)^{d-N\frac{(d-2)}{2}} \exp\left(\frac{N}{2} \int_\rho^1 \frac{dx}{x} \gamma_\phi(x) \right) \Gamma^{(N)}\left(\frac{k_i}{\rho\kappa}, \frac{m^2(\rho)}{\rho^2\kappa^2}, \lambda(\rho), 1 \right) \tag{6}$$

where $\lambda(1) = \lambda$, $m(1) = m$, and ρ is arbitrary. $m(\rho)$ and $\lambda(\rho)$, the running mass and coupling satisfy

$$\rho \frac{d\lambda(\rho)}{d\rho} = \beta \qquad \rho \frac{dm^2(\rho)}{d\rho} = \gamma_{\phi^2} m^2(\rho) \tag{7}$$

Equation (5) tells us how $\Gamma^{(N)}$ gets dressed by fluctuations between the scales κ and $\kappa + d\kappa$, in terms of parameters which get dressed according to (7). Integrating this equation tells us how $\Gamma^{(N)}$ gets dressed by fluctuations between the scales κ and $\kappa\rho$. This dressing induces a flow on $\mathcal{G}$. Equation (6) is the exact solution of an exact equation which is a result of an exact symmetry. The fixed point of the coupling λ, λ^* is given by the solution of $\beta = 0$. Now, we can use our freedom in choosing ρ to eliminate the variable $m(\rho)$ in (6) via the condition $m^2(\rho) = \rho^2\kappa^2$. At the fixed point λ^* one can solve the equation for $m(\rho)$ to find $\rho \sim \left(\frac{m}{\kappa}\right)^\nu$ where $\nu = (2 - \gamma^*_{\phi^2})^{-1}$, $\gamma^*_{\phi^2}$ being the value of γ_{ϕ^2} at the fixed point. Similarly, defining $\eta = \gamma^*_\phi$ one finds for instance

$$\Gamma^{(2)}(k = 0, m) \sim Am^{2\nu(2-\eta)} \qquad (8)$$

where A is some constant. Once again we emphasize that this is an exact result dependent only on the fact that a fixed point exists. The RG is not just about "improving perturbation theory". Of course, finding the fixed point and calculating A, ν and η is a different matter. In $d < 4$ dimensions for this model there are two known fixed points, the Gaussian fixed point $\lambda^* = 0$ and the Wilson-Fisher (WF) fixed point $\lambda^* \sim (4 - d)$. At the Gaussian fixed point $\nu = \frac{1}{2}$ and $\eta = 0$ whilst at the other e.g. in 3D $\nu = 0.630$ and $\eta = 0.031$.

Physically the importance of the fixed point for λ is the following. λ like all other quantities gets dressed as a function of scale and therefore changes its value. At the point $\lambda = \lambda^*$ the coupling becomes completely insensitive to dressing and therefore has essentially dropped out of the problem. Obviously $m = 0$ is a fixed point for the mass. As fixed points essentially define a theory finding them is one of the main tasks of field theory. Returning now to a non-perturbative aspect of renormalization; in (4) we could instead of differentiating the bare vertex function with respect to κ have differentiated the renormalized function with respect to Λ. This yields an equation analogous to (5). If one can find a fixed point of this equation then one can take the cutoff $\Lambda \to \infty$ and thereby recover a continuum theory.

The fact that there exist two fixed points for this theory means that one is really considering a class of field theories as a function of $x = \frac{\bar{\lambda}}{m^{4-d}}$. As $\bar{\lambda} \to 0$ one approaches the Gaussian theory, and as $m \to 0$ the WF fixed point. One crosses between them as a function of scale. The coupling $\bar{\lambda}$ is relevant in terms of RG flows with respect to the Gaussian fixed point. In other words a small perturbation from this fixed point induces a flow to larger length scales terminating at the WF fixed point. This is an example of crossover behaviour in field theory and describes a transition between qualitatively different DOF. For $x \ll 1$ the DOF are essentially non-interacting, whereas for $x \gg 1$ they are strongly interacting. The reader might legitimately enquire as to why, given that they are strongly interacting, one believes that perturbation theory can be used. This raises an important question: perturbation theory in terms of what coupling? In terms of $\bar{\lambda}$ straight perturbation theory breaks down as $m \to 0$ due to large dressings from the infrared (IR) regime as opposed to large dressings from the UV regime as was considered previously. The RG methodology tells you to ignore any differences between the UV and IR regimes. The essential problem is that of large dressings irrespective of whether the dressing arises from IR or UV fluctuations. Large dressings imply that one has used inadequate parameters to describe the physics, hence renormalization and the RG should be implemented. The correct parameter to perturb in is the running coupling constant which is a solution of the β function equation treated as a differential equation who's solution is a function of

x. The above is our first simple example of crossover behaviour in field theory. We would now like to proceed to other more pertinent examples showing some difficulties one encounters and their solution.

3.   Crossover Behaviour in Field Theory

One of the main themes we have tried to emphasize in the introduction is that the effective degrees of freedom of a physical system are scale dependent. Here we take a simple but physically relevant paradigm to show the difficulties involved in trying to describe a qualitative change in the DOF of a system. We will try to emphasize a physical approach, stating in general only results, leaving the details in our other papers [2]. Consider a Lagrangian

$$\mathcal{L} = \frac{1}{2}(\nabla\phi_B)^2 + \frac{1}{2}m_B^2\phi_B^2 + \sum_i \mu_B^i O_B^i \qquad (9)$$

where $\sum_i \mu_B^i O_B^i$ represents schematically a relevant or set of relevant operators that induce a crossover from a fixed point associated with $\mu^i = 0$ to some others. For the moment we specify neither the symmetry of the order parameter or the dimensionality of the system. Some examples of relevant operators are the following: i) for $d < 4$, Gaussian→WF fixed point as mentioned in the last section, $\mu_B^1 = \frac{\lambda_B}{4!}$, $O^1 = \phi_B^4$, $\mu_B^i(i \neq 1) = 0$; ii) quadratic symmetry breaking $(O(N) \to O(M))$, ϕ_B has an $O(N)$ symmetry, $\mu_B^1 = \frac{1}{2}\tau_B$, $O_B^1 = \sum_{i=1}^{M}(\phi_{iB})^2$, $\mu_B^2 = \frac{\lambda_B}{4!}$, $O_B^2 = \sum_{i=1}^{N}(\phi_{iB}^2)^2$; iii) uniaxial dipolar ferromagnets, where in Fourier space $\mu_B^1 = \frac{\alpha}{2}$, $O_B^1 = \frac{p_z^2}{p^2}\phi_B^2$, $\mu_B^2 = \frac{\lambda_B}{4!}$, $O_B^2 = \phi_B^4$. For the case of dimensional crossover one can determine the appropriate operators by Fourier transforming $\mathcal{L}$ with respect to the finite directions. One important common feature of the above is the introduction of an important new scale in each problem i.e. τ, the quadratic symmetry breaking term, α the strength of the dipole-dipole interactions and L the characteristic finite size scale. It is the existence of one or more new scales in a problem that makes a crossover much richer, more interesting and more complex than standard field theory. We call this generic length scale g. We also take this scale to be a physical scale and hence a RG invariant.

So, what does renormalization have to say about such systems? There is a widespread belief that renormalization just means getting rid of UV divergences. If we accept this belief and examine the above models one notices that the UV behaviour in these theories is independent of the parameter g, hence the UV divergences can be removed in a g independent way. We will give just one example of what happens if this philosophy is accepted. Consider $\lambda\phi^4$ on a manifold $S^1 \times R^3$ of size L. Using minimal subtraction gives for $mL \ll 1$ to one loop

$$\Gamma^{(2)} \to m^2\left(1 + \frac{\lambda}{32\pi^2}\ln\frac{m^2}{\kappa^2} + \frac{\lambda}{24m^2L^2} + O(\frac{\lambda^2}{m^3L^3})\right) \qquad (10)$$

Obviously the perturbative corrections are large in this regime, in fact in the limit $Lm \to 0$ these corrections become infinite. From the point of view of renormalization this is no different than the bare vertices in the L independent theory getting a large dressing due to fluctuations. Here we've done a renormalization but still the vertex has a large L dependent dressing. Why is that? In implementing minimal subtraction we have really made an assumption, that parameters associated with the $L = \infty$ system will provide a good description of the physics when $L \to 0$. In this limit the system is effectively 3D and so one can hardly expect 4D parameters to be adequate. The total breakdown in perturbation theory above is a reflection of this fact. 3D $\lambda\phi^4$ theory has completely different DOF to 4D $\lambda\phi^4$ theory.

The way out of this impasse is in many ways relatively simple — choose better renormalized parameters. Think back to the G-WF crossover discussed in the last section. The analog of the 4D theory there is the Gaussian theory and the analog of the 3D theory the WF fixed point theory. The analog of $\bar{\lambda}$ is L. L is a relevant parameter that causes a crossover from one fixed point to another. We managed to cope with the G-WF crossover, how so? Above we renormalized in an L independent way, the analog would be to renormalize in a $\bar{\lambda}$ independent manner. We could have certainly done this i.e renormalize the theory using only the counterterms appropriate for a Gaussian theory. For $\frac{\Lambda}{\kappa} \gg 1$ we would have found large dressings telling us that the Gaussian counterterms were not really sufficient to renormalize the theory. These large dressings occur because of the self interactions amongst the particles, because interacting DOF are qualitatively different to non-interacting ones. The correct thing to do was to choose renormalization conditions such as in (3) which were specified as functions of λ, i.e a good renormalization was dependent on λ the parameter that induces the crossover. In the case at hand we should therefore consider L dependent renormalization conditions such as

$$\Gamma^{(2)}(k = 0, m, \lambda, L, \kappa) = m^2 \qquad \Gamma^{(4)}(k = 0, m, \lambda, L, \kappa) = \lambda\kappa^{4-d} \qquad (11)$$

These conditions imply that the β function and anomalous dimensions are all functions of $L\kappa$ as well as λ, i.e the RG itself is L dependent. An L independent RG tells you how parameters are dressed in the theory by L independent fluctuations whereas an L dependent one tells how things are dressed by L dependent ones. In the real physical system it is manifestly obvious that the real fluctuations in the system are L dependent and that consequently conditions such as (11) will yield parameters which are a more faithful representation of the physics. The moral is: if the DOF of a system can qualitatively change as a function of scale then it is clearly better if one can derive a RG which can follow such a change.

It should be clear how to implement this philosophy more generally. For a crossover caused by a relevant parameter g, one should impose normalization conditions at an arbitrary value of g thereby obtaining a g dependent RG equation. In such a crossover one is interpolating between different conformal field theories i.e. different representations of the conformal groups associated with the limits $\frac{g}{m} \to 0$ and $\frac{g}{m} \to \infty$. Just as there are anomalous dimensions γ_ϕ and γ_{ϕ^2} which are characteristic of the conformal weights of the associated operators for a particular fixed point so one can define effective anomalous dimensions and critical exponents which are characteristic of the crossover system. Given that in d dimensions the dimension of the operator ϕ^4 is canonically $4 - d$ one can define an effective dimension d_{eff} via the relation $\frac{d\ln\Gamma^{(4)}}{d\ln m^2} = (4 - d_{eff} - 2\eta_{eff})\nu_{eff}$. What about the notion of a fixed point? For the system of size L true conformal symmetry is only realized in the limits $m \to 0$ and $Lm \to \infty$ which yields the d dimensional fixed point and $m \to 0$ $Lm \to 0$ which yields the $d - 1$ dimensional one. The equation $\beta = 0$ as an algebraic equation still has some meaning in these crossover systems. It does not, however, give a fixed point because the β function is now explicitly scale dependent through the variable $L\kappa$. If one thinks of the β function as being the velocity of the RG flow in the λ direction the value $\beta = 0$ is an equation that is satisfied only for a particular scale, not all scales as it should be for a true fixed point. The β function equation is a differential equation and should be integrated. However, one can in fact define an effective or "floating" fixed point in the following manner. Consider the β function generically as

$$\kappa \frac{d\lambda}{d\kappa} = \beta(\lambda, L\kappa) = -(4 - d)\lambda + a_1(L\kappa)\lambda^2 + a_2(L\kappa)\lambda^3 + O(\lambda^4) \qquad (12)$$

where a_1 and a_2 are known functions (see [3]). Define a new coupling $h = a_1\lambda$ to find

$$\kappa \frac{dh}{d\kappa} = -\varepsilon(L\kappa)h + h^2 + b(L\kappa)h^3 + O(h^4) \qquad (13)$$

where $\varepsilon(L\kappa) = 4 - d - \frac{d\ln a_1}{d\ln \kappa}$ and $b(L\kappa)$ is a combination of a_2 and a_1. Setting $\beta(h, L\kappa) = 0$ yields a solution $h^* \equiv h^*(L\kappa)$. This is the floating fixed point. As $L\kappa \to \infty$ it yields the d dimensional fixed point and as $L\kappa \to 0$ the $d - 1$ dimensional fixed point. Corresponding floating fixed points can be defined in all the crossover problems we have considered so far. The floating fixed point is the "small" parameter with which perturbation theory is implemented. A g dependent RG and a corresponding g dependent RG improved perturbation theory allow for complete perturbative access to the crossover. The main reason for this is that such a RG can interpolate between the qualitatively different DOF in the problem. As a specific example we will quote some one loop results [2] for the above finite size model. The fixed point coupling is $h^* = \varepsilon(\kappa L)$ where

$$4 - d_{eff} = \varepsilon(\kappa L) = 1 - \kappa \frac{d}{d\kappa} \ln(\sum_n (1 + \frac{4\pi^2 n^2}{L^2 \kappa^2})^{-\frac{3}{2}})$$

$$\gamma_{\phi^2}(h^*) = \frac{h^*}{3} \qquad \gamma_\phi(h^*) = 0$$

These functions all interpolate in a smooth way between their 4D and 3D values. $\varepsilon(\kappa L)$ is our "small" expansion parameter. It also yields (to first order) the effective dimensionality of the system. It is worth noting here that the sole requirement of finiteness of the correlation functions for all L is sufficient to determine the entire crossover.

So far we have outlined intuitively an approach to crossover behaviour and applied it to an interesting class of problems. Our considerations were governed by the RG flows of the parameters. The natural arena for such flows is $\mathcal{G}$. Rather than consider a particular crossover we would like to consider $\mathcal{G}$ more abstractly. This may also prove fruitful in cases where the relevant parameters are not a priori known.

4.   Geometry of $\mathcal{G}$

In this section we wish to begin an investigation of some of the geometrical structure that is inherent in the approach we are following. We will attempt to be as general as possible to begin with, and consequently somewhat vague. As was seen in the preceeding sections it was essential, if one wished to have a controlled perturbative expansion, to change from one set of parameters useful in one regime to a new set of parameters useful in another regime, for example the large mL and small mL regimes respectively. We are therefore working on a coordinate patch and changing coordinates on this patch. The immediate question would appear to be what space are we working on, i.e. a coordinate patch of what?

Examining the functional integral

$$Z[\mathcal{M}, \{\theta^i\}, \Lambda] = \int [D\phi]_\Lambda e^{-S[\mathcal{M}, \phi[\mathcal{M}], \{\theta^i\}, \Lambda]} \tag{14}$$

we see that it defines a map from the space, $\mathcal{F}$, parameterized by $(\mathcal{M}, \phi[\mathcal{M}], \{\theta^i\}, \Lambda)$ to a section of a line bundle over $\mathcal{G}$. $\mathcal{M}$ is the spacetime manifold, $\phi[\mathcal{M}]$ are fields on $\mathcal{M}$, $\{\theta^i\}$ are couplings between the fields and external sources and Λ plays the role of a regulator which will not be viewed as a true parameter of the theory, but rather as either a reflection of a true underlying lattice or a device to control any UV problems, and assist in the definition of the functional integral. We choose the set $\{g^i\}$ discussed in previous sections to be local coordinates on $\mathcal{G}$.

Earlier we saw that explicit calculations required a change of parameters, i.e. a change of coordinates on $\mathcal{G}$, from bare parameters (coordinates) to renormalized parameters (coordinates). If the object Z has any meaning it should have the same content in all coordinate systems. We will assume that Z is invariant under coordinate transformations on $\mathcal{G}$ and therefore is a scalar. Now, when one is looking at coordinate transformations, it is natural to examine what structure $\mathcal{G}$ posesses that can help one organize the analysis. Any structure $\mathcal{G}$ has must be induced from Z, or already exist in S. Ideally we would like our parameters to be related as simply as possible to the moments of the probability distribution, as these are generally the experimentally accessible objects. We will assume that $\mathcal{G}$ is a topological space with a differentiable structure and possibly isolated singularities, and that Z can be considered a differentiable function on $\mathcal{G}$ away from such special points. Thus if we consider an infinitesimal variation in S of the form dS, where d is the exterior derivative operator on $\mathcal{G}$, we get an induced change in Z. If the sources, masses etc. are position dependent then $\mathcal{G}$ is infinite dimensional and analogous to superspace, which would suggest that a mini-superspace may be useful. Mini-superspace in this context means restricting our considerations to a small finite dimensional subspace of $\mathcal{G}$. It is primarily this situation that will concern us here.

It is convenient for the following to work with the functional integral as a normalized probability distribution, which we can achieve by dividing by $Z = e^{-W}$. We therefore get a normalized functional integral

$$\int [D\phi]_\Lambda e^{W-S} = 1 \tag{15}$$

Because it is normalized and d is restricted to $\mathcal{G}$, we have

$$\int [D\phi]_\Lambda de^{W-S} = dW - <dS> = 0 \tag{16}$$

where $< A >$ means expectation value of A.

$$ds^2 = <(dW - dS) \otimes (dW - dS)> \tag{17}$$

defines a positive definite, symmetric, quadratic form on $\mathcal{G}$ arising from the positivity of the probability distribution or the convexity of the associated entropy functional. ds^2 plays the role of a metric on $\mathcal{G}$. An infinitesimal change in our parameters along some smooth curve in $\mathcal{G}$ defines a vector tangent to that curve and therefore we can express our metric as

$$g_{\mu\nu} = <\partial_\mu S \partial_\nu S> - \partial_\mu W \partial_\nu W \tag{18}$$

on the space satisfying $dW - <dS> = 0$. This metric is known as the Fisher information matrix [4] in probability theory and is used for comparing one probability distribution to another.

Let us begin with our most simple mini-superspace example, the Gaussian distribution, which corresponds to a free field theory in zero dimensions. We begin with a field ϕ coupled to an external source J described by the action

$$S[\phi, m^2, J] = \frac{1}{2}m^2\phi^2 + J\phi \tag{19}$$

$\mathcal{M}$ is now a single point and we have a coordinate patch on $\mathcal{G}$ with coordinates (J, m^2). The generator of connected correlation functions is

$$W[J, m^2] = -\frac{1}{2}\frac{J^2}{m^2} + \frac{1}{2}\ln[\frac{m^2}{2\pi}] \tag{20}$$

The condition $dW - < dS >= 0$ gives

$$-(< \phi > + \frac{J}{m^2})dJ - \frac{1}{2}(< \phi^2 > - \frac{J^2}{m^4} - \frac{1}{m^2})dm^2 = 0 \qquad (21)$$

The corresponding metric on using (21) is

$$ds^2 = \frac{1}{m^2}dJ^2 - \frac{2}{m^2}\frac{J}{m^2}dJ\,dm^2 + \frac{1}{m^2}((\frac{J}{m^2})^2 + \frac{1}{2}\frac{1}{m^2})(dm^2)^2 \qquad (22)$$

Note that this metric is not diagonal unless $J = 0$, however, a simple coordinate change allows us to diagonalize it, the appropriate choice of new coordinate being $\hat{\phi} = -\frac{J}{m^2}$ which is equivalent to starting with

$$S[\phi, \hat{\phi}, m^2] = \frac{1}{2}m^2\phi^2 - m^2\hat{\phi}\phi \qquad (23)$$

$$W[\hat{\phi}, m^2] = -\frac{1}{2}m^2\hat{\phi}^2 + \frac{1}{2}\ln[\frac{m^2}{2\pi}] \qquad (24)$$

the condition $dW - < dS >= 0$ now gives

$$m^2(< \phi > -\hat{\phi})d\hat{\phi} - \frac{1}{2}(< (\phi - \hat{\phi})^2 > -\frac{1}{m^2})dm^2 = 0 \qquad (25)$$

with metric

$$ds^2 = m^2 d\hat{\phi}^2 + \frac{1}{2}m^{-4}(dm^2)^2 \qquad (26)$$

Observe that if m^2 were negative this metric would not be positive definite and if $m^2 = 0$ it would be highly singular. This is connected to stability, unitarity and convexity of W. It is not difficult to verify that this metric (in either coordinate system) has scalar curvature $R = -\frac{1}{2}$. Before discussing the meaning of this let us see what happens in a more realistic field theoretic setting.

Consider a Gaussian model on a compact manifold $\mathcal{M}$ of volume L^d, where $d \le 4$ and

$$S[\phi, J, m^2, L, \Lambda] = \int_{\mathcal{M}}[\frac{1}{2}\phi(\Box + m^2)\phi + J\phi] \qquad (27)$$

J and m^2 can be position dependent, and in fact generically are on a curved $\mathcal{M}$. A coordinate transformation equivalent to above gives

$$S[\phi, \hat{\phi}, m^2, L, \Lambda] = \int_{\mathcal{M}}[\frac{1}{2}\phi(\Box + m^2)\phi - \phi(\Box + m^2)\hat{\phi}] \qquad (28)$$

with

$$W[\hat{\phi}, m^2, L, \Lambda] = -\frac{1}{2}\int_{\mathcal{M}}[\hat{\phi}(\Box + m^2)\hat{\phi}] + \frac{1}{2}Tr_\Lambda\ln[\frac{\Box + m^2}{\Lambda^2}] \qquad (29)$$

For simplicity we assume $\hat{\phi}$ and m^2 are constant on $\mathcal{M}$, consequently, treating Λ and L as constants, $\mathcal{G}$ is a 2D mini-superspace. Keeping Λ finite ensures we have no UV problems.

Examining the condition $dW = < dS >$ we obtain the equation

$$\int_{\mathcal{M}}[< (\Box + m^2)\phi > -(\Box + m^2)\hat{\phi}]d\hat{\phi} - \frac{1}{2}[\int_{\mathcal{M}} < (\phi - \hat{\phi})^2 > -Tr_\Lambda(\frac{1}{\Box + m^2})]dm^2 = 0$$

This expression is finite without a cutoff only for $d = 1$ where $Tr(\frac{1}{\Box + m^2})$ converges, and corresponds to the familiar situation of quantum mechanics. The metric induced on this 2D space is

$$ds^2 = \int_{\mathcal{M}}[m^2 d\hat{\phi}^2] + \frac{1}{2}Tr_\Lambda \frac{1}{(\Box + m^2)^2}(dm^2)^2 \tag{30}$$

This metric does not need a cutoff to be well-defined for $d < 4$, however for $d = 4$ $Tr(\frac{1}{\Box + m^2})^2$ is divergent and so our metric is not well-defined without Λ.

We can again look at the scalar curvature, which for the above metric is of the form $R = \frac{1}{4}Det^{-2}(g)\partial_{m^2}Det(g)$. Explicitly

$$R = -\frac{Tr_\Lambda(1 + \frac{\Box}{m^2})^{-3}}{\left(Tr_\Lambda(1 + \frac{\Box}{m^2})^{-2}\right)^2} + \frac{1}{2}\frac{1}{Tr_\Lambda(1 + \frac{\Box}{m^2})^{-2}} \tag{31}$$

For $d = 0$ this clearly reduces to the result for the Gaussian distribution. For $\mathcal{M} = (S^1)^d$, $d = 1, 2$ or 3, the cutoff can be taken to zero giving in Fourier space

$$R = -\frac{\sum_n(1 + (\frac{2\pi n}{mL})^2)^{-3}}{\left(\sum_n(1 + (\frac{2\pi n}{mL})^2)^{-2}\right)^2} + \frac{1}{2}\frac{1}{\sum_n(1 + (\frac{2\pi n}{mL})^2)^{-2}} \tag{32}$$

In the limit $mL \to 0$ the curvature reduces to the gaussian curvature $R = -\frac{1}{2}$ while in the limit $mL \to \infty$ it becomes

$$R = -\frac{1}{4}\frac{(2 - d)}{\Gamma(\frac{4-d}{2})}(\frac{m^2 L^2}{4\pi})^{-\frac{d}{2}} + \dots \tag{33}$$

This is a nice example of a crossover in the context of the geometry of $\mathcal{G}$. For $d = 2$, and 3 the corrections in (33) are exponentially small while for $d = 1$ they are power law. Interestingly for $2 < d < 4$ there is a crossover to positive curvature which requires R to pass through zero.

In a curved space setting for a conformally coupled free scalar field the formulae (30) and (31) for the metric and scalar curvature remain unchanged. We can consider a more general situation by including an additional coupling to the curvature, this will be associated with the mass term in a natural way. In the interacting case, treating λ

as constant, i.e. we look only at the curvature of $\mathcal{G}$ in the $\hat{\phi}$, m^2 plane, one would find that the curvature depended on the scaling variable Lm where L is the characteristic length scale of the geometry and m is now the dressed mass of equation (11) in this geometry. In the case of a totally finite geometry the RG is of interest as physically there is a maximum length scale in the problem. Hence the RG can only flow so far before it stops. We also note that if we take a finite temperature field theory [5] then $T = \frac{1}{L}$ and the considerations of section 3 undergo a corresponding translation, we therefore have a temperature dependent RG. In a real cosmological setting one can imagine including in a RG picture various other effects, such as curvature, to get a quite detailed picture of how the universe cooled from the big bang. Naturally one would also wish to generalize to the case of non-constant curvature where one needs to consider a position dependent RG. More discussion of these interesting matters in the context of cosmology and the early universe will be discussed elsewhere [5].

5.   Conclusions and Speculations

The main aim of this paper has been to try to stimulate thought along certain directions. There are certain problems that have remained intractable for many years now: the confinement problem in QCD and quantum gravity to name but two. We do not claim to have solutions to problems such as these. We do claim, however, that such theories exhibit certain key, common features, the chief one being that the DOF in the problem are radically different at different energy scales. We would also claim that if this metamorphosis could be understood then a quantitative understanding of the theory would probably follow.

The question of how systems behave under changes in scale is most naturally addressed using the field theoretic RG, a consequence of an exact symmetry. However, there are, as pointed out here, different, inequivalent representations of the RG. If one has a field theory parametrized by a set of parameters $P \equiv \{g^i\}$ corresponding to a point in $\mathcal{G}$ it might occur that different subsets of the parameters, relevant for describing the theory at different scales, are taken into one another by the RG flow on $\mathcal{G}$. If one's renormalization depends only on a subset of the parameters one is restricting one's flow to take place only in a subspace $\mathcal{T}$ of $\mathcal{G}$. The resultant RG, $RG_{\mathcal{T}}$, depends only on a subset K of the parameters and the RG flows take place only on $\mathcal{T}$. If any of the $P - K$ parameters are relevant in the RG sense then the true RG flows of the theory, $RG_{\mathcal{G}}$, thought of as true scale changes, will wish to flow off $\mathcal{T}$ into $\mathcal{G}$. However, the use of $RG_{\mathcal{T}}$ does not allow for such flows. Such a state of affairs would be shown up by the perturbative unreliability of the results based on $RG_{\mathcal{T}}$. If none of the parameters K are relevant then there should be no problem. However, one can only say what parameters are relevant when one knows the full fixed point structure of the theory! In principle it is obviously better to work with $RG_{\mathcal{G}}$. If a certain parameter was important then one has made sure that its effects are treated properly, and if it wasn't then that will come out of the analysis. There can be no danger, except for extra work, from keeping a parameter in, but there can be severe problems if it is left out. In the problems treated in this paper, although non-trivial, they were easy in the sense that the parameter space $\mathcal{G}$ was obvious. In the finite size case there were really 3 parameters m, λ and L. An L independent RG was equivalent to working on a 2 dimensional space which wasn't big enough to capture the physics. What about QCD, or gravity? After all, in QCD without fermions there appears to be only one parameter! There is another length scale in QCD, the confinement scale, however it is not manifest in the original Lagrangian, it comes out dynamically. This length scale is the analog of L. As we don't know how it really originates we arrive at a Catch 22 situation. Our suggestion in such cases would be the following: there are in most, if not all, of these type of problems important classical field configurations; instantons, monopoles, vortices etc. which are very important at one scale and not at another. What one should do is derive a RG which is explicitly dependent on such classical backgrounds just as we have shown in section 4 that one should have a RG that is explicitly dependent on one's background spacetime. This is contrary to the standard view which tries to make a clean split between the background (associated with IR effects) and renormalization of fluctuations (which are usually taken to be associated with UV effects). Although there may be scales where such an artificial split is sensible it will certainly be true that there will be scales where it manifestly is not. We hope it is clear from the above that when we talk about a parameter space it can be something quite complicated such as that of the standard model, a very pertinent example of crossover behaviour. We hope that we have convinced the reader that there is a lot to be said for developing a RG that can interpolate between different DOF.

We have considered here a class of problems that can be treated so as to yield perturbatively the full crossover behaviour. In section 4 we started to outline the most basic geometrical elements of a more general framework for treating crossovers. Our view was that a theory could be described by a set of moments of a probability distribution that was a function of a set of parameters. The idea was then to look at geometrical structures on $\mathcal{G}$ to see: i) whether some non-perturbative results could be gained in this way, and ii) whether through the geometry one could obtain a better, geometrical understanding of crossovers. It is obvious that in the more general setting we are at a very rudimentary stage indeed. We do believe however, that there are deep and important things to be learned from this approach.

The geometry we looked at in section 4 was ordinary Riemannian geometry based on a metric and a connection. There are many questions to be asked. For instance, is the connection we introduced the only relevant one? It would appear that symplectic and contact geometry also play an important role. There exists a symplectic form on the "phase" space composed of the $\{g^i\}$ and their Legendre transform conjugates which are expectation values of operators. There are also obvious connections with the trace anomaly that we will not go into here. From a more physical point of view one would imagine the intuition from lattice decimation could be extended to local decimations which would lead to a position dependent RG. In this setting the relevant geometry may be Weyl geometry. One may even speculate [6] on cosmological expansion as a form of natural decimation, where we are continuously decimating to scales larger and larger than the Planck scale, or equivalently we are following an RG flow further and further into the IR. One of the problems in GR is the origin of time. There is from the cosmological expansion of the universe a natural pinning of time to energy scale, is this an accident? It may be that the direction of time is due to gravity having an IR fixed point and that we are only observing its RG flow as time.

Acknowlegements: DOC is grateful to the Inst. theor. fys., Utrecht for travel support. CRS is grateful to FOM and DIAS for financial support. We would like to thank variously Rafael Sorkin, Gary Gibbons, Karel Kuchar and D. V. Shirkov for helpful conversations.

References

[1] Parisi G., *Statistical Physics Addison-Wesley, 1988.*

[2] O'Connor Denjoe, Stephens C. R., *Crossover Behaviour in Field Theory DIAS/Utrecht preprint (1992); Nucl. Phys.,* **B360** *(1991) 297-336.*

[3] O'Connor Denjoe, Stephens C. R., *Finite Size Scaling Functions to Two Loops. DIAS/Utrecht preprint (1992).*

[4] Amari S., *Differential-Geometric Methods in Statistics. Lecture Notes in Statistics Vol. 21, Springer Verlag (1985).*

[5] O'Connor Denjoe, Stephens C. R., Freire F., *Dimensional reduction and the non-triviality of* $\lambda\phi^4$ *theory at finite temperature. Utrecht/DIAS/Imperial preprint (1992); O'Connor Denjoe, Stephens C. R., The renormalization group in curved spacetime and finite size scaling. DIAS/Utrecht preprint (1992), to be published in Class. Quan. Grav. (1993).*

[6] Hu B. L., *Talk presented at Journees Relativistes, Amsterdam (1992).*

An Example of Indeterminacy in the Time-development of "Already Unified Field Theory": A Collision between Electromagnetic Plane Waves

R. Penrose[*][†]

Palmer Physical Laboratory

Princeton, New Jersey.

Abstract

An example is presented which points to a certain basic difficulty in the "already unified" approach to unified field theory. It is shown that one can construct a pair of solutions of the combined Einstein-Maxwell equations for which the two space-times are identical in the neighbourhood of an initial space-like hypersurface (and in fact they may also be identical at all earlier times), but the time-development of the equations leads to space-times which are essentially different in their futures. The construction of such examples requires the electromagnetic field to be null (or zero) in some regions. The example given here represents a collision between two gravitational-electromagnetic waves.

1. Introductory preamble

This paper was written in late 1959 or early 1960, while I was at Princeton University in the early part of my research career in general relativity. It was at a time when I knew Charlie Misner best, since he was also in Princeton then, and I learnt a great deal from him about issues of general relativity, such as the initial value problem etc. As far as I can recall, it was discussions with him, and also with John Wheeler, that led to the ideas described in this paper.

I had completed the paper, and gave it to John Wheeler for his comments. Unfortunately, unforseen circumstances intervened, and it was not until several months later

---

[*]NATO Science Visiting Fellow

[†]present address (1993): Mathematical Institute, Oxford, U.K.

that the paper resurfaced, at which time my own interests had moved elsewhere. The celebration of Charlie's 60th birthday seemed an ideal occasion on which to resurrect the paper, and I searched through old files in order to locate it. Fortunately I located almost all of it, but unfortunately page 3 of the typescript was nowhere to be found. I have done my best to reconstruct a substitute (positioned between the two dark horizontal lines in the article), although I am not really sure what equations (4) and (5) were. I do not think that this is of too great consequence, however. Apart from this insertion, and the completion of some references, I have left the original article completely unchanged.

As a final comment, in these introductory remarks, I should make reference to some work that appeared in two later papers[13,14]. Evidently, in 1959/1960, I was not aware that plane waves do not possess global Cauchy hypersurfaces[13], though I had some partial conception of the singularities that were likely to arise[14]. I hope that this paper, old though it is, is a worthy tribute to Charlie Misner's many original contributions to general relativity.

2.   Introduction

According to the reformulation of the combined Einstein-Maxwell theory put forward by Rainich[1] and by Misner and Wheeler[2] it is possible, under normal circumstances, to infer from the metric alone, all that is necessary for the physics of gravitation plus source-free electromagnetism. However, it will be shown here that in certain special cases the future space-time (*within* the domain of dependence) cannot be predicted from a knowledge of the metric in the neighbourhood of an initial space-like hypersurface alone, nor sometimes even from the whole past space-time metric. It follows, therefore, that in the "already unified" theory, no amount of allowable data on an initial hypersurface can be sufficient *in all cases* to determine the future uniquely. This situation is in sharp contrast with the analogous (Cauchy) problem for conventional general relativity, as the work of Foures-Bruhat[3] shows.

In order to present an example which has this indeterminacy property, it will be convenient first to discuss some rigorous solutions of th Einstein-Maxwell equations which are the curved-space analogues of the plane-wave solutions of Maxwell's equations.

3.   Plane Waves

A metric representing the general plane-polarized gravitational-electromagnetic plane wave has been given by Rosen[4]:

$$ds^2 = e^{\alpha}(dt^2 - dx^2) - e^{\gamma+\beta}dy^2 - e^{\gamma-\beta}dz^2 \tag{1}$$

where α, β, γ are functions of $t - x$. The condition

$$2\gamma'' + \gamma'^2 + \beta'^2 - 2\alpha'\beta' \leq 0 \qquad (2)$$

must also be imposed to ensure that the time-time component of the Ricci tensor is non-positive. With this condition satisfied, a Maxwell field can be constructed which satisfies, with the metric, the combined Einstein-Maxwell equations. The Maxwell field so constructed turns out to be necessarily null; that is, both its invariants vanish.

The necessary and sufficient condition for the Maxwell field to *vanish* for the metric (1) is that the equality sign in (2) should hold. The metric (1)

(reconstruction of essentials of missing page)

represents an entirely electromagnetic wave (i.e. vanishing Weyl tensor, cf. (7)) if

$$\beta'' + \beta'\gamma' - \beta'\alpha' = 0. \qquad (3)$$

We could, if desired, arrange that $\alpha = 0$ in (1) by rescaling the $t = x$ coordinate. However, α is retained here for later convenience.

There is also an alternative form[7,8] for this metric, namely

$$ds^2 = dudv + h_{ij}(u)x_i x_j du^2 - dx_i dx_i, \qquad (4)$$

where we require

$$h_{ii} \geq 0 \qquad (5)$$

to ensure that the time-time Ricci component is non-positive.

Bondi, Pirani and Robinson[9], following Robinson[7], showed (in the purely gravitational case) how the metric (1) could be used to describe a

---

non-singular "sandwich wave" which separates two regions of flat (Minkowskian) space. This situation is shown in Fig. 1. The space is exactly flat before the wave arrives and is again exactly flat after it has departed. The curved region is bounded by two (null) hyperplanes. Bondi, Pirani and Robinson considered only the purely gravitational case in their paper, but Rosen's metric (and also the metric (4)) allows the construction of similar sandwich waves where electromagnetic field is present in the curved region. A

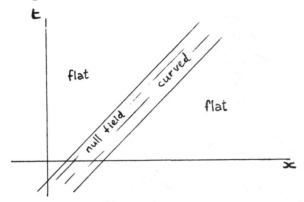

Fig. 1. Space-time picture of the sandwich wave.

simple way of constructing such an electromagnetic sandwich wave is to choose $\beta \equiv 0$, so that (3) is automatically satisfied, choose $\gamma \equiv 2(t-x)$ and $\alpha' \equiv \Omega+1$ where $\Omega \geq 0$ is a function of $t-x$ which is sufficiently smooth and which vanishes outside a certain region. Thus (2) is satisfied and satisfied with equality outside this region. Also (3) is satisfied everywhere so that outside this region the space-time is flat. The metric so chosen does not become singular, but the space-time so defined is not complete. To complete it, flat pieces with Minkowskian coordinates may be adjoined so as to overlap the flat regions of the metric (1) in the manner of Bondi, Pirani and Robinson. The transformations to Minkowskian coordinates take the form

$$t-x = \log(\tau - \xi), \ t+x = \frac{\tau^2 - \xi^2 - \eta^2 - \zeta^2}{a(\tau - \xi)}, \ y = \frac{\eta}{\tau - \xi}, \ z = \frac{\zeta}{\tau - \xi}, \quad (6)$$

where a is a suitable constant. The curved regions are bounded by null hyperplanes $(\tau - \xi = \text{constant})$ in the Minkowski spaces.

The tensor structure of such waves will be considered in Sec. IV.

4.   Collision between Sandwich Waves

Consider the following situation. Suppose two such electromagnetic sandwich waves A and B approach each other from infinity. Let S be a space-like (initial) hypersurface which intersects these waves before they collide and let A and B be the respective intersections of the waves A and B with S. (Fig. 2) It is clear that *whatever* initial value conditions are imposed on the *metric* in the neighbourhood of S, they will still be satisfied if two *independent* constant duality rotations[10] are applied to the electromagnetic fields in the neighbourhoods of A and B respectively. The space-time in the neighbourhood of S is unaffected by such duality rotations. However, at a later time, after the two waves have collided and are interfering, the space-time *will* be affected if *different* duality rotations of the electromagnetic field are applied to each of the two waves.

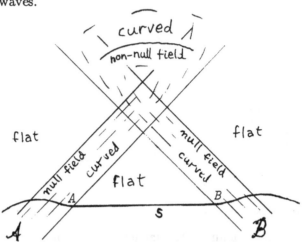

Fig. 2. Space-time picture of a collision between two sandwich waves A and B which approach each other from infinity. Before colliding the curved regions are bounded by flat characteristic hypersurfaces and the space-time between the waves is flat. When the waves recede again from each other, their non-linear interference causes the space-time between them to be curved and the chracteristic hypersurfaces which bound then on the outside also to be curved. The initial space-like hypersurface S may be chosen to be flat between the waves, but if so, it is necessarily non-flat outside the waves owing to the warping effect of the curvature in A and B.

One way of seeing this is to examine the Maxwell tensor in the region where the waves first overlap. Assuming the fields to be initially weak, the linear addition of the two fields will at first dominate any non-linear effects. But it can be easily seen that the sum of two differently duality rotated Maxwell tensors is not a duality rotation of the sum of the two original Maxwell tensors. Hence, if different duality rotations are applied to the two waves, the result must show up in the metric after they meet. (Only duality rotations do not affect the metric.) Also, in order to see that the space-time can be continued into the future, at least for a finite[11] time, it is sufficient to consider a space-like hypersurface passing through the (space-like 2-plane) region of first meeting of the waves. The Maxwell field and metric satisfies the required initial value conditions for this hypersurface and can therefore be continued for a finite time into the future (see Foures-Bruhat[3]).

Thus we have an example of two space-times which are identical in the neighbourhood of a certain space-like surface — and in fact they are also identical at all earlier times — but which are essentially different in their later time-developments. Both space-times admit Maxwell-type fields which satisfy, with the metric, the combined Einstein-Maxwell equations.

One might ask the question; how is this example compatible with the analysis given by Rainich and by Misner and Wheeler? All the necessary information about the Maxwell field ought to be contained in the metric, with no room for later indeterminacies. One answer to this question is that the Maxwell field in the whole initial region is *null* (i.e. both its invariants vanish). This is the case not treated by these authors. However, it seems to be possible to modify this example so that the *non-zero* Maxwell field is nowhere exactly null. This would seem to be the case if, instead of plane waves, radiation from two finite sources at a large distance apart had been used. The essential feature of such an example is that two regions of non-zero field are separated by a region where the Maxwell field vanishes. This region of zero field can be thought of as a special case of a null field, so it falls outside the scope of Rainich and Misner-Wheeler treatments.

It is of interest, however, to give an example which does *not* depend on two regions of non-zero Maxwell field being separated by a region of *zero* Maxwell field, but nevertheless indeterminacies arise in the future development of the space-time. Such an example (necessarily null in some parts of the space-time) can be given in terms of the plane wave collision described above, but it will be necessary first to examine the structure of the plane waves more closely. The plane waves exhibit a curious freedom with regard to duality rotations of the Maxwell field, which can only occur in the null case.

5. Tensor Structure of the Plane Waves

The Riemann tensor and Maxwell field tensor structure of the plane-waves can be inferred from some results given in Penrose[6]. The results are given there in a spinor form but they can easily be translated into a tensor form. At any point of a space-time in which the Einstein-Maxwell equations are satisfied, the Riemann tensor can be decomposed as follows[2]:

$$R_{\mu\nu\rho\sigma} = C_{\mu\nu\rho\sigma} + F_{\mu\nu}F_{\rho\sigma} + F^*_{\mu\nu}F^*_{\rho\sigma} \tag{7}$$

where $F_{\mu\nu}$ is the Maxwell field tensor defined in suitable "geometric" units (velocity of light = 1, gravitational constant = $1/(4\pi)$) and $F^*_{\mu\nu}$ its dual, defined by

$$F^*_{\mu\nu} = \frac{1}{2}(-g)^{\frac{1}{2}}F^{\alpha\beta}\epsilon_{\alpha\beta\mu\nu} \tag{8}$$

The tensor $C_{\mu\nu\rho\sigma}$ is Weyl's conformal tensor. It satisfies $C^\mu_{\nu\mu\sigma} = 0$ and can be thought of as describing the *purely gravitational* part of the curvature, the electromagnetic part having been subtracted out. Its dual $C^*_{\mu\nu\rho\sigma}$ is defined by

$$C^*_{\mu\nu\rho\sigma} = \frac{1}{2}(-g)^{\frac{1}{2}}C^{\alpha\beta}_{\mu\nu}\epsilon_{\alpha\beta\mu\nu} = \frac{1}{2}(-g)^{\frac{1}{2}}C^{\alpha\beta}_{\rho\sigma}\epsilon_{\alpha\beta\mu\nu} \tag{9}$$

the right and left duals being equal.

Consider the case of a plane-wave and choose, for convenience, a point O at which neithre the Maxwell field tensor $F_{\mu\nu}(O)$ nor the Weyl tensor $C_{\mu\nu\rho\sigma}(O)$ vanishes. Let X be any point of the manifold and let OX be a geodesic joining O to X. Let x^μ be a vector defined at O which is tangent to OX at O and whose length is the length fo OX (interpreted suitably if OX is null). Then the Maxwell field tensor and Weyl tensor at the point X are given (comparing the tensors at the two points by parallel transfer along OX) by

$$F_{\mu\nu}(X) = a\big(F_{\mu\nu}(O)\cos\theta - F^*_{\mu\nu}(O)\sin\theta\big) \tag{10}$$

$$C_{\mu\nu\rho\sigma}(X) = b\big(C_{\mu\nu\rho\sigma}(O)\cos\phi - C^*_{\mu\nu\rho\sigma}(O)\sin\phi\big), \tag{11}$$

a, b, θ, ϕ being real functions of $x^\lambda p_\lambda$ where p_λ is a null vector giving the direction of motion of the wave. The functions b determine the amplitudes and the functions θ, ϕ determine the planes of polarization of, respectively, the purely electromagnetic and the purely gravitational parts of the wave. Each of these four real functions may be chosen arbitrarily (apart from considerations, possibly, of continuity o differentiability) and these four functions determine the space–time[12]. Thus, to construct a sandwich wave, we merely choose for a and b, two functions which vanish outside a certain finite interval. If, as was the case with the particular sandwich wave constructed in II, the Weyl tensor vanishes everywhere we have $b \equiv 0$, so the point O need only be chosen where the Maxwell field does not vanish.

Only a, b and ϕ affect the geometry of the space-time. A change in the function $\theta(x^\lambda/p_\lambda)$ leads merely to a duality rotation of the Maxwell field at each point. This duality rotation is constant along each hypersurface $x^\lambda p_\lambda = $ constant, (these are null hyperplanes) but it is otherwise arbitrary. Thus a (continuous) duality rotation Θ may be performed which affects the trailing half of a sandwich wave only, the Maxwell field in the leading half being unaffected.

It may be mentioned here, in possing, that these duality rotations correspond to *actual* space rotations of the field tensor (rotation of plane of polarization). This is a characteristic feature of the null field and it shows why no scalar "complexion" can be defined for the null case. It may also be mentioned here that whereas the Maxwell field is null for these plane waves, so also, in a sense, is the Weyl tensor, that is (see Bondi *et al*[9], Penrose[6]) it is of Petrov type II with vanishing invariants.

6.   Indeterminacy with Non-zero Field

We are now in a position to consider again the space-time of Fig. 2 which represents a collision between plane waves. However, let us this time choose as the initial hypersurface a space-like hypersurface S' which intersects the two wave regions after they have collided but before the waves have completely overlapped. The situation is as shown in Fig. 3. Let A' and B' be the respective intersections of the waves $\mathcal{A}$ and

B with S' (A' and B' now overlap) and let C' be the intersection of a trailing portion C of the wave A with S'. Choose C so that, as in Section IV, a duality rotation Θ of the Maxwell field in A (prior to collision) may be performed which affects only the field in C and so does not affect the field in the rest of A. Make sure S' is chosen so that B' and C' are separated by a region of A'. Now the duality rotation Θ does not affect the space-time in the neighbourhood of S' but, as in the previous example, it *will* affect the later space-time when C and B have overlapped.

Thus, an example is provided with the proprety that the future space-time is not determined by the space-time in the neighbourhood of the initial hypersurface, but nevertheless no two regions of non-zero Maxwell field on this hypersurface are entirely separated by a region of zero Maxwell field. The essential feature of this example is that the regions B' and C' are separated by a region of *null* Maxwell field. It is also possible to arrange that the Maxwell field be *nowhere* zero on S', if desired, by letting the amplitudes of the waves tail off to infinity behind them.

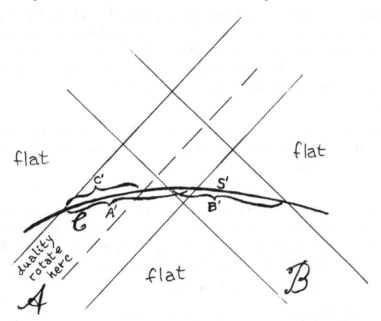

Fig. 3 Sandwich wave collision in more detail and with later initial hypersurface

7.   Acknowledgements

I am most grateful to Dr I. Robinson, Professor J.A. Wheeler and Dr W. Kundt for valuable discussions concerning various aspects of this paper. I am grateful also to K. P. Tod for some assistance in connection with the missing p.3.

[1] G.Y. Rainich, Trans. Am. Math. Soc. **27**, 106 (1925).

[2] C.W. Misner and J.A. Wheeler, Ann. Phys. **2**, 525 (1957).

[3] Y. Foures-Bruhat, J. RAtional Mech. Anal. **5**, 951 (1956).

[4] N. Rosen, Phys. Z. Sowjet. **12**, 366 (1937).

[5] H. Takeno, Tensor **8**, 59 (1958).

[6] R. Penrose, Ann. Phys. **10**, 171-201 (1960).

[7] I. Robinson, Report to the Royamont Conference (1959).

[8] I. Robinson, Report to the Eddington Group, Cambridge, 1956, H W Brinkman, Proc. Natl. Acad. Sci. (U.S.) **9** (1923), 1.

[9] H. Bondi, F.A.E. Pirani and I. Robinson, Proc. Roy. Soc. London **A251**, 519 (1959).

[10] A duality rotation (see Misner and Wheeler[2]) of the Maxwell field tensor $F_{\mu\nu}$ is given by $F_{\mu\nu} \to F_{\mu\nu} \cos\theta - F_{\mu\nu}^* \sin\theta$ where $F_{\mu\nu}^*$ is the dual of $F_{\mu\nu}$ (see eq. (8)). Thus, in particular, a duality rotation through 90° interchanges (apart from a sign) the electric and magnetic fields. The Maxwell stress energy tensor is invariant under any duality rotation (and under no other transformation) of $F_{\mu\nu}$.

[11] It will be shown in another paper, however, that after the collision the wave fronts necessarily become eventually singular. It seems likely that the space-time may eventually become singular also.

[12] In Penrose[6] these functions were taken to be analytic. However, this is inessential since any continuous function of a real variable can be smoothly approximated arbitrarily closely by an analytic function. Alternatively, C^∞ piecewise analytic functions may be used. Such functions have all the generality that is required here.

[13] R. Penrose, Revs. Mod. Phys. **37** 215-220 (1965)

[14] K.A. Khan and R. Penrose, Nature **229** 185-186 (1971)

Non-static Metrics of Hiscock-Gott Type

A.K. Raychaudhuri *

Abstract

Considering some non-static metric of Hiscock-Gott type, the standard formula for the angle deficit is recovered. However, the result may not hold in more general cases.

The gravitational field of cosmic gauge strings was first investigated by Vilenkin[1] who used the linearized field equations. This approach is clearly faulty inside the body and the immediate neighbourhood of the string. An exact solution for an infinitely long string of finite thickness and uniform density was obtained independently by Hiscock[2] and Gott[3]. Their solution ran as follows - inside the string the metric was

$$ds^2 = dt^2 - d\rho^2 - dz^2 - a^2 \, \sin^2\left(\frac{\rho}{a}\right) \, d\phi^2 \quad (\rho \leq \rho_0) \tag{1}$$

and outside,

$$ds^2 = dt^2 - dr^2 - dz^2 - b^2 r^2 d\phi^2 \quad (r \geq r_0) \tag{2}$$

With the metric (1), (i.e., inside the string)

$$T_0^0 = T_2^2 = \frac{1}{8\pi a^2}$$

$$T_1^1 = T_3^3 = 0$$

while outside $T_\nu^\mu = 0$. The continuity conditions for the two metrics at $\rho = \rho_0 (r = r_0)$ are satisfied if one sets

$$br_0 = a \, \sin\left(\frac{\rho_0}{a}\right)$$

and

$$b = \cos\left(\frac{\rho_0}{a}\right) \tag{3}$$

---

*Relativity and Cosmology Centre, Physics Department, Jadavpur University, Jadavpur, Calcutta 700 032, India

It is somewhat interesting to note that $\rho_0 \neq r_0$, the coordinates ρ and r are related by

$$r = \rho - \rho_0 + a \tan \left(\frac{\rho_0}{a} \right)$$

The metric (2) is reduced to the Minkowski form

$$ds^2 = dt^2 - dr^2 - dz^2 - r^2 d\,\bar{\phi}^2$$

if one sets $\bar{\phi} = b\phi$. Thus while ϕ runs from 0 to 2π. $\bar{\phi}$ goes from 0 to $2\pi b$ which in view of 3(b) is less than 2π. Thus one talks of an angle deficit of $2\pi(1 - b)$. Again one has for the mass per unit length of the string,

$$\mu = \int_0^{\rho_0} \int_0^{2\pi} T_0^0 \, a \, \sin \left(\frac{\rho}{a} \right) \, d\rho d\phi$$

$$= \frac{1}{4} \left[1 - \cos \left(\frac{\rho_0}{a} \right) \right] = \frac{1}{4} (1 - b)$$

The angle deficit is thus simply $\delta\phi = 8\pi\mu$.

However whether this simple and important result will remain valid in case one considers the string placed in a non-static universe seems not quite clear. Indeed, expanding universe solutions, in general, are not cylindrically symmetric and we have not investigated the general problem. We merely consider some non-static metrics which are of the Hiscock-Gott type and the outside space is still a vacuum.

For strings $(T_0^0 - \frac{1}{2}T)$ vanishes, hence from the Raychaudhuri equation, the expansion will be controlled merely by the shear, if the relevant vector be geodetic and hypersurface orthogonal. This makes it plausible that we may have an identical temporal behaviour in the string and the surrounding vacuum.

Again the Hiscock-Gott metric admits six Killing vectors[4], we take a simplifying assumption that the four Killing vectors pertaining to the space sections are still existent in the non-static case. We are thus led to the metrics

$$ds^2 = dt^2 - Bdz^2 - A \left(d\rho^2 + a^2 \, \sin^2(\frac{\rho}{a})d\phi^2 \right) \text{ for } \rho < \rho_0$$

$$= dt^2 - Bdz^2 - A \left(d\rho^2 + b^2 r^2 d\phi^2 \right) \text{ for } r > r_0$$

We get a string like object in vacuum with

$$T_2^2 = T_0^0 = \frac{1}{8\pi a^2 A}$$

and $A = t^{4/3}, B = t^{-2/3}$. (Another choice of A and B namely $A = $ const, $B = t^2$ also gives a string like field but this metric can be reduced to the static form by the transformation

$$\bar{t} = t \cosh z, \; \bar{z} = t \sinh z$$

The space sections of the vacuum metric ($t = \text{constant} = t_0$ say) again assume the Minkowski form

$$ds^2 = dt^2 - d\bar{z}^2 - d\bar{r}^2 - \bar{r}^2 d\bar{\phi}^2$$

with

$$\bar{z} = z t_0^{-1/3}, \; \bar{r} = r t_0^{2/3}, \; \bar{\phi} = b\phi$$

so that the angle deficit is again

$$2\pi \left(1 - \cos(\frac{\rho_0}{a})\right)$$

The mass per unit length in the present case is

$$\mu = \int_0^{t^{1/3}} \int_0^{\rho_0} \int_0^{2\pi} T_0^0 \, t_0 \, dz \, a \, \sin(\frac{\rho}{a}) d\rho \, d\phi = \frac{1}{4}\left(1 - \cos(\frac{\rho_0}{a})\right)$$

Thus while the relation $\delta\phi = 8\pi\mu$ remains frozen, the string starting from a cigar like singular form at $t = 0$, goes over to a pancake form as $t \to a$.

The metrics we have considered are admittedly of a very special form and it is hazardous to guess that the results are of general validity but it seems rather difficult to consider more general cases.

References

[1] A. Vilenkin, *Phys.Rev.D*, **23**, 852 (1981)

[2] W.A. Hiscock, *Phys.Rev.D*, **31**, 3288 (1985)

[3] G.R. Gott, *Astrophys.J*, **288**, 422 (1985)

[4] A.K. Raychaudhuri, *Phys.Rev.D*, **41**, 3041 (1990)

Non-Standard Phase Space Variables, Quantization, and Path Integrals, or Little Ado about Much

MICHAEL P. RYAN, JR.

Instituto de Ciencias Nucleares   Universidad Nacional Autónoma de México
A. Postal 70-546
México 04510 D.F.
MEXICO

SERGIO HOJMAN

Departamento de Física   Facultad de Ciencias
Universidad de Chile
Casilla 653
Santiago
CHILE

1. INTRODUCTION

In this article we want to describe a long-term research project that the authors plan to carry out over a period in the immediate future. We would like to outline the basic ideas of the project and give a few preliminary calculations the bear on the validity of our ideas as well as some speculations on where the research will lead. We chose this volume to do this because it occured to us that each of the components of our plan lies in a field that Charlie has touched on at some point in his varied career. These components are the Hamiltonian formulation of the gravitational field equations, path integral quantization, and quantum cosmology. Such a wide-ranging list shows how much we all owe to Charlie as a scientist and we hope that our efforts will demonstrate our debt to his lead in these fields.

The genesis of our paper lies in a long-term concern that both of us have had about the structure of quantum mechanics that has been the subject of years of blackboard discussions between us (the speculations of one of us [M.R.] go all the way back to fondly-remembered discussions with Charlie and other Charlie-students in Maryland). The focal point of our ideas has always been the cookbook nature of quantum mechanics as based on a Hamiltonian action functional for the classical equations of motion and canonical commutation relations derived from the Poisson bracket relations of the variables in the action. Perhaps the main among many objections to this procedure are that not all classical equations of motion are derivable from an action principle and the well-known fact that not all canonical phase-space variables (q_a, p_a) give a reasonable quantum theory when simple quantization procedures are applied to them. In the quantum mechanics of mechanical systems and some fields experiment has given enough guidance to be able to sidestep many of these difficulties, but in the quantization of the gravitational field there are no such experiments, and it is much more important to have a well-formulated quantum theory that is as free from ambiguities as possible.

One of us (S.H.) has spent some time studying the Lagrangian and Hamiltonian formulations of classical systems with an eye toward quantization, and especially the quantization of the gravitational field. The other has studied quantum cosmologies as models for quantum gravity. We have already combined these studies in a paper with Dario Núñez on the quantization of Bianchi-V cosmological models using non-standard Hamiltonian techniques[1]. The project we have in mind has several objectives and in this article we plan to give examples from quantum cosmology of several of the possible lines of investigation that we hope to pursue.

Basically, we hope to give a better foundation to the idea of quantization using non-standard Hamiltonian techniques by studying the possible meaning of using very general phase-space variables as a basis for quantization. In the current paper we will not use the Heisenberg picture (although this is not excluded as a future possiblity), but will instead confine ourselves to Schrödinger and path-integral methods. One of the problems with Schrödinger quantization in non-standard phase space variables is that some operators are realized as derivatives, and it is vital to consider the function space on which these derivatives operate. As Bargmann[2] showed in the context of phase-space variables related to the usual momentum and coordinate variables of the harmonic oscillator by a complex "canonical" transformation, this function space need not be a Hilbert space, but that one needs only demand that the space be capable of being mapped onto a Hilbert space in a meaningful way. While we are not in a position to develop a general mapping scheme here, we will mention possible forms that such a mapping might take.

With these goals in mind we will, in Sec. 2, sketch the problem of quantization using non-standard phase-space variables, with some emphasis on the difficulties of Schrödinger quantization. In Sec. 3 we will use a Bianchi type-I cosmological model as a model for quantum gravity and quantize it by means of BRST path integral methods using a simple set of non-standard phase space variables. Because of the speculative nature of the article we will not carry any of our discussions too far, but only attempt enough to give the broad outline of a picture we hope to fill in in the future.

2. PHASE SPACE VARIABLES IN QUANTUM MECHANICS.

The usual formulation of quantum mechanics begins with an action principle

$$S = \int_{t_0}^{t_1} L(q_a, \dot{q}_a, t)dt = \int_{t_0}^{t_1} [p_a \dot{q}_a - H(p_a, q_a)]dt, \tag{2.1}$$

which in principle gives the classical equations of motion when one varies S with respect to $(q_a, \dot{q}_a)$ or (p_a, q_a). The standard procedure is to convert the phase space variables (q_a, p_a) into operators $(\hat{q}_a, \hat{p}_a)$ and convert the Poisson bracket relations $\{q_a, q_b\} = 0 = \{p_a, p_b\}$, $\{q_a, p_b\} = \delta_{ab}$ into operator commutator relations $[\hat{q}_a, \hat{q}_b] = 0 = [\hat{p}_a, \hat{p}_b]$, $[\hat{q}_a, \hat{p}_b] = i\delta_{ab}$. One also takes any function $O(q_a, p_b)$ and converts it to an operator $\hat{O}(\hat{q}_a, \hat{p}_b)$ which has (modulo factor -ordering problems) the same functional form as O. The most improtant function for the quantum dynamics of the system is the Hamiltonian operator $\hat{H}(\hat{q}_a, \hat{p}_b)$. In this article we will concern ourselves mainly with the Schrödinger picture, where the momentum or coordinate variables are realized as derivative operators on a function space consisting of functions of eigenvalues of the other operators and the time t. Note that we say "function space" rather than Hilbert space; this is deliberate.

This procedure leads to a Schrödinger equation

$$\hat{H}\Psi(q_a, t) = i\frac{\partial \Psi}{\partial t}(q_a, t) \tag{2.2}$$

[or the equivalent for a momentum space wave function, $\Phi(p_a, t)$]. If the solutions to (2.2) form a Hilbert space, we can define a positive- definite probability density $\rho(q_a, t) = \Psi^* \Psi(q_a, t)$ on the space of solutions.

This simple-minded precis of quantum theory ignores a large number of well-known problems, some of which have at least tentative solutions, and we will assume the reader is familiar with these. The quantization of constrained systems, the quantum theory of relativistic systems, and factor ordering difficulties are among them. The problems we plan to discuss in most detail are basically difficulties associated with the process of Schrödinger quantization we outlined above. It is well known (and usually cheerfully ignored) that the process is only valid for equations of motion that admit an action principle of the form (2.1), and the complete procedure requires a Hamiltonian form of the action, and that the phase space formulation where one realizes operators as derivatives is only supposed to be valid for a restricted class of phase space coordinates.

One of the best known examples of the failure of straightforward Schrödinger quantization is the motion of a particle in a central potential. In Cartesian coordinates we have

$$S = \int [p_x \dot{x} + p_y \dot{y} + p_z \dot{z} - \{\frac{1}{2}(p_x^2 + p_y^2 + p_z^2) + V(x^2 + y^2 + z^2)\}]dt \qquad (2.3)$$

and if we realize $\hat{p}_x, \hat{p}_y, \hat{p}_z$ as $-i\partial_x, -i\partial_y, -i\partial_z$, then we arrive at a correct Schrödinger equation, $i\partial\Psi/\partial t = \hat{H}\Psi$. If we make the canonical transformation to spherical coordinates, $x, y, z \to r, \theta, \phi, \quad p_x, p_y, p_z \to p_r, p_\theta, p_\phi$, the action becomes

$$S = \int [p_r \dot{r} + p_\theta \dot{\theta} + p_\phi \dot{\phi} - \{\frac{1}{2}(p_r^2 + \frac{p_\theta^2}{r^2} + \frac{p_\phi^2}{r^2 \sin^2 \theta}) + V(r)\}]dt, \qquad (2.4)$$

and if we try to realize the momentum operators $\hat{p}_r, \hat{p}_\theta, \hat{p}_\phi$ as $-i\partial_r, -i\partial_\theta, -i\partial_\phi$, the resulting Schrödinger equation is "incorrect", at least given a naive interpretation of the resulting wave function $\Psi(r, \theta, \phi, t)$. Since it can be shown that any infinitesimal canonical transformation is equivalent to a unitary transformation acting on the corresponding operators, is is only transformations such as those to spherical coordinates that cannot be built up from infinitesimal transformations that will cause such problems.

In this article we would like to speculate on possible solutions to these problems. We have discussed problems that can be broken down into three main points: 1)How can one quantize a system described by any phase space variables? 2) How can one quantize systems whose equations of motion do not come from an action principle?; and 3)Can one quantize a system in the Schrödinger representation in non-standard phase-space variables? In order to make at least some steps toward answering these questions, we will begin with an outline of the symplectic approach to the canonical formulation of the equations of motion.

In principle one needs a first order formulation of the equations of motion of a system described by a set of variables x^a, $a = 1, \cdots, 2n$. It is usual to suppose that half of these variables are coordinates and half are velocities or momenta. The equations of motion are

$$\dot{x}^a = f^a(x^b). \qquad (2.5)$$

A canonical set of equations that reproduces these is

$$\{x^a, x^b\} = J^{ab}; \qquad (2.6a)$$

$$\dot{x}^a = \{x^a, H\} = J^{ab}\frac{\partial H}{\partial x^b} = f^a, \qquad (2.6b)$$

where for consistency of the equations of motion J^{ab} must obey the Jacobi identity,

$$J^{ab}{}_{,d}J^{dc} + J^{bc}{}_{,d}J^{da} + J^{ca}{}_{,d}J^{db} = 0. \qquad (2.7)$$

The symplectic structure matrix J^{ab} is usually assumed to have the form

$$J^{ab} = \begin{bmatrix} 0 & I_N \\ -I_N & 0 \end{bmatrix}, \tag{2.8}$$

where I_N is the $n \times n$ unit matrix, so that half of the x^a are momenta and the other half configuration variables. Of course, one must still be able to find a Hamiltonian H that gives (2.6b).

In an unpublished work reported by Dyson[3], Feynman recognized that the form of J^{ab} given by (2.8) is very restrictive and that relaxing this requirement leads to easier solution of certain problems, but dropped the idea because he felt it was unphysical. Recently Hojman and Shepley[4] have attempted to use Feynman's extension of symplectic theory to find new methods of quantization. They have shown that there exists a Hamiltonian H for any system of the form (2.6), and a J^{ab} can be constructed for such a system if one knows $2n$ constants of the motion C_i, ($2n-1$ of which do not depend explicitly on time), that is knows them explicitly as functions of the coordinates (a fairly strong requirement, equivalent to knowing the full classical solution). This J^{ab} may be constructed by summing elements of the basic form[5]

$$J_1^{ab} = \lambda(x^c)\varepsilon^{ab\mu_1\cdots\mu_{2n-2}}C_{1,\mu_1}\cdots C_{2n-2,\mu_{2n-2}}, \tag{2.9}$$

where $C_{,\mu} \equiv \frac{\partial C}{\partial x^\mu}$, $\varepsilon^{ab\mu_1\cdots\mu_{2n-2}}$ is the $2n$-index Levi-Civita symbol, and $\lambda(x^c)$ is a function of the phase space coordinates that will be explained below. This J^{ab} satisfies the Jacobi identity. The $C_1\cdots C_{2n-1}$ are time independent constants of the motion. The Hamiltonian is defined as $H = C_{2n-1}$, along with $C_{2n} = t + d_{2n}$, where d_{2n} is time independent. This can always be achieved by a change of coordinates. It is easy to realize that $\lambda(x^c)$ may always be chosen so that $J_1^{ab}\frac{\partial H}{\partial x^b} = f^a$. There is considerable freedom in this formalism in selecting the Hamiltonian H. The main advantage of this formulation is that it allows one to find Hamiltonians for systems that do not admit an action principle (an example that will be mentioned below is the set of Class B Bianchi models when the isometries are imposed directly in the Hilbert action). A disadvantage (which we feel may not be a true disadvantage) is that the Hamiltonian is not unique.

Perhaps the best way of characterizing this concept is to imagine that the form of J^{ab} given in (2.8) is the phase space equivalent of a "Minkowski metric", and that phase space "general coordinate transformations" as opposed to ordinary canonical transformations, the equivalent of Lorentz transformations, will naturally change the metric form while preserving the equations of motion (2.6). Notice that one can make very general phase space transformations, including transformations to sets of variables that are all constants of the motion, the equivalent of the procedure that leads to the Hamilton-Jacobi formulation of classical mechanics.

In the future we are planning to use these ideas to attempt to build up a quantum theory and to apply it to a list of problems. Perhaps one of the most interesting possibilities will be its application to quantum gravity. We have already made a first attempt, applying it to the quantization of Bianchi Type V models in quantum cosmology.

There are essentially three paths we could take. One would be the use of the Heisenberg picture, since the equations of motion (2.6) are written in a form that is particularly amenable to this representation. While we plan to investigate this line later, our interest in quantum gravity has, given the direction that field has taken lately, led us to consider first the two other major methods of quantization, both of which are a little more difficult in the context of Eqns. (2.6). One is Schrödinger quantization and the other is path integral quantization. We will discuss the difficulties of Schrödinger quantization below, while path integral quantization will be the subject of Sec. 3. The symplectic structure language that we use is reminiscent of that used in geometric quantization[6], although the aims and methods are different, and only canonical transformations ("Lorentz") transformations are allowed there. It is possible that there may be some natural points of contact between our ideas and that theory.

In principle, Schrödinger quantization is not difficult, at least mechanically. We must select half of the operators $\hat{x}^a$ as multiplication operators so the "wave function" will be a function of the eigenvalues of these operators and, in principle, time. The rest of the operators will be realized as derivative operators with respect to their "conjugates". We can then construct a Schrödinger equation

$$i\frac{\partial \Psi}{\partial t}(x^a, t) = \hat{H}(x^a, -i\partial_{x^a})\Psi(x^a, t). \tag{2.10}$$

There are two main problems here, one obvious and the other a bit more subtle. One is that when J^{ab} is no longer of the form (2.8) it becomes a tricky problem to decide which variables are conjugate to which. In the examples we have considered so far this problem can be handled, but we do not yet have a general procedure. The second problem is: To what function space do the solutions $\Psi(x^a, t)$ belong? There is no guarantee that they form a Hilbert space, and this is often considered a fatal defect. However, the original Bargmann formulation for the harmonic oscillator[2], that is the germ of the Ashtekar variables[7], addressed exactly this question. The complex "canonical" transformation $\bar{\Pi} \equiv \frac{1}{\sqrt{2}}(p + iq)$, $X \equiv \frac{1}{\sqrt{2}}(q + ip)$ leads to an action (total derivatives dropped)

$$S = \int [\bar{\Pi}\dot{X} + iX\bar{\Pi}]dt, \tag{2.11}$$

which can be quantized in the Schrödinger representation by realizing $\bar{\Pi}$ as $-i\partial/\partial X$.

This leads to a very simple Schrödinger equation of the form

$$\hat{H}\Psi(X,t) = -X\frac{\partial\Psi}{\partial X} = i\frac{\partial\Psi}{\partial t}, \tag{2.12}$$

first given by Fock[8]. What Bargmann showed was that the function space upon which $\partial/\partial X$ operates is not a Hilbert space, but that there exists a kernel $K(q,X)$ which maps the space of functions $F(X)$ of solutions to (2.12) onto the usual Hilbert space $H(q)$, that is

$$\psi(q,t) = \int K(q,X)\Psi(X,t)dX, \tag{2.13}$$

where dX means $dpdq$. This, of course, means that wave functions $\Psi(q_a, p_a, t)$ need not be wave functions in the usual sense. However, we have achieved simpler equations (note that [2.13] is much simpler than the usual harmonic-oscillator Schrödinger equation) at the cost of finding the kernel K that maps the new wave functions into useful wave functions that have the properties usually associated with them. In the case of the harmonic oscillator, finding K is relatively easy, but for more general phase space variables it may become quite difficult. We plan to investigate this in the future.

In a first attempt to apply some of our ideas to quantum gravity, the authors and Dario Núñez quantized Bianchi type V cosmological models as quantum cosmologies with one-sixth the logarithm of the determinant of the three-metric as an internal time[1]. The Einstein equations in this case do not result from the variation of the Hilbert action restricted by symmetry, but we were able to find a Hamiltonian H in the sense of Eqns. (2.6) and use it for Schrödinger quantization of the theory. We used a simple-minded mapping of the resulting wave function to one that could be used to construct a Hilbert space rather than attempt to construct a Bargmann-like kernel. We argued that given the confused state of the current interpretation of the wave function of the universe, that our solution has as much claim as any other to be a viable candidate for such a wave function.

We will not give the details here of the Bianchi V calculation, but will retreat to a simpler model, the Bianchi type I cosmologies. In the section that follows we will go on to the next part of the program outlined in the Introduction and apply BRST path-integral quantization to these models in non-standard phase space coordinates. We will again use a simple mapping of the wave function we obtain rather than attempt to develop here the more complicated integral mapping outlined above.

3. PATH INTEGRAL QUANTIZATION OF BIANCHI TYPE I MODELS.

Part of our plans for future research are aimed at path-integral quantization of gravity using non-standard phase space variables. In general this will be difficult to achieve since Eqs. (2.6) do not not necessarily admit an action which can be calculated for different histories. In the present article we will do a preliminary calculation in order to show how such a quantization might be expected to come out using a diagonal Bianchi type I cosmology where the results of standard path-integral quantization are known and the Hilbert action calculated for the model is a valid action for its equations of motion.

The form of the metric is[9]

$$ds^2 = -N^2 d\alpha^2 + e^{2\alpha} e_{ij}^{2\beta} dx^i dx^j, \tag{3.1}$$

where $\beta_{ij} = \text{diag}(\beta_+ + \sqrt{3}\beta_-, \beta_+ - \sqrt{3}\beta_-, -2\beta_+)$, and $\frac{1}{6}\ln g = \frac{1}{6}\ln[\det(g_{ij})] = \alpha$ is taken as an internal time. The action can be reduced to the form

$$I = \int [p_+\dot{\beta}_+ + p_-\dot{\beta}_- - p_\alpha - \tilde{N}(-p_\alpha^2 + p_+^2 + p_-^2)]d\alpha, \tag{3.2}$$

where $\cdot \equiv d/d\alpha$ and $\tilde{N}$ is a normalized lapse function such as that used by Berger and Voegli[10]. Variation of this action with respect to $p_\pm$, $\beta_\pm$, p_α and $\tilde{N}$ along with the equation $\dot{p}_\alpha = 0$ give the full set of equations for the model. Our non-standard variable set will be $p_\pm$, $\beta_\pm^{(0)}$, K, and $C = -p_\alpha^2 + p_+^2 + p_-^2$, where the $\beta_\pm^{(0)}$ are constants of motion that are the initial values of $\beta_\pm$, i.e. $\beta_\pm(\alpha = 0)$, $p_\pm$ are unchanged, C is the non-linear combination given above, and K is an extension of the phase space that we need to obtain the full equations of motion. In terms of these variables an action that gives the correct classical equations of motion is

$$I = \int [p_+\dot{\beta}_+^{(0)} + p_-\dot{\beta}_-^{(0)} + C\dot{K} - \tilde{N}C]d\alpha. \tag{3.3}$$

Varying I with respect to $p_\pm$, $\beta_\pm^{(0)}$, C, K and $\tilde{N}$ we find

$$\dot{p}_\pm = \dot{\beta}_\pm^{(0)} = 0, \tag{3.4a}$$

$$\dot{C} = \dot{K} = 0, \qquad C = 0. \tag{3.4b}$$

The classical map between these variables and $\beta_\pm$, p_α ($p_\pm$ being unchanged) gives the usual relations

$$p_\pm = p_\pm^{(0)} = \text{const.}, \qquad p_\alpha = \sqrt{p_+^{(0)2} + p_-^{(0)2}}, \qquad \beta_\pm = \beta_\pm^{(0)} + \frac{p_\pm^{(0)}\alpha}{\sqrt{p_+^{(0)2} + p_-^{(0)2}}}. \tag{3.5}$$

The BRST path integral quantization of this system is relatively easy to carry out. What we would like to do is use a different coordinate system on the space of paths than that generated by skeletonization. That is, we would like to expand the possible paths in Fourier series, something that is rarely done except for the harmonic oscillator, but, of course, is possible for any continuous path.

For BRST quantization we have to extend the phase space to a set of both normal and anticommuting (ghost) variables. As is usual in phase space path-integral quantization, one must treat "momentum" and "coordinate" variables differently, and the decision about how to do this is somewhat of an art. In the Fourier-series formulation one way to handle this is to decide on the type of Fourier series to be used for each variable. We will not discuss this problem in detail here, but it will become obvious in the type of series we choose. The phase space is extended to include Π, a momentum conjugate to $\tilde{N}$ and ρ, $\bar{\rho}$, c, $\bar{c}$, four anticommuting functions of α, which is all we need since there is only one constraint, $C = 0$ and one gauge fixing. We will take the proper time gauge, so the gauge function $\chi(p_\pm, \beta_\pm^{(0)}, K, C, \tilde{N}, \alpha) \equiv \dot{\tilde{N}}$ is zero, so the gauge-fixing potential Φ is simply $\bar{\rho}\tilde{N}$, while the BRST charge Ω has the form $\Omega = cC + \rho\Pi$ [11]. With all of these choices the Batalin-Fradkin-Vilkovisky[12] action for the system described by (3.3),

$$I_B = \int [p_+\dot{\beta}_+^{(0)} + p_-\dot{\beta}_-^{(0)} + C\dot{K} - \tilde{N}C + \bar{\rho}\dot{c} + \bar{c}\dot{\rho} + \Pi\dot{\tilde{N}} - \{\Phi,\Omega\}]d\alpha, \qquad (3.6)$$

becomes

$$I_B = \int [p_+\dot{\beta}_+^{(0)} + p_-\dot{\beta}_-^{(0)} + C\dot{K} - \tilde{N}C + \bar{\rho}\dot{c} + \bar{c}\dot{\rho} + \Pi\dot{\tilde{N}} - \bar{\rho}\rho]d\alpha. \qquad (3.7)$$

The Fourier series we need for each of the variables which describe paths between some value of the variables at $\alpha = 0$ and other values at $\alpha = T$ are

$$\Pi = \sum_{n=1}^{\infty} \Pi_n \sin\left(\frac{n\pi\alpha}{T}\right); \qquad \tilde{N} = N_0 + \sum_{n=1}^{\infty} N_n \cos\left(\frac{n\pi\alpha}{T}\right); \qquad (3.8a)$$

$$c = \sum_{n=1}^{\infty} c_n \sin\left(\frac{n\pi\alpha}{T}\right); \qquad \rho = \rho_0 + \sum_{n=1}^{\infty} \rho_n \cos\left(\frac{n\pi\alpha}{T}\right); \qquad (3.8b)$$

$$\bar{c} = \sum_{n=1}^{\infty} \bar{c}_n \sin\left(\frac{n\pi\alpha}{T}\right); \qquad \bar{\rho} = \bar{\rho}_0 + \sum_{n=1}^{\infty} \bar{\rho}_n \cos\left(\frac{n\pi\alpha}{T}\right); \qquad (3.8c)$$

$$\beta_\pm^{(0)} = \beta_{\pm c}^{(0)}(\alpha) + \sum_{n=1}^{\infty} \beta_\pm^{(0)n} \sin\left(\frac{n\pi\alpha}{T}\right); \qquad K = K_c(\alpha) + \sum_{n=1}^{\infty} K^{(n)} \sin\left(\frac{n\pi\alpha}{T}\right); \qquad (3.8d)$$

$$p_\pm = p_\pm^{(0)} + \sum_{n=1}^{\infty} p_\pm^{(n)} \cos\left(\frac{n\pi\alpha}{T}\right); \qquad C = C^{(0)} + \sum_{n=1}^{\infty} C^{(n)} \cos\left(\frac{n\pi\alpha}{T}\right), \qquad (3.8e)$$

where the Fourier coefficients of the anticommuting variables are anticommuting numbers, and, in principle functions with subscript c are the classical solutions for those variables. Since the variables that appear in (3.8d) have classical solutions that are constant, $\beta_{\pm c}^{(0)}$ and K_c would be constants. These classical solutions give unphysical results, so we will use a linear solution for each of them connecting $\beta_{\pm 0}^{(0)}$, K_0 at $\alpha = 0$ with $\beta_{\pm 1}^{(0)}$, K_1 at $\alpha = T$, or

$$\beta_{\pm c}^{(0)} = \frac{1}{T}(\beta_{\pm 1}^{(0)} - \beta_{\pm 0}^{(0)})\alpha + \beta_{\pm 0}^{(0)}; \qquad K_c = \frac{1}{T}(K_1 - K_0)\alpha + K_0. \qquad (3.9)$$

The ghost part of the action $I_{gh} = \int_0^T [\bar{p}\dot{c} + \bar{c}\dot{p} - \bar{\rho}\rho]d\alpha$ becomes

$$I_{gh} = \sum_{n=1}^{\infty} [\bar{\rho}_n c_n \frac{n\pi}{2} - \bar{c}_n \rho_n \frac{n\pi}{2} - \bar{\rho}_n \rho_n T/2] - \bar{\rho}_0 \rho_0 T. \qquad (3.10)$$

The ghost part of the propagator $\int \prod_n dc_n \prod_n d\bar{c}_n \prod_n d\rho_n \prod_n d\bar{\rho}_n d\rho_0 d\bar{\rho}_0 e^{iI_{gh}}$ can be shown, using Berezin integration (with the normalization $\int \theta d\theta = 1$)[13], to be $T[-i\prod_n(\frac{n^2\pi^2}{4})]$. The rest of the action, I_A is

$$I_A = \int_0^T [\{p_+^{(0)} + \sum_{n=1}^{\infty} \cos\left(\frac{n\pi\alpha}{T}\right)\}\{(\frac{1}{T})(\beta_{+1}^{(0)} - \beta_{+0}^{(0)}) + \sum_{n=1}^{\infty} \frac{n\pi}{T}\beta_+^{(n)} \cos\left(\frac{n\pi\alpha}{T}\right)\}+$$

$$+\{p_-^{(0)} + \sum_{n=1}^{\infty} p_-^{(n)} \cos\left(\frac{n\pi\alpha}{T}\right)\}\{(\frac{1}{T})(\beta_{-1}^{(0)} - \beta_{-0}^{(0)}) + \sum_{n=1}^{\infty} \frac{n\pi}{T}\beta_-^{(n)} \cos\left(\frac{n\pi\alpha}{T}\right)\}+$$

$$+\{C^{(0)} + \sum_{n=1}^{\infty} C^{(n)} \cos\left(\frac{n\pi\alpha}{T}\right)\}\{(\frac{1}{T})(K_1 - K_0) + \sum_{n=1}^{\infty} \frac{n\pi}{T}K^{(n)} \cos\left(\frac{n\pi\alpha}{T}\right)\}-$$

$$-\sum_{n=1}^{\infty} \Pi_n \sin\left(\frac{n\pi\alpha}{T}\right) \sum_{m=1}^{\infty} N_m \frac{m\pi}{T} \sin\left(\frac{m\pi\alpha}{T}\right) -$$

$$-\{N_0 + \sum_{n=1}^{\infty} N_n \cos\left(\frac{n\pi\alpha}{T}\right)\}\{C^{(0)} + \sum_{n=1}^{\infty} C^{(n)} \cos\left(\frac{n\pi\alpha}{T}\right)\}]d\alpha$$

$$= p_+^{(0)}(\beta_{+1}^{(0)} - \beta_{+0}^{(0)}) + \sum p_+^{(n)}\beta_+^{(n)} \frac{n\pi}{2} + p_-^{(0)}(\beta_{-1}^{(0)} - \beta_{-0}^{(0)})+$$

$$+\sum_{n=1}^{\infty} p_-^{(n)}\beta_-^{(n)} \frac{n\pi}{2} + C^{(0)}(K_1 - K_0) + \sum_{n=1}^{\infty} C^{(n)}K^{(n)} \frac{n\pi}{2}+$$

$$+\sum_{n=1}^{\infty} \Pi_n N_n \frac{n\pi}{2} - N^{(0)}C^{(0)}T + \sum_{n=1}^{\infty} N_n C^{(n)} \frac{n\pi}{2}. \qquad (3.11)$$

Integrating e^{iI_A} over Π_n (we will always use the Liouville measure, e.g. for Π_n, $d\Pi_n/2\pi$) from $-\infty$ to $+\infty$ gives an infinite product of delta functions of the form $\delta(\frac{n\pi}{2}N_n)$ which when integrated over the N_n removes the last term of (3.11) at the cost of a term $\prod_n(1/n\pi^2)$ as an overall factor. Similar integrations over $C^{(n)}$, $p_\pm^{(n)}$ and subsequently $K^{(n)}$ and $\beta_\pm^{(n)}$ remove all of the summations at the cost of three more $\prod_n(1/n\pi^2)$ factors. The integration over $N^{(0)}$ gives $\delta(C^{(0)}T) = (1/T)\delta(C^{(0)})$ which removes the factor of T in the ghost integration and also gets rid of the $C^{(0)}(K_1 - K_0)$ term. The final form of the propagator is

$$< \beta_{\pm1}^{(0)}, K_1, T \mid \beta_{\pm0}^{(0)}, K_0, 0 > = -i\prod_n \left(\frac{n^2\pi^2}{4}\right)\left\{\left(\prod_\ell \frac{1}{\ell\pi^2}\right)^4\right\} \times$$

$$\times \int_{-\infty}^{\infty} \frac{dp_+^{(0)}}{2\pi} \frac{dp_-^{(0)}}{2\pi} e^{ip_+^{(0)}(\beta_{+1}^{(0)} - \beta_{+0}^{(0)})} e^{ip_-^{(0)}(\beta_{-1}^{(0)} - \beta_{-0}^{(0)})}$$

$$= N\delta(\beta_{+1}^{(0)} - \beta_{+0}^{(0)})\delta(\beta_{-1}^{(0)} - \beta_{-0}^{(0)}). \tag{3.12}$$

The constant normalization N can be treated as one over the Jacobian of the transformation from skeletonization coordinates to Fourier-series coordinates on the space of paths as Feynman does for the case of the harmonic oscillator.

Notice that the Fourier series coordinates are simpler to use than skeletonization for linear equations of motion because the coefficients are nicely grouped to give δ-functions that kill terms on further integration. Unfortunately, this simplicity disappears for more complicated motions unless one can calculate explicitly the Fourier series for the motion and integrate easily over the resulting coefficients.

The propagator above can be compared to the Green function for the Schrödinger quantization of the system, and they agree, since the solution for $\Psi(\beta_\pm^{(0)}, K, \alpha) = \Psi(\beta_\pm^{(0)})$, that is, an arbitrary function of $\beta_\pm^{(0)}$ that is independent of time. This wave function is perhaps the ultimate in frozen dynamics, but, of course, this fact means little until one knows how to map such a wave function into a true Hilbert space function. In principle one must develop a kernel such as those mentioned in Sec. 2 to map $\Psi(\beta_\pm^{(0)})$ to $\psi(\beta_\pm, \alpha)$, but we will leave this to future work. Here we will appeal to an argument similar that used in our Bianchi-V work, where we argued that the $\partial/\partial t$ that appears in the Schrödinger equation is a partial derivative that implies holding certain variables constant[1]. However, since the variables $\beta_\pm^{(0)}$ depend on α implicitly, the function Ψ can depend on α through this dependence. That is, since $\beta_\pm^{(0)} = \beta_\pm - (p_\pm^{(0)}/\sqrt{p_+^{(0)2} + p_-^{(0)2}})\alpha$, Ψ can be written as

$$\Psi = \Psi\left(\beta_\pm - \frac{p_\pm^{(0)}\alpha}{\sqrt{p_+^{(0)2} + p_-^{(0)2}}}\right). \tag{3.13}$$

In fact, the product of eigenstates of $p_\pm$ with eigenvalue $p_\pm^{(0)}$ becomes

$$\Psi = e^{i(p_+^{(0)}\beta_+ + p_-^{(0)}\beta_- - \sqrt{p_+^{(0)2} + p_-^{(0)2}}\,\alpha)}, \tag{3.14}$$

which is the solution found by Charlie in his original study of quantum cosmology[9].

As we mentioned in the Introduction, we have not attempted to carry this path integral formulation too far. We have only tried to give the flavor of the quantization of the gravitational field using non-standard phase space variables by presenting this simple model. Path integral quantization in non-standard phase space variables still faces many problems, especially for systems that do not come from a variational principle, where even the definition of the process will require new ideas, and it may even be impossible to achieve a consistent theory.

REFERENCES

1. S. Hojman, D. Núñez, and M. Ryan, *Phys. Rev. D* **45**, 3523 (1992).
2. V. Bargmann, *Comm. Pure and App. Math.* **14**, 2960 (1961).
3. F. Dyson, *Am. J. Phys.* **58**, 209 (1990).
4. S. Hojman and L. Shepley, *J. Math. Phys.* **32**, 142 (1991)
5. S. Hojman, in press
6. See, for example, D. Simms and N. Woodhouse, *Lectures on Geometric Quantization* (Springer, Berlin, 1977);
   N. Woodhouse, *Geometric Quantization* (Oxford U. P., Oxford, 1981).
7. A. Ashtekar *et al.*, *New Perspectives in Canonical Gravity* (Bibliopolis, Naples, 1988).
8. V. Fock, *Z. Phys.* **49**, 339 (1928).
9. C. Misner, *Phys. Rev.* **186**, 1319 (1969).
10. B. Berger and C. Voegli, *Phys. Rev. D* **32**, 2477 (1985).
11. See J. Guven and M. Ryan, *Phys. Rev. D* **45**, 3559 (1992), and references therein.
12. L. Fradkin and G. Vilkovisky, CERN Report TH 2332 (1977);
    I. Batalin and G. Vilkovisky, *Phys. Lett.* **69B**, 309 (1977).
13. F. Berezin, *The Method of Second Quantization* (Academic Press, New York, 1966).

The Present Status of the Decaying Neutrino Theory

D.W Sciama [*]

Abstract

The present status of the decaying neutrino theory is reviewed. Three recent developments are highlighted:

(a) the proposal that the dark matter in rich clusters of galaxies is mainly baryonic

(b) the implications of the hypothesis that decay photons are the solution of the C^o/CO ratio problem

(c) the strong supporting evidence from recent observations with the Hubble Space Telescope.

1.   Introduction

It gives me great pleasure to contribute an article in honour of Charles Misner. I still have vivid memories of his great impact on us during his visit to Cambridge in 1966-7. His combination of physical insight, mathematical power and forceful originality took us all by storm. That was the year the mixmaster universe was born. It was also the year of neutrino viscosity in the early universe (Misner 1968). In memory of that exciting time I write here about some (possible) recent uses of neutrinos in cosmology and in astrophysics.

2.   The Decaying Neutrino Theory

If neutrinos have non-zero rest mass one would expect a more massive neutrino type ν_1 to decay into a photon and a less massive neutrino type ν_2:

$$\nu_1 \rightarrow \gamma + \nu_2$$

[*]SISSA and ICTP, Trieste and Department of Physics, Oxford. Institute of Astronomy, Cambridge

Conservation of energy and momentum in the decay tells us that the energy E_γ of the photon in the rest frame of ν_1 is given by

$$E_\gamma = \frac{1}{2} m_1 (1 - \frac{m_2^2}{m_1^2})$$

It is very likely that $\frac{m_2^2}{m_1^2} \ll 1$. In that case we would have the simple result

$$E_\gamma = \frac{1}{2} m_1.$$

Cowsik (1977) pointed out that even if the lifetime for this decay is much longer than the age of the universe, the photon flux from the cosmological distribution of neutrinos pair created in the hot big bang might be appreciable, simply because of the great length of the line of sight through the universe. This point has led to a considerable literature which I will not go into here. As I write this article I am completing a book which gives a detailed discussion of this question and is entitled Modern Cosmology and the Dark Matter Problem.

The point of departure for this article was my suggestion (Sciama 1990a) that decay photons from dark matter neutrinos in our galaxy might be responsible for the otherwise puzzling widespread ionisation of hydrogen in the intersteller medium. The key point here is that the ionisation potential of hydrogen is 13.6 eV, so we require

$$E_\gamma > 13.6 \mathrm{eV}$$

and therefore

$$m_1 > 27.2 \mathrm{eV}.$$

This gives us the first of the many coincidences in this subject because (a) the Tremaine-Gunn phase space constraint for neutrinos to be bound in the galaxy leads to essentially the same lower limit (b) the cosmological density ρ_{ν_1} of ν_1 neutrinos is such that

$$m_1 = 93 \Omega_{\nu_1} h^2 \ \mathrm{eV}$$

where as usual $\Omega_{\nu_1} = \rho_{\nu_1}/\rho_{crit}$ and h is the Hubble parameter given by

$$H_0 = 100h \ \mathrm{km./sec}/Mpc, \quad (\frac{1}{2} < h < 1)$$

If $\Omega_{\nu_1} \sim 1$ and $h \sim \frac{1}{2}$ (which would be compatible with the observed age of the universe) we would have

$$m_1 \sim 23 \ \mathrm{eV}$$

Thus if h is only slightly greater than $\frac{1}{2}$ we would again have

$$m_1 > 27.2 \ \mathrm{eV},$$

and the decay photon would be able to ionize hydrogen.

Encouraged by these coincidences I evaluated the decay lifetime τ needed to account for the observed electron density in the interstellar medium ($n_e \sim 0.03 cm^{-3}$)and obtained

$$\tau \sim 2 \times 10^{23} \text{ secs.}$$

This choice of lifetime leads to another coincidence. Let us evaluate the intergalactic ionising flux F of decay photons from the cosmological distribution of neutrinos of density n_{ν_1}. A naive estimate would be

$$F \sim \frac{n_{\nu_1}}{\tau} \frac{c}{H_0}.$$

With $n_{\nu_1} \sim 100 \ cm^{-3}$, $\tau \sim 2 \times 10^{23}$ secs and $c/H_0 \sim 2 \times 10^{28}$ cm we would obtain

$$F \sim 10^7 \text{ photons cm}^{-2}\text{sec}^{-1}.$$

At first sight this result is a disaster because there is an observational upper limit on F, namely

$$F <\sim 10^6 \text{ cm}^{-2} \text{ sec}^{-1}.$$

We can try to solve this problem by using the fact that the decay photons will be red shifted as they propagate, and once their energy is reduced below 13.6 eV they will no longer contribute to the ionising flux. This effect is simple to include. If one writes

$$E_\gamma = 13.6 + \epsilon \ eV,$$

one has

$$F = \frac{n_{\nu_1}}{\tau} \frac{c}{H_0} \frac{\epsilon}{13.6}$$

We can therefore solve our problem if

$$\epsilon <\sim 1,$$

so that

$$E_\gamma <\sim 14.6 \ eV.$$

Hence

$$E_\gamma = 14.1 \pm 0.5 \ eV,$$

and

$$m_1 = 28.2 \pm 1 \ eV.$$

This gives us a new coincidence since this value of m_1 leads to $\Omega_\nu \sim 1$ for a low value of h. Also the resulting value of F would account for the otherwise puzzling ionisation of the intergalactic medium and Lyman α clouds.

We can do even better than this if we use recent observational data to argue that the decay photons must also be able to ionise nitrogen (Sciama 1992). This requires that

$$E_\gamma > 14.53 \text{ eV}.$$

Remarkably, this is just compatible with our upper limit and would lead to

$$E_\gamma = 14.565 \pm 0.035 \text{ eV},$$

and

$$m_1 = 29.13 \pm 0.07 eV.$$

Thus the mass of the decaying neutrino $(?\nu_\tau)$ would be pinned down to better than 1%. Also if we assume that $\Omega = 1$ when the baryon density is included then we obtain (Sciama 1990b)

$$H_0 = 56 \pm 0.5 \text{km/sec} Mpc$$

so that the Hubble constant would be pinned down to 1%.

With this background we are now ready to consider three recent developments in these ideas.

3. The Nature of the Dark Matter in Clusters of Galaxies

This decaying neutrino theory is widely regarded as having been disproved by the failure of Davidsen et al (1991) to observe a decay line from dark matter neutrinos in the rich cluster of galaxies A665. However, since then Hughes and Tanaka (1992) have published detailed observations of the x-ray emission from the hot gas in this cluster and have derived model distributions for the density of the galaxies, the gas and the dark matter. From this one can show (Sciama, Persic and Salucci 1992, 1993) that the dark matter is more concentrated to the centre of the cluster than are either the galaxies or the gas. A similar concentration is found in other clusters both from the x-ray data and the gravitational lens data. I find it rather pleasing that general relativity can be used to help demonstrate this excess concentration of the dark matter.

It seems unlikely that neutrinos would be more concentrated than baryons in clusters, since the neutrinos are non-dissipative (unlike in Misner's discussion of the early universe). It is more likely that most of the dark matter in this case is baryonic. One possibility is that a cooling flow in the hot gas ends up as faint low mass stars (Thomnas and Fabian 1990). The absence of an observable decay line would then be explained.

By contrast the dark matter in our galaxy has a more extended distribution than the visible matter. This would fit in well with the hypothesis that this dark matter is in the form of neutrinos, but does not demand it. In any case I believe that these considerations make clear that the decaying neutrino theory is still viable.

4. Ultra Violet Radiation in Dense Molecular Clouds

It has been known for a long time that the abundance ratio of atomic carbon to carbon monoxide in the interiors of dense molecular clouds in the galaxy is typically 10^5 times greater than would be expected from simple equilibrium models. This is referred to as the C^o/CO ratio problem. In these models the flux of ultraviolet radiation in the interiors of the clouds is very low because of their great opacity, and so the dissociation rate of CO is also very low. One class of proposed solutions to this problem involves mechanisms for increasing the ultra violet flux in the interiors. These proposals have recently been reviewed by Sorrell (1992). They involve complicated mechanisms which are difficult to assess.

By contrast, decaying neutrinos situated in the clouds would provide a straightforward mechanism, as was pointed out by Tarafdar (1991). This proposal has recently been updated by Sciama (1993a). A crucial role in Tarafdar's analysis is played by the opacity of the clouds to the decay photons, since this controls the resulting flux of these photons. The clouds typically have particle densities in the range $10^3 - 10^5$ cm^{-3}, and the vast majority of these particles is in the form of H_2 molecules. The flux of decay photons would only be sufficient to solve the C^o/CO ratio problem if these photons are unable to dissociate the H_2 molecules, so that the opacity of the clouds would be relatively small. Here enters a new coincidence. The photodissociation continuum of H_2 has a threshold at 14.68 eV. Hence we require that

$$E_\gamma < 14.68eV.$$

This constraint is essentially the same as our previous one, which followed from considering the integalactic ionising flux ($E_\gamma <\sim 14.6eV$). The new constraint is more stringent in the sense that the threshold involved has been accurately measured, whereas the previous constraint involved more uncertain quantities (the decay lifetime and the observational limit on the intergalactic flux). If our nitrogen and C^o/CO assumptions are both correct we would then have the essentially exact constraints

$$14.53 < E_\gamma < 14.68eV,$$

and (if $m_2 << m_1$)

$$29.06 < m_1 < 29.36eV.$$

The consequences for the value of the Hubble constant would then be essentially the same as before.

5. Recent HST Observations of the Electron Density in the Interstellar Medium

Strong support for the neutrino decay theory can be derived (Sciama 1993b) from the analysis by Spitzer and Fitzpatrick (1993) of their observations of a galactic halostar HD93521 using the Hubble Space Telescope. They observed the ultraviolet absorption spectrum of several regions along the line of sight to this star. One of the absorbing species was singly ionised carbon in an excited state, CII*. The excitation is probably due to electron collisions and so leads to the value of the electron density n_e in each absorbing region.

In their analysis Spitzer and Fitzpatrick also used the 21 cm emission observations of Danly et al (1992) which showed that the atomic hydrogen column density makes each region opaque to Lyman continuum photons. They concluded that the ionised and neutral gas are intimately mixed up together, in constrast to the standard view that these two components of the interstellar medium are located in separate regions (the warm ionised medium or WIM and the warm neutral medium or WNM).

If this is correct we face the same type of problem as we did in the previous section with the molecular clouds. In this case we are dealing with an excess of hydrogen-ionising photons in an opaque region rather than an excess of CO dissociating photons. Again decaying neutrinos come to the rescue. In fact it was related opacity problems which led to the decaying neutrino theory in the first place, but we have here the first clear observational evidence that the free electrons (which were previously known to exist from $H\alpha$ and pulsar dispersion data) are located in opaque regions.

Further support for the theory comes from the values of n_e derived by Spitzer and Fitzpatrick. They found for the slowly moving absorbing regions (which would be unaffected by shock waves) that n_e is the same in each region to their observing precision of 10%. This is just what would be expected for opaque regions in the decaying neutrino theory, and it was predicted in the original paper (Sciama 1990a). The reason is simple. Since every decay photon produces an ionisation locally, to be followed by a recombination, one has the ionisation equilibrium equation

$$\frac{n_\nu}{\tau} = \alpha n_e^2,$$

where α is the appropriate recombination coefficient. In limited regions of the galaxy n_ν will be nearly constant, and n_e will depend on the gas temperature T (via $\alpha(T)$) only as $T^{0.37}$. Moreover in warm regions, such as those observed by Spitzer and Fitzpatrick, the temperature is fairly constant. Hence n_e should be nearly constant, as is observed.

Moreover, if one assumes that carbon is undepleted in these regions (for which there is some evidence (Sciams 1993b), one finds for the four slowly moving regions n_e = 0.055 cm^{-3}. This is just the same as the values found for three other regions from an analysis of the pulsar dispersion measures of three pulsars with accurately known (parallactic) distances (Reynolds 1990, Sciama 1990c). One would not expect such constancy for other ionisation mechanisms. Future HST measurements should provide a stringent test of these ideas.

6.   Conclusions

I conclude from this discussion that the evidence in favour of the decaying neutrino theory is rather strong. One should therefore take seriously its other implications. These include the early reionisation of the universe and the resulting suppression of anisotropies in the cosmic microwave background on small angular scales (Scott, Rees and Sciama 1991), and attempts to derive the required decay lifetime from elementary particle physics (Roulet and Tommasini 1991, Gabbiani, Masiero and Sciama 1991).

I am grateful to the Italian Ministry of Universities and Scientific and Technological Research for their financial support.

References

R. Cowsik, Phys. Rev. Lett. **39**, 784, 1977.

L. Danly et al., Ap. J. Suppl. **81** 125, 1992.

A. F. Davidsen et al., Nature, 351, 128, 1991.

F. Gabbiani, A. Masiero and D. W. Sciama, Phys. Lett. 259**B**, 323, 1991.

J. P. Hughes and K. Tanaka, Ap. J. 398, 62, 1992.

C. W. Misner, Ap. J. 151, 431, 1968.

R. J. Reynolds, Ap. J. 348, 153, 1990.

E. Roulet and D. Tommasini, Phys. Lett. 256**B**, 218, 1991.

D. W. Sciama, Ap. J. 364, 549, 1990a.

D. W. Sciama, Phys. Rev. Lett. **65**, 2839, 1990b.

D. W. Sciama, Nature, 346, 40, 1990c.

D. W. Sciama, Int. Journ. Mod. Phys. D.1, 161, 1992.

D. W. Sciama, to be published, 1993a.

D. W. Sciama, to be published, 1993b.

D. W. Sciama, M. Persic and P. Salucci, Nature, 358, 718, 1992.

D. W. Sciama, M. Persic and P. Salucci, PASP, Jan. issue, 1993.

D. Scott, M. J. Rees and D. W. Sciama, Astron. and Astrophys. 250, 295, 1991.

W. Sorrell, Comments Astrophys. 16, 123, 1992.

L. Spitzer and E. L. Fitzpatrick, Ap. J., in press, 1993.

S. P. Tarafadar, MNRAS, 252, 55P, 1991.

P. A. Thomas and A. C. Fabian, Mon. Not. R.A.S. 246, 156, 1990.

Exploiting the Computer to Investigate Black Holes and Cosmic Censorship

STUART L. SHAPIRO AND SAUL A. TEUKOLSKY

Center for Radiophysics and Space Research, Cornell University, Ithaca, NY 14850, USA.

ABSTRACT

We describe a method for the numerical solution of Einstein's equations for the dynamical evolution of a *collisionless* gas of particles in general relativity. The gravitational field can be arbitrarily strong and particle velocities can approach the speed of light. The computational method uses the tools of numerical relativity and N-body particle simulation to follow the full nonlinear behavior of these systems. Specifically, we solve the Vlasov equation in general relativity by particle simulation. The gravitational field is integrated using the $3 + 1$ formalism of Arnowitt, Deser, and Misner. Our method provides a new tool for studying the cosmic censorship hypothesis and the possibility of naked singularities. The formation of a naked singularity during the collapse of a finite object would pose a serious difficulty for the theory of general relativity. The hoop conjecture suggests that this possibility will never happen provided the object is sufficiently compact ($\lesssim M$) in all of its spatial dimensions. But what about the collapse of a long, nonrotating, prolate object to a thin spindle? Such collapse leads to a strong singularity in Newtonian gravitation. Using our numerical code to evolve collisionless gas spheroids in full general relativity, we find that in all cases the spheroids collapse to singularities. When the spheroids are sufficiently compact the singularities are hidden inside black holes. However, when the spheroids are sufficiently large there are no apparent horizons. These results lend support to the hoop conjecture and appear to demonstrate that naked singularities can form in asymptotically flat spacetimes.

1 INTRODUCTION

Despite many years of effort, very few analytic solutions of Einstein's equations that are applicable to realistic situations are known. Even these are simplified by imposing a high degree of symmetry. The goal of numerical relativity is to devise methods for solving Einstein's equations for general spacetimes, albeit numerically. The field was launched by the pioneering paper of May and White (1966), who followed the collapse of a spherical star to a black hole. Since then, several research groups around the world have expended considerable effort and computer resources to extend this enterprise.

Charles Misner, renowned for his mathematical insights and analytic prowess in general relativity, is in fact one of the founders of the field of numerical relativity. The May and White paper actually solved the equations provided by Misner and Sharp (1964). While suitable for many problems involving spherical gravitational collapse, the equations do not generalize to more general problems. Once again, Misner had the answer. All workers in the field today know the classic paper of Arnowitt, Deser, and Misner (1962), which introduced the 3+1 decomposition of spacetime. This decomposition allows one in principle to integrate Einstein's equations forward in time for arbitrary spacetimes, given data on some initial time slice. This procedure is very natural for computer solution, and is analogous to the numerical approach used in hydrodynamics, plasma physics, and so on, for solving time-dependent problems.

Another tack that is currently being pursued in numerical relativity is to evolve data not from an initial spacelike slice, but rather from null surfaces. This approach has the potential advantage of tracking outgoing gravitational radiation more easily. While it might not be so widely known, Misner was also one of the founders of this second approach to numerical relativity. The paper by Hernandez and Misner (1966) was the first to propose null coordinates for the numerical integration of Einstein's equations. They wrote down the complete set of equations for the collapse of a spherical star to a black hole, and pointed out that these coordinates would prevent the formation of trapped surfaces and avoid the associated singularity during the integration. Choosing a suitable set of coordinates that allow one to integrate Einstein's equations arbitrarily far into the future without encountering singularities is still at the heart of modern research in numerical relativity.

For several years we have been developing the tools of numerical relativity to explore the dynamical behavior of collisionless matter. One of our goals is to learn how to solve Einstein's equations on the computer, and collisionless matter provides a particularly simple matter source. The dynamical equations for collisionless particles are simply the geodesic equations, which are ODEs and are considerably easier to solve than the equations for hydrodynamic matter, which are PDEs. In contrast to hydrodynamic matter, collisionless matter is not subject to shocks or other discontinuities which present computational difficulties unrelated to solving Einstein's equations.

Basically we are solving relativistic Vlasov equation for the matter coupled to Einstein's equations for the gravitational field. The method is mean-field particle simulation scheme: the distribution function is sampled by a large number of particles to get the stress-energy tensor $T_{\mu\nu}$. This stress-energy tensor is the source for the the field equations. Once they have been solved for the metric, the particles are moved for a small time step along geodesics of this background metric, and $T_{\mu\nu}$ is recalculated.

The whole procedure is repeated to follow a complete evolution.

A second goal of our study is to explore the dynamical fate and evolution of relativistic star clusters, and their possible astrophysical importance. For example, we have shown how a sufficiently relativistic star cluster collapses to form a supermassive black hole. Since supermassive black holes are likely to be the engines that power quasars and AGNs, the question of their origin is an important unsolved problem that our work addresses. In this paper, we will not deal with this and other astrophysical issues related to our work. Instead, we will focus on some of the fundamental issues connected with general relativity, rather than with astrophysical applications.

As in hydrodynamics, the simplest problems to tackle first are those with lots of symmetry and thus fewer degrees of freedom. Even in spherical symmetry, the dynamical evolution of a star cluster is not trivial. One must choose a good set of coordinates that allow the evolution to be calculated without the appearance of coordinate singularities. More importantly, when there are black holes, one wants coordinates that avoid the accompanying physical singularities, while continuing to follow the evolution outside the black holes. There are no recipes for choosing good coordinates in general. However, the spherical problem is essentially solved (Shapiro and Teukolsky 1985a,b,c, 1986, 1988). One can start with an arbitrary distribution of particle velocities and positions and follow the cluster's evolution, even for cases in which the velocities are close to the speed of light and the gravitational fields become arbitrarily strong. Black hole formation and the growth of the event horizon can be tracked with reasonable accuracy.

Fully three-dimensional problems with no symmetry whatsoever are beyond the reach of present computational resources and algorithmic developments. The current focus is on handling nonspherical problems in axisymmetry. These problems allow one to study two new qualitative features absent in spherical symmetry: rotation and the production of gravitational waves. Also, as we shall see below, they allow us to investigate some fundamental issues of general relativity including cosmic censorship and the formation of naked singularities.

2 COSMIC CENSORSHIP AND NAKED SINGULARITIES

It is well-known that general relativity admits solutions with singularities, and that such solutions can be produced by the gravitational collapse of nonsingular, asymptotically flat initial data. The *cosmic censorship hypothesis* (Penrose 1969) states that such singularities will always be clothed by event horizons and hence can never be visible from the outside (no naked singularities). If cosmic censorship holds, then there is no problem with predicting the future evolution outside the event horizon. If it does not hold, then the formation of a naked singularity during collapse would be a crisis for general relativity theory. In this situation, one cannot say anything precise about the future evolution of any region of space containing the singularity since new information could emerge from it in a completely arbitrary way.

Are there guarantees that an event horizon will always hide a naked singularity? No definitive theorems exist. Proving the validity of cosmic censorship is perhaps the most outstanding problem in the theory of general relativity. Until recently, possible counter-examples (see, e.g. Goldwirth *et al.* 1989) have all been restricted to spherical symmetry and typically involve shell crossing, shell focusing, or self-similarity. Are these singularities an accident of spherical symmetry?

Very little is known about nonspherical collapse in general relativity. In the absence of concrete theorems, Thorne (1972) has proposed the *hoop conjecture*: Black holes with horizons form when and only when a mass M gets compacted into a region whose circumference in *every* direction is $C \lesssim 4\pi M$. If the hoop conjecture is correct, aspherical collapse with one or two dimensions appreciably larger than the others might then lead to naked singularities.

For example, consider the Lin-Mestel-Shu instability (Lin *et al.* 1965) for the collapse of a nonrotating, homogeneous spheroid of collisionless matter in Newtonian gravity. Such a configuration remains homogeneous and spheroidal during collapse. If the spheroid is slightly oblate, the configuration collapses to a pancake, while if the spheroid is slightly prolate, it collapses to a spindle. While in both cases the density becomes infinite, the formation of a spindle during prolate collapse is particularly worrisome. The gravitational potential, gravitational force, tidal force, kinetic and potential energies all blow up. This behavior is far more serious than mere shell-crossing, where the density alone becomes momentarily infinite. For collisionless matter, prolate evolution is forced to terminate at the singular spindle state. For oblate evolution the matter simply passes through the pancake state, but then becomes prolate and also evolves to a spindle singularity.

Does this Newtonian example have any relevance to general relativity? We already know that *infinite* cylinders do collapse to singularities in general relativity, and, in accord with the hoop conjecture, are not hidden by event horizons (Thorne 1972; Misner *et al.* 1973). But what about *finite* configurations in asymptotically flat spacetimes?

Previously, we constructed an analytic sequence of momentarily static, prolate and oblate collisionless spheroids in full general relativity (Nakamura *et al.* 1988). We found that in the limit of large eccentricity the solutions all become singular. In agreement with the hoop conjecture, extended spheroids have no apparent horizons. Can these singularities arise from the collapse of nonsingular initial data? To answer this, we have performed fully relativistic dynamical calculations of the collapse of these spheroids, starting from nonsingular initial configurations (Shapiro and Teukolsky 1991*a*, *b*).

We find that the collapse of a prolate spheroid with sufficiently large semi-major axis leads to a spindle singularity without an apparent horizon. Our numerical computations suggest that the hoop conjecture is valid, but that cosmic censorship does not hold because a naked singularity may form in nonspherical relativistic collapse.

2.1 Collapse of Collisionless Spheroids

We followed the collapse of nonrotating prolate and oblate spheroids of various initial sizes and eccentricities. We use maximal time slicing and isotropic spatial coordinates in axisymmetry. The metric is

$$ds^2 = -\alpha^2 dt^2 + A^2(dr + \beta^r dt)^2 + A^2 r^2(d\theta + \beta^\theta dt)^2 + B^2 r^2 \sin^2\theta d\phi^2. \qquad (1)$$

The key equations, derived in $3 + 1$ form following Arnowitt, Deser, and Misner (1962), are presented in Shapiro and Teukolsky (1992a). The matter particles are instantaneously at rest at $t = 0$ and the configurations give exact solutions of the relativistic initial-value equations (Nakamura *et al.* 1988). In the Newtonian limit, these spheroids reduce to homogeneous spheroids. When they are large (size $\gg M$ in all directions) we confirm that their evolution is Newtonian (Lin *et al.* 1965; Shapiro and Teukolsky 1987).

We constructed a number of geometric probes to diagnose the evolving spacetime. We tracked the Brill mass and outgoing radiation energy flux to monitor mass-energy conservation. To confirm the formation of a black hole, we probed the spacetime for the appearance of an apparent horizon and computed its area and shape when it was present. To measure the growth of a singularity, we computed the Riemann invariant $I \equiv R_{\alpha\beta\gamma\delta} R^{\alpha\beta\gamma\delta}$ at every spatial grid point. To test the hoop conjecture, we computed the minimum equatorial and polar circumferences outside the matter.

Typical simulations were performed with a spatial grid of 100 radial and 32 angular zones, and with 6000 test particles. A key feature enabling us to snuggle close to singularities was that the angular grid could fan and the radial grid could contract to follow the matter.

Figure 1 shows the fate of a typical prolate configuration that collapses from a highly compact and relativistic initial state to a black hole. Note that in isotropic coordinates a Schwarzschild black hole on the initial time slice would have radius $r = 0.5M$, corresponding to a Schwarzschild radius $r_s = 2M$.

Figure 2 depicts the outcome of prolate collapse with the same initial eccentricity but from a larger semi-major axis. Here the configuration collapses to a spindle singularity at the pole without the appearance of an apparent horizon. (We searched for both a single global horizon centered on the origin as well as a small disjoint horizon around the singularity in each hemisphere.) The spindle consists of a concentration of matter near the axis at $r \approx 5M$.

Figure 3 shows the growth of the Riemann invariant I at $r = 6.1M$ on the axis, just outside the matter. Before the formation of the singularity, the typical size of I at any exterior radius r on the axis is $\sim M^2/r^6 \ll 1$. With the formation of the spindle singularity, the value of I rises without bound in the region near the pole.

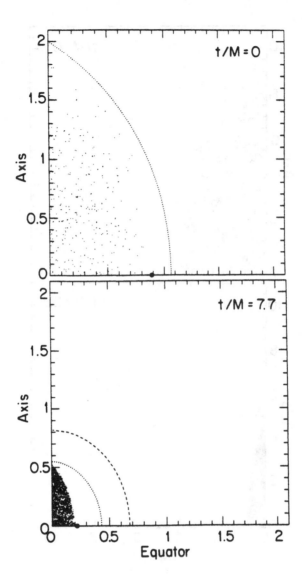

Figure 1 Snapshots of the particle positions at initial and late times for prolate collapse. The positions (in units of M) are projected onto a meridional plane. Initially the semi-major axis of the spheroid is $2M$ and the eccentricity is 0.9. The collapse proceeds nonhomologously and terminates with the formation of a spindle singularity on the axis. However, an apparent horizon (dashed line) forms to cover the singularity. At $t/M = 7.7$ its area is $A/16\pi M^2 = 0.98$, close to the asymptotic theoretical limit of 1. Its polar and equatorial circumferences at that time are $C^{AH}_{pole}/4\pi M = 1.03$ and $C^{AH}_{eq}/4\pi M = 0.91$, At later times these circumferences become equal and approach the expected theoretical value 1. The minimum exterior polar circumference is shown by a dotted line when it does not coincide with the matter surface. Likewise, the minimum equatorial circumference, which is a circle, is indicated by a solid dot. Here $C^{min}_{eq}/4\pi M = 0.59$ and $C^{min}_{pole}/4\pi M = 0.99$. The formation of a black hole is thus consistent with the hoop conjecture.

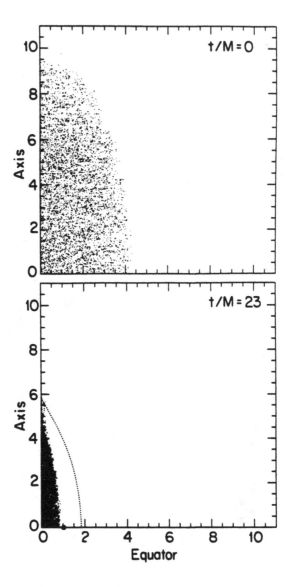

Figure 2 Snapshots of the particle positions at the initial and final times for prolate collapse with the same initial eccentricity as Figure 1 but with initial semi-major axis equal to $10M$. The collapse proceeds as in Figure 1, and terminates with the formation of a spindle singularity on the axis at $t/M = 23$. The minimum polar circumference is $C_{\text{pole}}^{\min}/4\pi M = 2.8$. There is no apparent horizon, in agreement with the hoop conjecture. This is a good candidate for a naked singularity, which would violate the cosmic censorship hypothesis.

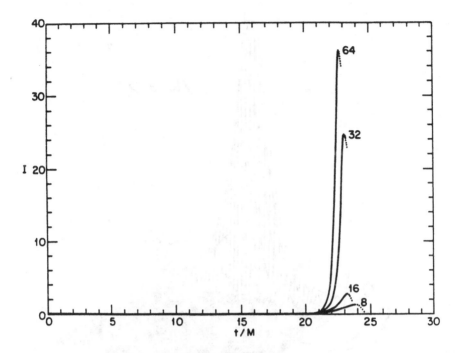

Figure 3   Growth of the Riemann invariant I (in units of M^{-4}) versus time for the collapse shown in Figure 2. The simulation was repeated with various angular grid resolutions. Each curve is labeled by the number of angular zones used. We use dots to show where the singularity has caused the code to become inaccurate.

The maximum value of I determined by our code is limited only by the resolution of the angular grid: the better we resolve the spindle the larger the value of I we can attain before the singularity causes the code (and spacetime!) to break down. Unlike shell-crossing singularities, where I blows up in the matter interior whenever the matter density is momentarily infinite, the singularity also extends *outside* the matter beyond the pole at $r = 5.8M$ (Fig. 4). In fact, the peak value of I occurs in the vacuum at $r \approx 6.1M$. Here the *exterior* tidal gravitational field is blowing up, which is not the case for shell crossing. *The absence of an apparent horizon suggests that the spindle is a naked singularity.*

When our simulation terminates, I along the axis falls to half its peak value at $r \approx 4.5M$ inside the matter and $r \approx 6.7M$ outside the matter. The singularity is not a point. Rather it is an extended region which includes the matter spindle, but grows most rapidly in the vacuum exterior above the pole. A $t = $ constant slice has a spatial metric $ds^2 = A^2 dr^2 + A^2 r^2 d\theta^2 + B^2 r^2 \sin^2 \theta \, d\phi^2$. In flat space $A = B = 1$. At the

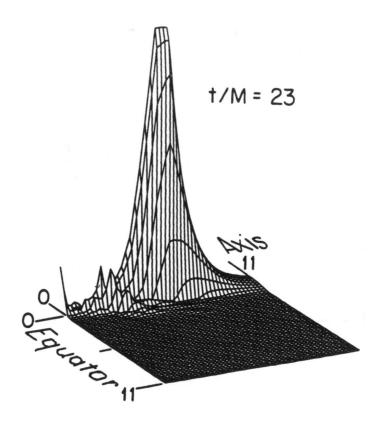

Figure 4 Profile of I in a meridional plane for the collapse shown in Figure 2. For the case of 32 angular zones shown here, the peak value of I is $24/M^4$ and occurs on the axis just outside the matter.

termination of the simulation these quantities have a modest maximum value $A \approx B \approx 1.7$, which occurs at the origin. They decrease monotonically outwards, reaching unity at large distances. However, it is their second derivatives that contribute to I and these blow up. While A and B steadily grow with time, I diverges much more rapidly. The behavior is similar to the logarithmic divergence of the metric in the analytic prolate sequence of Nakamura *et al.* (1988). We emphasize that the above characterization of the singularity and the behavior of the metric is dependent on the time slicing and may be different for other choices of time coordinate. In principle the spindle singularity might first occur at the center rather than the pole with a different time slicing.

The absence of an apparent horizon does not necessarily imply the absence of a global event horizon, although the converse is true. Recently Wald and Iyer (1991) have emphasized this point by showing that even Schwarzschild spacetime can be sliced with nonspherical slices that approach arbitrarily close to the singularity without any trapped surfaces. Because such singularities cause our numerical integrations to terminate, we cannot map out a spacetime arbitrarily far into the future, which would be necessary to completely rule out the formation of an event horizon. However, we do not think this is at all likely: for collapse from an initially compact state (Fig. 1), outward null geodesics turn around near the singularity. For collapse from large radius, by contrast, (Fig. 2) outward null geodesics are still propagating freely away from the vicinity of the singularity up to the time our integrations terminate. It is an interesting question for future research whether any time slicing can be found which will be more effective in snuggling up to the singularity without actually hitting it[1]. Such a slicing would enable one to confirm that all outward null geodesics propagate to large distances.

Further evidence for the nakedness of the singularity is the similarity of the spindle singularity to the infinite cylinder naked singularity. In both cases the proper length of a given segment of matter along the axis grows slowly, while its proper circumference and surface area shrink to zero much more rapidly. Also, the singularity is an extended region along the axis and not just a point.

We have also followed the collapse of an initially oblate configuration with the same initial eccentricity and semi-major axis as Figure 2. Following pancaking, it overshoots, becomes prolate and forms a black hole. At the time our integrations terminate, we find that $C_{\text{pole}}^{\text{min}} = C_{\text{eq}}^{\text{min}} = 0.85(4\pi M)$.

All of the above results are consistent with the hoop conjecture. When black holes form, the minimum polar and equatorial circumferences satisfy $C^{\text{min}} \lesssim 4\pi M$. Conversely, when naked singularities form the minimum polar circumference is much bigger than this value. In all cases where an apparent horizon forms, its area satisfies to within numerical accuracy $A \leq 16\pi M^2$, as required theoretically (see, e.g., Eardley 1975). In every case we find that gravitational radiation carries away a negligible fraction ($\ll 1\%$) of the total mass-energy by the time a black hole or naked singularity forms.

3 CONCLUSIONS

Numerical simulations of relativistic star clusters are a useful tool for probing the nature of spacetimes with strong gravitational fields. We have applied this tool to investigate cosmic censorship and the possible formation of naked singularities.

We have presented numerical evidence that the hoop conjecture is a valid criterion for the formation of black holes during nonspherical gravitational collapse. We have also found numerical candidates for the formation of naked singularities from nonsingular initial configurations. These examples are in contrast with any cases of singularities

---

[1] Maximal slicing apparently does not hold back the formation of prolate spindle singularities. For prolate spheroids the Newtonian potential diverges only logarithmically as the eccentricity $\rightarrow 1$, which may explain why α does not plummet precipitously near a spindle.

which may arise during spherical collapse. There the exterior spacetime is always the Schwarzschild metric and the Riemann invariant is always exactly $48M^2/r_s^6$, which is finite outside the matter. In spherical collapse the singularities can thus only occur inside the matter. Here the singularities extend above the pole into the vacuum exterior. These examples suggest that the unqualified cosmic censorship hypothesis cannot be valid.

While the matter treated here has kinetic pressure, it is collisionless, not fluid. We do not regard the collisional properties of the matter as crucial: First, the formation of naked singularities should not depend on the particular details of the fundamental interactions affecting matter at high densities. The gravitational field equations alone should be sufficient to rule out naked singularities, at least in the vacuum exterior, for true cosmic censorship. Second, collisional effects may even accelerate the formation of singularities via relativistic "pressure regeneration" (Misner *et al.* 1973) There is at least one tentative numerical example of prolate *fluid* collapse (adiabatic index $\Gamma \to 2$) that appears to be evolving to a singular state without the formation of an apparent horizon (Nakamura and Sato 1982; Nakamura *et al.* 1987).

It is not impossible that naked singularities qualitatively similar to the ones here may even arise in vacuum spacetimes. Since the sequence of momentarily static spheroids (Nakamura *et al.* 1988) proved to be a predictor of the singularities found in the dynamical calculations, we have been motivated to seek similar sequences of pure vacuum inital data. We have constructed two sequences characterized by long prolate concentrations of mass-energy: linear strings of black holes, and Brill waves with characteristic widths much less than their lengths (Abrahams *et al.* 1992). We find once again that the surrounding gravitational tidal field diverges for limiting members of these sequences, but that no "common" apparent horizons occur when the configurations are sufficiently long. It would be interesting to employ these solutions as initial data in dynamical evolutions.

The collisionless matter simulations described so far have no angular momentum. The presence of angular momentum could prevent an infinitesimally thin spindle singularity from forming on the axis. Recently, Apostolatos and Thorne (1992) have shown that, as in Newtonian theory, an infinitesimal amount of rotation is sufficient to prevent the formation of a singularity in the relativistic collapse of an *infinite* dust cylinder. This still leaves open the possibility that collapsing configurations of finite size can collapse to singularities in the presence of rotation. Recall that a small amount of angular momentum does not prevent the formation of a singularity when a Kerr black hole forms.

To explore this question, we have recently used our code to study the collapse of rotating collisionless spheroids (Shapiro and Teukolsky 1992b). The spheroids are initially prolate and consist of equal numbers of co- and counter-rotating particles, as in the infinite case treated by Apostolatos and Thorne (1992). Although individual particles are rotating, the spheroid has no net angular momentum. This restriction greatly simplifies the spacetime—the metric still has the same form as Equation (1). We find that rotation significantly modifies the evolution when it is sufficiently large. Imploding configurations with appreciable rotation ultimately collapse to black holes. However, for small enough angular momentum, our simulations cannot at present distinguish rotating from nonrotating collapse: spindle singularities appear to arise without apparent horizons. Hence it is possible that even spheroids with some angular momentum may form naked singularities.

Acknowledgments
This research was supported in part by NSF grants AST 91-19475 and PHY 90-07834 and NASA grant NAGW-2364 at Cornell University. Computations were performed on the Cornell National Supercomputer Facility.

References

Abrahams, A. M., Heiderich, K., Shapiro, S. L., and Teukolsky, S. A. 1992 *Phys. Rev. D*, in press.

Apostolatos, T. A., and Thorne, K. S. 1992 *Phys. Rev. D*, in press.

Arnowitt, R., Deser, S., and Misner, C. W. 1962 In *Gravitation: an Introduction to Current Research* (ed. L. Witten), p. 227. New York: John Wiley.

Eardley, D. M. 1975 *Phys. Rev. D*, **12**, 3072.

Goldwirth, D. S., Ori, A., and Piran, T. 1989 In *Frontiers in Numerical Relativity* (ed. C. R. Evans, L. S. Finn, and D. W. Hobill), p. 414. Cambridge University Press.

Hernandez, W. C., and Misner, C. W. 1966 *Astrophys. J.*, **143**, 452.

Lin, C. C., Mestel, L., and Shu, F. H. 1965 *Astrophys. J.*, **142**, 1431.

May, M. M., and White, R. H. 1966 *Phys. Rev*, **141**, 1232.

Misner, C. W., and Sharp, D. H. 1964 *Phys. Rev*, **136**, B571.

Misner, C. W., Thorne, K. S., and Wheeler, J. A. 1973 *Gravitation*, San Francisco: Freeman.

Nakamura, T., Oohara, K., and Kojima, Y. 1987 *Prog. Theor. Phys. Suppl.*, **90**, 57.

Nakamura, T., and Sato, H. 1982 *Prog. Theor. Phys.*, **68**, 1396.

Nakamura, T., Shapiro, S. L., and Teukolsky, S. A. 1988 *Phys. Rev. D*, **38**, 2972.

Penrose R. 1969 *Rivista del Nuovo Cimento*, **1** (Numero Special), 252.

Shapiro, S. L., and Teukolsky, S. A. 1985a *Astrophys. J.*, **298**, 34.

—— 1985b *Astrophys. J.*, **298**, 58.

—— 1985c *Astrophys. J. Lett.*, **292**, L41.

—— 1986 *Astrophys. J.*, **307**, 575.

—— 1987 *Astrophys. J.*, **318**, 542.

—— 1988 *Science*, **241**, 421.

—— 1991a *Phys. Rev. Lett.*, **66**, 994.

—— 1991b *Amer. Sci.*, **79**, 330.

—— 1992a *Phys. Rev. D*, **45**, 2739.

—— 1992b *Phys. Rev. D*, **45**, 2206.

Thorne, K. S. 1972 In *Magic Without Magic: John Archibald Wheeler* (ed. J. Klauder), p. 1. San Francisco: Freeman.

Wald, R. M., and Iyer, V. 1991 *Phys. Rev. D*, **44**, R3719.

Misner Space as a Prototype for Almost Any Pathology

Kip S. Thorne *

Abstract

Misner space (i.e. Minkowski spacetime with identification under a boost) is a remarkably rich prototype for a variety of pathologies in the structure of spacetime—some of which may actually occur in the real Universe. The following examples are discussed in some detail: traversable wormholes, violation of the averaged null energy condition, chronology horizons (both compactly and noncompactly generated) at which closed timelike curves are created, classical and quantum instabilities of chronology horizons, and chronology protection.

1. Introduction

In August 1965, at the *Fourth Summer Seminar on Applied Mathematics* at Cornell University, Charles Misner gave a lecture titled "Taub-NUT space as a counterexample to almost anything" [Mis67]. Near the end of his lecture, Misner introduced an exceedingly simple spacetime that shares some of Taub-NUT's pathological properties. This spacetime has come to be called *Misner space.*

Twenty-three years later, when I became intrigued by the question of whether the laws of physics forbid traversable wormholes and closed timelike curves, and if so, by what physical mechanism the laws prevent them from arising, Bob Wald and Robert Geroch reminded me of Misner space and its relevance to these issues. Since then, I and others probing these issues have found Misner space and its variations to be fertile testing grounds and powerful computational tools.

In this paper, I shall describe the remarkable pathologies that Misner space encompasses, and what we have learned about wormholes and closed timelike curves with

---

*Theoretical Astrophysics, California Institute of Technology, Pasadena, California 91125. This paper and the author's recent research on this topic were supported by Caltech's Feynman Research Fund, but not by the author's NSF physics grant, since NSF has forbidden the author to spend any of his grant money on wormhole or closed-timelike-curve issues. The writing of this paper was completed at the Institute for Theoretical Physics, Santa Barbara California, with support from NSF Grant PHY89-04035.

the aid of Misner space. My discussion will be quite elementary, requiring little more than a basic understanding of special relativity. For a somewhat more sophisticated discussion, but one that precedes the recent new insights, see the classic book of Hawking and Ellis [HaE73]

2.   Misner Space

Everything in this section was known to Misner in 1965 [Mis67] and is contained in Hawking and Ellis [HaE73], though they describe it in more sophisticated language and equations than I shall use.

Go into your bedroom (in flat, Minkowski spacetime), and identify the right and left walls with each other, so if you walk into the right wall, you find yourself emerging from the left wall without having encountered any matter. Then set the right wall into motion toward the left with a speed β (in units of the speed of light). The result is Misner space. Put differently, Misner space is Minkowski spacetime, closed up in one spatial dimension, and contracting along its closed dimension at a rate $d(\text{circumference})/dt = -\beta$.

Denote by (t, x, y, z) the Lorentz coordinates of a reference frame that is at rest with respect to the left wall, with the left wall at $x = 0$ and the right wall initially at $x = L$. Set $t = 0$ at the moment the right wall begins to move. Place a clock on the wall (left or right, it doesn't matter since they are physically identical), and denote by τ the proper time it ticks. Then, depicted in a spacetime diagram, Misner space will look like Figure 1a.

Notice that the clock's world line (thick line) appears twice in the diagram, once on the left wall, and again on the right. The events $\tau = 0$ on the left and on the right are physically identical, and similarly for all other τ. On the left, $\tau = t$; on the right, $\tau = t/\sqrt{1 - \beta^2}$ because of special relativistic time dilation. This time dilation leads inexorably to the creation of closed timelike curves (CTCs). For example, if one departs from the wall at the event $\tau = 3$ and travels rightward along the timelike curve labeled "CTC", one reaches the wall again at the same event $\tau = 3$ where one began the trip.

The dashed null line labeled *chronology horizon* is the location at which CTCs first arise. To the past of the chronology horizon (the *chronal region*) there are no CTCs; everywhere in the diagram to the future of the chronology horizon (the *non-chronal region*) there exist CTCs.

There are two generic families of timelike geodesics (straight, timelike world lines) in the chronal region, a "leftward family" and a "rightward family". Each geodesic in the leftward family (for example the line L in Fig. 1b) passes unscathed through the chronology horizon and into the non-chronal region. Each line in the rightward family (for example R in Fig. 1b) encircles Misner space rightward an infinite number of

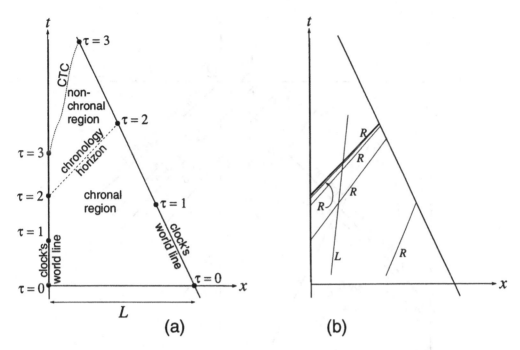

Figure 1. (a) Misner space depicted in a spacetime diagram. (b) Two timelike geodesics, L and R, in Misner space.

times. If, during its n'th trip around Misner space, it travels at speed v_n relative to the Lorentz coordinates, then its speed relative to the wall at the end of that trip is (by the standard special relativistic law for composing velocities) $v_{n+1} = (v_n + \beta)/(1 + v_n\beta)$, which is greater than v_n. This v_{n+1} must also be the speed with which it leaves the wall at the beginning of trip $n + 1$, and thus also its speed relative to the Lorentz coordinates during trip $n + 1$. Thus, on each successive trip, the rightward geodesic is boosted to a higher speed. Finally, after an infinite number of trips through the wall but a finite amount of proper time, it is moving at the speed of light and has asymptoted to the chronology horizon. (This and most other statements made in this paper can be verified fairly easily by elementary, special relativistic calculations.)

The seemingly pathological rightward geodesics are *not* a set of measure zero. Rather, as should become evident below, half the geodesics in the chronal region are of the rightward type and half of the leftward type.

What can possibly happen to a rightward family of observers, who move along the rightward geodesics, after they asymptote to the chronology horizon? They have lived only a finite amount of proper time. They have not encountered any spacetime curvature and thus cannot have been killed by infinite tidal forces. So where do they go? As Misner showed in his seminal lecture [Mis67], they pass through a chronology horizon of their own (distinct and different from that of the leftward observers), and

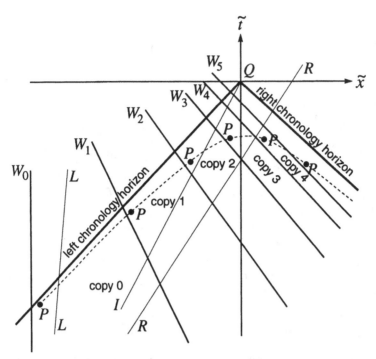

Figure 2. Minkowski spacetime in the role of the covering space for Misner space; $\tilde{t}$ and $\tilde{x}$ are Lorentz coordinates that coincide with t and x in copy 0 of Misner space, but are boosted relative to t and x in all other copies.

into a non-chronal region of their own (distinct and different from the leftward one).

This pathological behavior can be understood using the *covering space* of Misner space [HaE73]; Figure 2. This covering space is constructed by lining up a sequence of copies of Misner space, side by side, each one boosted by speed β relative to the last one. The copies of the (physically irrelevant) wall are labeled W_0, W_1, W_2, etc. in Figure 2; and each Misner space is labeled "copy 1", "copy 2", etc. A representative event P in Misner space is shown in each of the copies. There is actually an infinite number of copies of Misner space and of the wall and of the point P, with the high-order copies asymptoting to the rightward chronology horizon. The typical leftward geodesic L and typical rightward geodesic R are shown in the covering space, along with the two distinct chronology horizons through which they pass.

From this covering space it should be clear that there is a complete symmetry between the leftward observers and the rightward ones. Just as the leftward observers, as they near their leftward chronology horizon, see the rightward observers circle around and around Misner space an infinite number of times, approaching the speed of light, so also the rightward observers, as they near their rightward chronology horizon, must see the leftward ones circle infinitely and approach the speed of light.

Intriguingly, there is a third family of timelike geodesics and observers that are intermediate between the leftward and rightward ones. This family consists of geodesics that hit the covering space's origin (the event Q in Fig. 2); an example is the geodesic I. Note that in the covering space there is only one copy of the event Q, whereas there is an infinite number of copies of every other event, e.g. P. This presumably is related to the following remarkable pathology. Although the event Q exists in the covering space, it does *not* exist in Misner space. There is no way to include it in the spacetime, if one insists that the spacetime be a manifold and one includes the chronal region, and both the left and right chronology horizons, and both the left and right non-chronal regions. Those regions cannot be meshed smoothly with Q; and the impossibility of meshing makes Misner space geodesically incomplete: the intermediate geodesics (e.g. I) all terminate after finite proper time just before reaching the non-existent event Q. As pathological as this may seem, it would have been much more pathological if the terminating geodesics were not a set of measure zero.

The intermediate geodesics can be used as the time lines for a coordinate system that treats leftward and rightward geodesics on an equal footing. This coordinate system (T, X, y, z) is related to the Lorentz coordinates $(\tilde{t}, \tilde{x}, y, z)$ of the covering space (Figure 2) by

$$\tilde{t} = T \cosh X , \quad \tilde{x} = T \sinh X ; \qquad (1)$$

and correspondingly, the metric in this coordinate system is

$$ds^2 = -dT^2 + T^2 dX^2 + dy^2 + dz^2 . \qquad (2)$$

The boost-related points that are identified to produce Misner space (e.g. the points P in Figure 2) all are on the same hyperboloid $\tilde{t}^2 - \tilde{x}^2 = \text{constant}$, and therefore are all at the same T, y, z, but different X. The boost that takes one of the P's into the next one is simply a displacement in X by $\tanh^{-1} \beta$. Therefore, the n'th copy of P is at $X_n = X_0 + n \tanh^{-1} \beta$, and *Misner space can be regarded as the space of Equation (2) with X periodic with period $\tanh^{-1} \beta$.*

As seen by the intermediate observers, who sit at fixed (X, y, z), the leftward observers circle leftward around Misner space an infinite number of times as they approach their chronology horizon, and similarly for the rightward observers. The intermediate observers never see either family, leftward or rightward, reach its chronology horizon, because, as they watch and wait, they all come crashing together and cease to exist just before non-event Q, i.e. just before the "moment" $T = 0$ of the two chronology horizons.

3.   The Classical and Quantum Instability of Misner Space

The infinite relative circling of leftward and rightward geodesics produces severe instabilities in Misner space. Anything (dust particles, electromagnetic waves, gravitons, etc.) moving in the rightward manner (e.g. along the geodesic R) will become infinitely energetic as seen by the leftward observers, as they approach their leftward chronology horizon; and the resulting infinite energy density presumably must act back on the spacetime, via Einstein's equations, to change it radically. Change it how? Nobody knows for sure, but most likely to prevent the creation of CTCs at the leftward horizon, i.e. to *enforce chronology protection* (a phrase coined by Hawking [Haw92]). Similarly, anything moving in the leftward manner will become infinitely energetic on the rightward horizon, and probably act back to prevent the creation of CTCs there.

One might think it possible to stabilize the spacetime along one of the chronology horizons, e.g. the leftward one, by ensuring that nothing (no particles, no waves, no gravitons, ...) in Misner space moves in the opposite direction (rightward). However, there is one sort of thing that cannot be so controlled: Vacuum fluctuations of quantum fields. So long as the "universe" (Misner space) is truly closed up in the x-direction and truly shrinking in size, there is no way to prevent vacuum fluctuations of very short wavelength from traveling rightward and thereby piling up on themselves at the leftward chronology horizon. The pileup makes each rightward mode of any quantum field appear more than once at the same location in spacetime near the leftward chronology horizon. The shorter the mode's wavelength, the nearer the horizon it must approach for this to happen, but even a mode of arbitrarily short wavelength will pile up on itself an arbitrarily large number of times when it approaches arbitrarily close to the horizon. When this pileup begins, the mode's half quantum also piles up on itself, thereby endowing the mode with more than a single half quantum; and, as a result, when one renormalizes the mode's energy one winds up with a finite amount rather than zero. This finite energy in each rightward mode gives rise to a diverging renormalized energy density as one approaches the leftward chronology horizon—as Bill Hiscock and Deborah Konkowski showed by a rigorous "point-splitting" calculation, when they were in Misner's research group in the early 1980s [HiK82].

Thus, although Misner space is a solution to Einstein's classical vacuum field equations, it cannot be a solution to the equations of *semi-classical gravity* in which spacetime is treated as classical and all matter fields are quantized; in semiclassical gravity the quantum fields probably distort the spacetime geometry away from that of Misner, as one approaches the chronology horizon, and thereby probably enforce chronology protection.

4.   Misner Space with Spherical Walls: a Traversable Wormhole

If you give the walls of Misner space vanishing relative velocity $\beta = 0$ and change them from flat planes to spheres, you will obtain the simplest example of a traversable

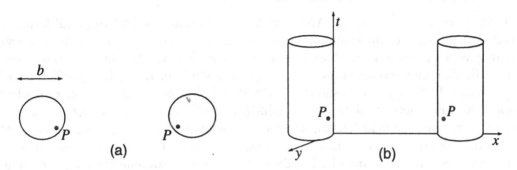

Figure 3. (a) A wormhole in Euclidean space. (b) A wormhole with both mouths at rest in Minkowski spacetime, and with synchronous time connection.

wormhole.

More specifically: extract two balls of radius b from 3-dimensional Euclidean space, and identify their surfaces so when you enter the surface of the right ball, you find yourself emerging from the surface of the left ball; the result is the wormhole of Figure 3a. For concreteness and simplicity, make the identification via reflection of the balls in the plane that is half way between them, so for example the point P appears at the locations shown in the figure. In Minkowski spacetime, this wormhole is obtained by identifying the two world tubes swept out by the two balls, as in Figure 3b, with events at the same Lorentz time t identified ("synchronous identification"), for example the events P shown in the figure.

There is spacetime curvature on the wormhole's coinciding spherical walls (which are called its *mouths*). A key feature of this curvature can be deduced by noticing that a bundle of light rays that enter the right mouth radially converging must emerge from the left mouth radially diverging (because of spherical symmetry). This means that the mouths act like a diverging lens with focal length equal to $2b$. The equations of null-ray propagation [Wal84] reveal that the focusing or defocusing of a bundle of rays is governed by the integral along the null ray of the Ricci curvature tensor—i.e., according to Einstein's equations, by the integral of the stress-energy tensor $T_{\alpha\beta}$. For the mouths to defocus rays, this integral must be negative: $\int T_{\alpha\beta} l^\alpha l^\beta d\zeta < 0$, where $l^\alpha = dx^\alpha/d\zeta$ is the tangent vector to the null ray and ζ is its affine parameter. A negative value of this integral is (by definition) a violation of the *averaged null energy condition* (ANEC).

4.1.   ANEC Violation

What is true of this special wormhole turns out to be true quite generally [MTY87]: to hold any wormhole open so it can be traversed by timelike and null curves, one must thread it with material that violates ANEC.

Do the laws of physics permit ANEC-violating material to exist? Since all forms of material, no matter how exotic, are made, ultimately, from quantum fields, the laws that govern the answer are those of quantum field theory. Motivated by this, Gunnar Klinkhammer several years ago undertook a study of the ANEC predictions of quantum field theory. He was able to show that in Minkowski spacetime, a massless scalar field can never violate ANEC [Kli91], and this has since been generalized to the electromagnetic field [Fol92]. However, as soon as one changes the topology of spacetime or curves it, the answer can change: ANEC can then be violated, at least in some cases. Klinkhammer [Kli91] discovered the first example of such a violation: Misner space with zero-velocity, flat walls—i.e., Minkowski spacetime closed up in one dimension and not shrinking. The left and right walls are very similar to two electrically conducting plates; and just as conducting plates distort the vacuum fluctuations of the electromagnetic field between themselves in such a way as to endow the vacuum field with tension and negative renormalized energy density (the *Casimir effect* [Cas48]), so also Misner space's identified walls distort the vacuum fluctuations, giving them tension and negative energy density. This means that the field has a negative renormalized value of $T_{\alpha\beta}l^\alpha l^\beta$ for l^α directed between the plates. Since the material of any real plate has a positive value of $T_{\alpha\beta}l^\alpha l^\beta$ that vastly exceeds the field's negative value between the plates, the Casimir setup for real plates does not violate ANEC. By contrast, the flat walls of Misner space are imaginary; they have no physical reality, so they contribute nothing to $T_{\alpha\beta}l^\alpha l^\beta$. The only contributions are from the field, those contributions are everywhere negative, and therefore ANEC is violated everywhere in the quantum-field vacuum of zero-velocity, flat-walled Misner space: $\int T_{\alpha\beta}l^\alpha l^\beta d\zeta < 0$.

A more recent study by Wald and Yurtsever [WaY91] has shown that in 3 + 1 dimensions (but not in 1 + 1), spacetime curvature can interact with the vacuum fluctuations of quantum fields to produce violations of ANEC. However, it is not yet known whether the specific distributions of curvature that occur in wormholes can ever produce such ANEC violations, nor (most importantly) whether they can do so in a self-consistent way, so the wormhole's curvature and quantum fields together produce a renormalized stress-energy tensor $T_{\alpha\beta}$ that in turn, through the Einstein equations, produces the curvature.

4.2.   Wormholes with Moving Mouths

Thus far we have kept the wormhole's spherical mouths at rest relative to each other. Next, set the right mouth in motion at uniform speed β toward the left one; Figure 4a. This motion must produce CTCs in just the same manner as in flat-walled Misner space. One can easily see this by noting that along the mouths' symmetry axis (x-axis), the spacetime structure is identical to that of flat-walled Misner space, i.e. it

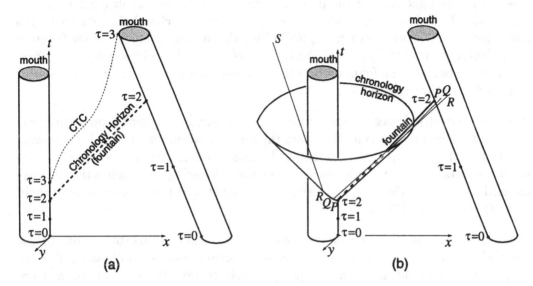

Figure 4. (a) A wormhole with the right mouth moving toward the left; cf. Figure 1a. (b) The fountain and chronology horizon of such a wormhole spacetime.

is that of Figures 1 and 2.

Though unimportant on the axis, the spherical shape of the wormhole mouths is extremely important elsewhere: it causes the mouths to behave like a diverging lens (as we saw above), and this in turn radically alters some key pathologies of Misner space:

First, the mouths' diverging-lens action deflects rightward-propagating geodesics away from the symmetry axis (e.g. the geodesic $PQRS$ in Fig. 4b), thereby preventing them from circling around Misner space an infinite number of times—unless they are directed precisely along the symmetry axis. This converts the family of infinitely circling rightward geodesics from a set of finite measure in flat-walled Misner space to a set of zero measure in spherical-walled Misner space (i.e. in the wormhole spacetime); and correspondingly, it destroys the rightward chronology horizon and non-chronal region. The wormhole spacetime, therefore, is endowed with only one chronology horizon and one non-chronal region: the leftward one [MTY87]. The (zero-measured) rightward geodesics that move along the symmetry axis are analogous to the intermediate geodesics of flat-walled Misner space: they asymptote to the chronology horizon (actually to the smoothly closed null geodesic on the horizon labeled "fountain" in Fig. 4), and there, after an infinite number of circuits through the wormhole but finite proper time, they terminate without ever experiencing infinite spacetime curvature.

Second, the mouths' diverging-lens action radically changes the geometry of the chronology horizon. No longer is it a flat, null surface. Now, roughly speaking, it is the future light cone of the point P that lies on the wormhole's left mouth in Figure 4b. This chronology horizon (location of onset of closed timelike curves) is generated by null geodesics (e.g. $PQRS$ in Fig. 4b) that all emerge from the fountain (asymptoting to it in the past) [MTY87, FMN90]. It is said to be a *compactly generated horizon* because the fountain on which its null generators originate is a compact region of spacetime [Haw92].

Third, the mouths' diverging lens action causes classical, rightward-propagating particles and waves to *defocus* as they pass through the wormhole, and this defocusing counteracts their pileup at the chronology horizon; if the mouths are far enough apart when the chronology horizon is reached, then the defocusing overwhelms the velocity-induced pileup, and the energy densities of the rightward particles or waves remain finite at the horizon [MTY87].

When Mike Morris, Ulvi Yurtsever, and I realized this remarkable ability of the mouths' sphericity to stabilize the chronology horizon in classical physics [MTY87], we presumed it would do so also in quantum field theory. We were wrong, as Kim, Frolov, I and others [KiT91, Fro91] later discovered via point-splitting calculations of a quantized scalar field's renormalized stress-energy tensor near the chronology horizon. The diverging lens effect weakens the growth of the renormalized energy density as one approaches the chronology horizon, but does not prevent it from growing infinitely as the horizon is reached. A heuristic explanation lies in the fact that the closer one is to the horizon, the shorter the wavelengths of the modes that are beginning to pile up on themselves, and therefore the larger the number of piling up modes; in the limit as one reaches the chronology horizon, their number becomes infinite. Although the diverging-lens effect makes each mode pile up on itself less vigorously than in flat-walled Misner space, and thereby makes each mode's contribution to the renormalized energy density finite as one reaches the horizon rather than infinite, because an infinite number of modes contribute, the total energy density still grows infinitely at the horizon.

It turns out that spherical-walled Misner space (the wormhole spacetime) is a prototype for the most general spacetime with a *compactly generated* chronology horizon, in the following sense. Hawking [Haw92] has shown that in the generic case, the horizon possesses one or more fountains (smoothly closed null geodesics); and it seems almost certain, though nobody has proved it firmly, that all the horizon's generators emerge from these fountains in the same manner as in Figure 4b. Moreover, Kim and I [KiT91] and Klinkhammer [Kli92] have shown that modes of quantum fields pile up on themselves at every location on a generic chronology horizon in much the same manner as in Figure 4b, thereby producing divergent vacuum polarization.

Does this mean that vacuum fluctuations always prevent the creation of closed time-like curves in a compact region of spacetime? Perhaps, but we are not at all sure. The energy density diverges sufficiently slowly, as one nears the chronology horizon, that before it can act back on the spacetime geometry via the Einstein equations, quantum gravity effects might invalidate our analysis. [KiT91, Tho91, Haw92].

5. Grant Space, or Displaced Misner Space

James Grant [Gra93] has recently found a simple generalization of Misner space, in which closed timelike curves are created at a *non-compactly generated chronology horizon*:

Go into your bedroom (in flat, Minkowski spacetime), and identify the right and left walls with each other, with a horizontal displacement of one wall relative to the other by some distance a, and with the right wall moving toward the left at speed β. The result is *Grant space* (or, one might like to say, *displaced Misner space*). Put differently, Grant space is Minkowski spacetime, closed up in the x-direction, with a displacement $\Delta y = a$ of all x-directed geodesics when they travel around the space once, and with the space contracting along its x-direction at a rate $d(\text{circumference})/dt = -\beta$.

It should be clear that the lateral displacement $\Delta y = a$ does not alter the existence of two families of geodesics in spacetime, rightward and leftward, nor of the leftward and rightward chronology horizons and leftward and rightward non-chronal regions, as depicted in Figures 1b and 2. What is changed in Figure 2 is that the identified points P in the covering space's successive copies of Grant space are displaced relative to each other by an amount $\Delta y = a$ into the paper. Correspondingly, the generators of the leftward (or rightward) chronology horizon (null geodesics traveling in the x-direction) do not close on themselves; rather, they begin at $\tilde{t} = \tilde{x} = y = -\infty$ and travel rightward to $\tilde{t} = \tilde{x} = 0$, $y = +\infty$, where they leave the chronology horizon. Moreover, the chronology horizon possesses no fountains (no smoothly closed null geodesics from which generators spring); and no closed timelike curve passes through any event on the chronology horizon.

On the other hand, arbitrarily close to each event on the horizon there are closed timelike curves. This can be understood using Figure 5, which is the x-t portion of the covering space. Event P is in the leftward non-chronal region and arbitrarily close to the leftward chronology horizon. Figure 5 shows many copies of P, each related to the preceding one by a boost with speed β and a displacement into the paper by a distance a. P can be connected to itself by many geodesics, with each one C_n circling around Grant space a different number of times, n. Geodesics C_{10} and C_{14} are shown in the figure. Since C_{14} has twice as long a temporal duration $\Delta \tilde{t}$ in the covering space as C_{10}, but only makes 40% more trips around Grant space and thus has only a 40% longer extent in the (not depicted) y direction, $dy/d\tilde{t}$ is $1.4/2.0 = 0.7$ as large on C_{14} as on C_{10}. As one goes to an ever larger number n of traversals, $dy/d\tilde{t}$ gets ever smaller (as one can readily show by a detailed calculation), so that eventually, for a sufficiently large number of traversals, $dy/d\tilde{t}$ is small enough for the geodesic to

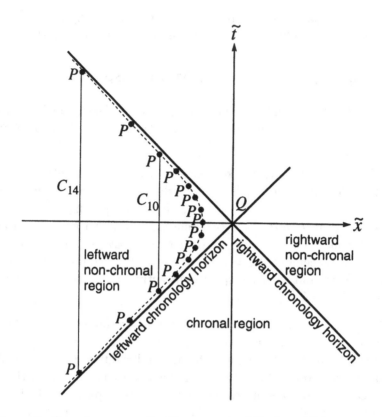

Figure 5. Covering space for Grant space (displaced Misner space).

be timelike.

Since there are closed (i.e. self-intersecting) timelike geodesics arbitrarily close to the chronology horizon, there must also be self-intersecting null geodesics arbitrarily close to the chronology horizon. Vacuum fluctuations that travel around these geodesics superimpose on themselves, thereby producing nonzero renormalized energy density. Since an infinite number of arbitrarily high-frequency modes can do this with each one producing a finite renormalized energy density, the resulting total renormalized energy density diverges as one approaches the chronology horizon. The details of the divergence have been computed by Grant [Gra93] using point splitting techniques; as one might expect, the divergence is much weaker than near the fountain of a compactly generated horizon. Whether the divergence is strong enough to enforce chronology protection is not yet known.

Grant [Gra93] has argued that Grant space is the same as Gott's two-moving-cosmic-strings spacetime [Got91], but as yet the precise mapping from one to the other has not been spelled out. This is but one of many features of the generalizations of Misner space that remain to be explored.

References

[Cas48]   H. B. G. Casimir, *Proceedings of the Koninklijke Nederlandse Akademie van Wetenschappen, Series B (1948)*, Vol. 51, p. 793.

[Fol92]   A. Folacci, *Averaged null energy condition for electromagnetism in Minkowski spacetime. Physical Review D (1993)*, Vol. 47, pp. 2726–2729.

[FMN90] J. Friedman, M. S. Morris, I. D. Novikov, F. Echeverria, G. Klinkhammer, K. S. Thorne, and U. Yurtsever, *Cauchy problem in spacetimes with closed timelike curves. Physical Review D (1990)*, Vol. 42, pp. 1915–1930.

[Fro91]   V. P. Frolov, *Vacuum polarization in a locally static multiply connected spacetime and a time-machine problem. Physical Review D (1991)*, Vol. 43, pp. 3878–3894.

[Got91]   J. R. Gott, *Closed timelike curves produced by pairs of moving cosmic strings—exact solution. Physical Review Letters (1991)*, Vol. 66, pp. 1126–1129.

[Gra93]   J. D. E. Grant, *Cosmic strings and chronology protection. Physical Review D (1993)*, in press.

[Haw92] S. W. Hawking, *The chronology protection conjecture. Physical Review D (1992)* Vol. 46, pp. 603–611.

[HaE73] S. W. Hawking and G. F. R. Ellis, *The large scale structure of space-time.* Cambridge University Press, 1973.

[HiK82] W. A. Hiscock and D. A. Konkowski, Quantum vacuum energy in Taub-NUT-type cosmologies. *Physical Review D (1982)* Vol. 26, pp. 1225–1230.

[KiT91] S.-W. Kim and K. S. Thorne, *Do vacuum fluctuations prevent the creation of closed timelike curves? Physical Review D (1991)* Vol. 43, pp. 3929–3947.

[Kli91]   G. Klinkhammer, *Averaged energy conditions for free scalar fields in flat spacetime. Physical Review D (1991)* Vol. 43, pp. 2542–2548.

[Kli92]   G. Klinkhammer, *Vacuum polarization of scalar and spinor fields near closed null geodesics. Physical Review D (1992)* Vol. 46, pp. 3388–3394.

[Mis67] C. W. Misner, *Taub-NUT space as a counterexample to almost anything. Relativity Theory and Astrophysics I. Relativity and Cosmology.* Edited by J. Ehlers. American Mathematical Society, Providence, RI, 1967.

[MTY87] M. S. Morris, U. Yurtsever, and K. S. Thorne, *Wormholes, time machines, and the weak energy condition. Physical Review Letters (1988)* Vol. 61, 1446–1449.

[Tho91]   K. S. Thorne, *Do the laws of physics permit closed timelike curves?* Annals of the New York Academy of Sciences (1991), Vol. 631, 182–193.

[Wal84]   Section 9.2 of R. M. Wald, *General Relativity.* University of Chicago press, 1984.

[WaY91]   R. Wald and U. Yurtsever, *General proof of the averaged null energy condition for a massless scalar field in 2-dimensional curved spacetime.* Physical Review D (1991) Vol. 44, pp. 403–416.

Relativity and Rotation

C. V. Vishveshwara *

1. Introduction

It was an honour to be a student of Charles Misner. And it was a pleasure to work with him. It is once again an honour and a pleasure to write an article in celebration of his sixtieth birthday.

The phenomenon of rotation generates interesting physical effects and involves intriguing basic concepts. One of the earliest classic examples of this is Newton's water-pail experiment. Within the framework of the general theory of relativity rotation displays rather unusual features. These are incorporated, for instance, in the spacetime structure such as that of a rotating black hole. Dragging of inertial frames and the occurrence of the ergosphere, to name two examples, are the outcome of the rotation built into the spacetime. Rotational effects also show up in the characteristics of particle motion and associated phenomena like gyroscope precession. These effects are revealed in an elegant manner by the invariant geometrical description of particle trajectories following the directions of spacetime symmetries, assuming that the spacetime under consideration admits such symmetries. In three dimensions, this geometrical characterization involves the specification of the curvature κ and torsion τ of the curve as functions of some parameter that varies along the curve, normally the arc-length. This can be extended to higher dimensions including time and the relevant geometrical parameters would then be $\kappa, \tau_1, \tau_2...\tau_{n-1}$ for n-dimensions. These parameters fit naturally into the Frenet-Serret formalism that is well known to the geometers. When the formalism is applied to timelike integral curves of space symmetries many interesting features emerge. In the major part of this article we shall briefly review the results that have been obtained in this area of investigation. Details of proofs and computations as well as extended discussions of results may be found in the references cited in the text.

We shall also discuss Killing and Killing-Yano tensors and their relevance to geodetic angular momentum. Finally we shall indicate some of the other topics that are related to the phenomenon of rotation in general relativity.

*Raman Research Institute, Bangalore 560 080 India. **Present Address:** Indian Institute of Astrophysics, Bangalore 560 034, India.

2.   Frenet-Serret Formalism

Consider a time-like curve parametrised by the four dimensional proper length s (or proper time τ). At each point along the curve one can define a characteristic orthonomal tetrad $e_{(i)}^a$ the Frenet-Serret tetrad (Latin indices range from 0 to 3). Here a and (i) are respectively the tensor and the tetrad indices. We suppress the tensor indices whenever convenient and no confusion may arise. The four-velocity u^a of the particle following the trajectory is identified with $e_{(0)}^a$: $u^a \equiv e_{(0)}^a$. Denoting by a dot the covariant derivative with respect to s, one has the relations

$$\begin{aligned}
\dot{e}_{(0)} &= \kappa e_{(1)} \\
\dot{e}_{(1)} &= \kappa e_{(0)} + \tau_1 e_{(2)} \\
\dot{e}_{(2)} &= -\tau_1 e_{(1)} + \tau_2 e_{(3)} \\
\dot{e}_{(3)} &= -\tau_2 e_{(2)}
\end{aligned} \tag{1}$$

In the above equations κ, τ_1, τ_2 are respectively the curvature, the first and the second torsions. They are in general functions of s and their specification completely characterises the curve. These geometric parameters and the tetrad are obtained from the physics of the situation, for instance as in the case of motion of charged particles[1].

3.   Charged Particle Motion in a Homogeneous Electromagnetic Field

The Lorentz equation governing the motion of a particle of mass m and charge e is given by

$$m\dot{u}^a = e\bar{F}_b^a u^b \tag{2}$$

where $\bar{F}_{ab}$ is the electromagnetic field tensor. Equivalently,

$$\dot{e}_{(0)} = F e_{(0)} \tag{3}$$

where $F = \frac{e}{m}\bar{F}$. The tensor F_{ab} is antisymmetric:

$$F_{ab} = -F_{ba} \tag{4}$$

Further, if the electromagnetic field is homogenous. it is covariantly constant by definition,

$$F_{ab;c} = 0 \tag{5}$$

Condition (5) implies the weaker condition

$$\dot{F}_{ab} \equiv F_{ab;c}\, u^c = 0 \tag{6}$$

As a consequence of conditions (4) and (6) several interesting results follow. These results are also valid for particle motion along symmetry directions, i.e. for Killing trajectories, since the equations of motions in this case turn out to be formally the same as for charges moving in homogeneous electromagnetic fields. We shall therefore consider the equations governing the Killing trajectories first and then discuss the results common to both cases.

4. Killing Trajectories

Consider a timelike Killing vector field ξ^a. A particle moving along this direction will have the four-velocity

$$u^a \equiv e^a_{(0)} = e^\psi \xi^a \tag{7}$$

with the condition $u^a u_a = 1$ so that

$$e^{2\psi} = \left(\xi^b \xi_b \right)^{-1} \tag{8}$$

From the Killing equation $\xi_{a;b} + \xi_{b;a} = 0$, one can show that $\psi_{,a} \xi^a = 0$. Then it follows that

$$\dot{e}^a_{(0)} = e^\psi \dot{\psi} \xi^a + e^{2\psi} \xi^a_{\ ;b} \xi^b = F^a_{\ b} e^b_{(0)} \tag{9}$$

where $F_{ab} \equiv e^\psi \xi_{a;b}$

We note that F_{ab} is an antisymmetric tensor and equation (9) has the same form as the Lorentz equation. Further,

$$\dot{F}_{ab} = 0 \tag{10}$$

as a consequence of the equation

$$\xi_{a;b;c} = R_{abcd} \xi^d \tag{11}$$

Therefore we have exactly the same conditions (4) and (6) that govern the motion of charged particles in a homogeneous electromagnetic field. Thus the formal similarity· between the two cases has been established. We now discuss the results that are common to these two cases.

5. Results

1. The geometric parameters κ_1, τ_1, τ_2 are constants along the trajectories, i.e. $\dot{\kappa} = \dot{\tau}_i = 0$.

2. Each vector of the tetrad satisfies the Lorentz equation $\dot{e}_{(i)} = F e_{(i)}$.

3. The parameters κ_1, τ_1, τ_2 and the tetrad $e_{(i)}$ can be expressed in terms of $e_{(0)}$ and F^n, where $(F^n)_{ab} = F^a_{\ e}...F^p_{\ b}$ (n terms). For instance, n has the highest value of 6 for τ_2. Thus the expressions for κ_1, τ_1 and τ_2 are progressively complicated. However, two very simple relations emerge that lend themselves to interesting physical interpretation.

4. Define the dual

$$*F^{ab} = \frac{1}{2\sqrt{-g}} \epsilon^{abcd} F_{cd} \tag{12}$$

Then $\alpha \equiv \frac{1}{2}\mathrm{Tr}F^2, \beta \equiv \frac{1}{2}\mathrm{Tr}F *F$ are essentially the Lorentz invariants $(E^2 - H^2)$ and $(E \cdot H)$ respectively in the electromagnetic case. One can show by straightforward algebra that the following relations hold.

(a.1) $\kappa^2 - \tau_1^2 - \tau_2^2 = \alpha$ and (a.2) $\kappa\tau_2 = \beta$

(b) Define $\omega^a \equiv \tau_1 e_{(3)}{}^a + \tau_2 e_{(1)}^a$. Then, $\omega^a = *F^a{}_b e_{(0)}{}^b$ (13)

Note that $*F e_{(0)}$ is the magnetic field seen by the particle in the electromagnetic case, while it is the vorticity of the Killing trajectories denoted by $\bar{\omega}^a$ when the particle motion is along the symmetry direction. In general ω^a defined as above is not proportional to the vorticity $\bar{\omega}^a$. However, the two are equivalent for the Killing trajectories. From the point of view of physical interpretation, relation (a.1) is connected to the structure of black holes, whereas relation (b) is associated with the precession of gyroscopes.

Let us first elucidate relation (a.1). Consider the family of surfaces $\Sigma : \xi_a \xi^a = constant$ that are analogous to equipotentials. The surface normal is given by

$$n_a = \left(\xi^b \xi_b\right)_{;a} = -2e^{-2\psi} \kappa e_{(1)a} \tag{14}$$

which is proportional to the acceleration or equivalently to the gravitational field. From equation (13) we have

$$\bar{\omega}^a \bar{\omega}_a = \tau_1^2 + \tau_2^2 \tag{15}$$

Therefore, (a.1) can be recast as

$$-\bar{\omega}^a \bar{\omega}_a + \frac{1}{4}n^a n_a = +\frac{1}{2}\left(\xi_{b;c}\xi^{b;c}\right)\xi^a\xi_a \tag{16}$$

This relation is well-known in black hole theory[2]. It tells us that the event horizon, which is a null surface ($n^a n_a = 0$) is a Killing horizon ($\xi^a \xi_a = 0$) provided the vorticity is null on it ($\bar{\omega}^a \bar{\omega}_a = 0$). In the case of Schwarzschild (non-rotating) black hole this is true for the global timelike Killing vector $\xi \equiv \xi^a \frac{\partial}{\partial x^a} = \frac{\partial}{\partial t}$, since $\bar{\omega}$ is 0 to begin with. In the Kerr spacetime (or rotating black hole metric), this happens for $\xi = \frac{\partial}{\partial t} + \Omega_H \frac{\partial}{\partial \psi}$ where Ω_H is the angular velocity of the black hole, which is constant. The event horizon is, therefore, a Killing horizon as well.

Let us consider relation (b). A Fermi transported triad orthogonal to $e_{(0)}$ obeys the equation $\dot{f}^i = \kappa f_k e^k_{(1)} e^i_{(0)}$. Such a triad represents three mutually orthogonal gyroscopes transported along the curve. These rotate with angular velocity $-\omega^a$ with respect to the Frenet-Serret triad. In the case of Killing trajectories the latter can be shown to be Lie dragged and the rotation of the gyroscopes is given by $\bar{\omega}^a$, the vorticity associated with the Killing congruence. This precession can be obtained by computing $\tau_1, \tau_2, e_{(1)}$ and $e_{(3)}$ in terms of ξ_a and $\xi_{a;b}$.

On the basis of these calculations several examples of gyroscope precession can be studied. Suppose the Killing vector we are considering is of the form $\xi = \frac{\partial}{\partial t} + \Omega \frac{\partial}{\partial \phi}$, where Ω is the angular velocity. We have assumed that t and ϕ are symmetry directions and taken Ω to be a constant. Some of the interesting cases of gyroscope precession are as follows. In flat spacetime for arbitrary Ω one finds Thomas precession. In the Schwarzschild spacetime, we can carry out the computations for arbitrary Ω, obtaining Fokker-de Sitter-Thomas precession. If we then make Ω the Keplerian angular speed $(M/R^3)^{\frac{1}{2}}$ for circular orbit of radius R around mass M, we get the geodetic or Fokker-de Sitter precession. The same procedure can be carried out in the Kerr spacetime getting the Schiff correction, which depends on a the angular momentum per unit mass of the black hole. Furthermore, gyroscope precession is non-zero even for a stationary observer with $\Omega = 0$. We may remark that all computations can be elegantly performed without any approximation using the present formalism.

6. Generalization to Higher Dimensions

The foregoing considerations can be generalized in a straightforward manner to higher dimensional spacetimes[3]. This would be relevant to the study of the intrinsic geometry of Killing trajectories and rotational effects in higher dimensional solutions of Einstein's equations or their generalizations. It is also related to the higher dimensional black hole theory. The formalism has been worked out in detail for 5 and 6 dimensions. The structure of the Frenet-Serret equations is the same as before. However, we have to add $e_{(4)}$ and τ_3 for 5 dimensions and again $e_{(5)}$ and τ_4 for 6. We have $F_{ab} = e^\psi \xi_{a;b}$ as before, but its dual will be a tensor of rank (D-2), eg. $^*F^{abc} = \frac{1}{\sqrt{-g}} \epsilon^{abcde} F_{de}$ for D=5. Consequently, the formalism has to be modified and some of the results have to be obtained by methods that are different from those employed in the case of D=4. Using the projection operator $h_{ab} = g_{ab} - e_{(0)a}e_{(0)b}$, define

$$V_{ab} \equiv h^c_a \, h^d_b \, e_{(0)c;d} \tag{17}$$

Then the vorticity tensor $\omega_{ab} \equiv V_{[ab]}$ where the square brackets denote antisymmetrization. Define also $e^{2\psi}\Omega^{ab\cdots} = \,^*F^{abc\cdots p}e_{(0)p}$. Then one finds that $\Omega^{ab\cdots}\Omega_{ab\cdots}$ is proportional to $\omega^{ab}\omega_{ab}$.

As we have noted already, the expressions for the geometric parameters become increasingly complicated. Nevertheless, for both D=5 and 6, it can be shown that the simple relation

$$\kappa^2 - \sum_i \tau_i^2 = \alpha \tag{18}$$

is still valid. This can be recast into a formula analogous to equation (16):

$$\frac{(-1)^{D-2}}{(D-3!)}\,\Omega^{ab..}\Omega_{ab..} + \frac{1}{4}n^a\,n_a = (\xi^a\,\xi_a)\,\left(\xi_{b;e}\,\xi^{b;e}\right) \tag{19}$$

which is the black hole equation in higher dimensions. Whether this is true for arbitrary D is an open question. One would also like to see further elucidation of the significance and the properties of the tensor $\Omega^{ab...}$.

7. Use of Rotating Coordinates

Rotating coordinate systems have been used to study geodesics and gyroscope precession in an interesting manner[4]. This method can be extended to the Frenet-Serret analysis of Killing trajectories as well.

In a static coordinate system, we have the Newtonian equation for the acceleration of a particle

$$\vec{a} = -\nabla\phi_g, \tag{20}$$

where ϕ_g is the gravitational potential. Going over to a coordinate system rotating about the z−axis with constant angular speed Ω so that $\phi \equiv \phi' + \Omega t$, the equation of motion is transformed to

$$0 = -\nabla(\phi_g + \phi_{cf}) \tag{21}$$

where ϕ_{cf} is the centrifugal potential. From this condition one arrives at the Keplerian angular speed for circular orbit immediately

$$\phi_g = -GM/r \text{ and } \phi_{cf} = -\frac{1}{2}\Omega^2 r^2 \text{ so that } \Omega = \sqrt{GM/r^3} \tag{22}$$

In general relativity, the same transformation can be adopted. A particle originally moving in a circular orbit with angular speed Ω will appear static in the new rotating coordinates. If $g_{00} \sim (1 + 2\phi_g)$, it will go over to $[1 + 2(\phi_g + \phi_{cf})]$ and the condition $g_{00,i}^{new} = 0$ yields the geodesic equation. The analogy is precise in the case of Schwarzschild metric. Gyroscope precession can also be computed more simply by referring to the rotating coordinate system.

In the case of the Frenet-Serret formalism, a similar procedure is carried out. We can transform to a rotating coordinate system so that the orbiting particle appears stationary. Using the new g_{00} and $g_{0\mu}$, the parameters κ, τ_i and e_i can be computed. The precession of gyroscope can then be studied under different circumstances[5].

8. Generalization to Quasi-Killing Trajectories

The formalism we have developed can be naturally extended to what we may term as quasi-Killing trajectories[5]. These are timelike curves following the direction defined by a vector field χ^a given by

$$\chi^a = \xi_{(0)}^a + \sum_i \omega_{(i)}(x^b)\, \xi_{(i)}^a \qquad (23)$$

where $\xi_{(0)}$ and $\xi_{(i)}$ are Killing vectors, $\omega_{(i)}(x^b)$ are scalar functions such that the Lie derivatives of $\omega_{(i)}$ with respect to χ vanish, i.e. $\underset{\chi}{\mathcal{L}}\, \omega_{(i)} = 0$. If the trajectory is to represent the world-line of a particle, then χ^a has to be timelike. All previous results go through, provided we make the identification.

$$F_b^a = e^\psi \left[\xi_{(0);b}^a + \sum_i \omega_{(i)}\, \xi_{(i);b}^a\right] \qquad (24)$$

with $e^{2\psi} = \left(\chi^b\chi_b\right)^{-1}$. This is the consequence of the fact that F_{ab} thus defined satisfies the conditions $F_{ab} = -F_{ba}$ and $\dot{F}_{ab} = 0$.

A well-known example of this type of trajectories is the irrotational or hypersurface orthogonal or globally stationary congruence in the Kerr spacetime associated with the Zero Angular Momentum Observers (ZAMO)[6],[7]. For these,

$$\chi = \xi + \omega\eta = \frac{\partial}{\partial t} + \omega\frac{\partial}{\partial \phi} \qquad (25)$$

with $\omega = -\frac{\xi\cdot\eta}{\eta\cdot\eta}$. The vorticity of the χ-congruence is zero. However, the gyroscope precession with respect to the Frenet-Serret triad is given by $^*Fe_{(0)}$ which is non-zero, since F differs from the vorticity by terms containing derivatives of Ω.

9. Killing and Killing-Yano Tensors

We have so far concentrated on the geometry of Killing trajectories and its relation to rotational effects. On the other hand, as is well known, Killing vectors lead to constants of motion in the case of particles following godesics. If a Killing vector involves rotational symmetry, the associated constant of motion can be identified with a specific component of angular momentum of the particle. This is once again related to the rotational effects inherent to the particle motion. Analogous to Killing vectors there can exist tensors, the Killing and Killing-Yano tensors, that are also connected with the angular momentum of geodesic motion and hence to rotational effects. The Kerr spacetime is an important example of this phenomenon.

Killing and Killing-Yano tensors admitted by the Kerr spacetime are related to the angular momentum of geodesics as has been mentioned already. They also play an important role in the separability of geodesic and wave equations. The timelike and rotational Killing vectors of the Kerr spacetime engender geodesic constants of motion identified respectively with energy and the z-component of angular momentum.

Remarkably enough, the total separability of the geodesic equations revealed the existence of an additional quadratic constant of motion $K = K_{ab}p^a p^b$, where p^a is the geodesic four-momentum[8]. The symmetric Killing tensor K_{ab} satisfies the equation

$$K_{(ab;c)} = 0 \qquad (26)$$

Soon after this discovery, considerable amount of work was done on the construction of the Killing tensor in type-D spacetimes with particular attention to Kerr and Kerr-Newman spacetimes[9]. In the Kerr spacetime expressed in the usual Boyer-Lindquist coordinates, with l^a and n^a as the repeated null directions ($n \cdot l = 1 = -m \cdot \bar{m}$), the tensor has the form

$$K_{ab} = 2(r^2 + a^2 \cos^2 \theta)l_{(a}n_{b)} - r^2 g_{ab} \qquad (27)$$

Let us consider the interpretation of the Killing tensor[10]. If we set $a = 0$ we obtain K_{ab} for the Schwarzschild and the flat spacetimes and it is easy to show that $K^{ab} = L_x^a L_x^b + L_y^a L_y^b + L_z^a L_z^b$, where L_x, L_y and L_z are the rotational Killing vectors about x, y and z axes and satisfy the commutation relations $[L_x, L_y] = -L_z$ etc. Thus in this case K is the square of the angular momentum. Can K^{ab} of the Kerr spacetime be expressed analogously in terms of three vectors (only L_z being Killing) having the angular momentum commutation relations? The answer is no. The closest approach to angular momentum interpretation of K^{ab} is achieved via the Killing-Yano tensor which is the 'square root' of the Killing tensor.

The Killing-Yano tensor f_{ab} has the properties $f_{ab} = -f_{ba}$, $f_{a(b;c)} = 0$ and $f_{ab} = f_a{}^c f_{cb}$. Then $J^a = f^a{}_b p^b$ is parallely propagated along the geodesic ($J^a{}_{;c}p^c = 0$). In the Kerr spacetime we have $f_{ab} = a \cos \theta l_{[a}n_{b]} + ir \, m_{[a}\bar{m}_{b]}$.

The close relation of f_{ab} to the flat spacetime angular momentum can be strikingly demonstrated by transforming to the Kerr-Schild coordinates (x^0, x, y, z) in which the metric takes on the form $g_{ab} = \eta_{ab} + 2H(x, y, z)l_a l_b$, where η_{ab} is the Lorentz metric. In these coordinates the Killing-Yano tensor has a very simple form.

The non-vanishing components of f^{ab} are

$$f^{03} = a = -f^{30}; f^{12} = z = -f^{21}; f^{13} = -y = -f^{31}; f^{23} = x = -f^{32} \qquad (28)$$

Except for $f^{03} = a$ it is identical to the flat space angular momentum tensor $f_{ab} \equiv \epsilon^{abcd}x_c t_d$, where t^a is the timelike Killing vector. Further $J^0 = a, J^x = yp_z - zp_y$ etc.

Let us note further similarities shared by the f^{ab} of Kerr and flat spacetimes.

(a) 'Sphere' : The surface on which the angular momentum operators act is given by

$$\frac{1}{2}f_{ab}f^{ab} = x^2 + y^2 + z^2 = \text{constant} \qquad (29)$$

The 'radial direction' can be defined by

$$r_i = \frac{1}{2}\epsilon_{ijkl}\, t^j f^{kl} = {}^*f_{ij}t^j = (0, x, y, z) \qquad (30)$$

which turns out to be the gradient to the 'sphere'.

(b) 'Angular Momentum Operators' : Define the vectors $X_a \equiv x_{,a}$ $Y_a \equiv y_{,a}$ and $Z_a \equiv z_{,a}$. Then $L_x^a = f^{ab} X_b, L_y^a = f^{ab} Y_b$ and $L_z^a = f^{ab} Z_b$ can be identified with angular momentum operators with the usual commutation relations.

(c) 'First Order Precession of Angular Momentum' : Define $\dot{L} = (L^a p_a)^{;b} p_b = \left(\underset{L}{\mathcal{L}} g^{ab} \right) p_a p_b$. In the Schwarzschild and flat spacetimes $\dot{L} = 0$ since each of L^a is a Killing vector. In the case of Kerr, evaluating $\underset{L}{\mathcal{L}} g^{ab}$ to first order in a one finds

$$\underset{y}{L^{(a;b)}_x} = \pm \frac{Ma}{r} l^{(a} \underset{x}{L^{b)}_y} \tag{31}$$

implying

$$\underset{y}{\dot{L}_x} = \pm \omega \underset{x}{L_y} \; ; \; \omega = Mar^{-3} l^a p_a \; ; \; \dot{L}_z = 0 \tag{32}$$

displaying the first order precession of the angular momentum about the z-axis.

We have already seen how $F_{ab} \equiv e^{\psi} \xi_{a;b}$ resembles the electromagnetic field tensor. In fact $\xi_{a;b}$ by itself satisfies source-free Maxwell's equations in vacuum spacetimes. Similarly, the Killing-Yano tensor has an 'electromagnetic structure' of its own. From the properties of f_{ab} one can show

$$f^{ab}_{;b} = 0 \; ; \; {}^* f^{ab}_{;b} = 4\pi j^a \tag{33}$$

with $j^a = 3t^a/4\pi$ where t^a is the global timelike Killing vector. Further ${}^* f^{ab}$ resembles a Maxwell field and can be expressed in terms of a four-potential A_a,

$$ {}^* f_{ab} = \nabla_{[a} A_{b]} \tag{34}$$

Imposing the background symmetries on A_a,

$$\underset{t}{\mathcal{L}} A_a = \underset{L_z}{\mathcal{L}} A_a = 0 \; , \tag{35}$$

we can show that

$$ {}^* f_{ab} t^b = \nabla_a (A_b t^b) \tag{36}$$

This is the usual relation between the electric field and the electrostatic potential. In the Kerr spacetime one finds

$$A_a = \left[\frac{1}{2} \left(x^2 + y^2 + z^2 \right), ay, -ax, o \right] \tag{37}$$

Thus A_a is composed of the sphere ($A_o = A_a t^a =$ constant surfaces) and L_z. As we have already seen the electric field ${}^* f^a_b t^b$ is obviously along the 'radial direction' normal to the equipotential surface. The close connection between Killing-Yano tensors and Maxwell fields has been further developed by Carter[11].

10.　Related Topics and Conclusion

There are several areas that are related to the foregoing considerations. Some of them involve the use of the Frenet-Serret formalism and implicitly incorporate rotational effects. We shall only mention these topics here. Stationary motions in flat spacetimes, i.e., motions along different combinations of Killing directions, have been classified on the basis of the Frenet-Serret parameters of the trajectories of motion. Coordinate systems with the time direction oriented along any such trajectory have been obtained and the corresponding metrics of the spacetime determined. Particle production in some of these non-inertial frames has been investigated thereby displaying the influence of rotation or linear acceleration on this process. It would be interesting to extend this procedure to non-flat spacetimes admitting suitable Killing vectors[12].

The Frenet-Serret characterization of Killing trajectories is useful in the classifiction of spacetimes also. For instance, the Perjes classification of zero Simon tensor vacuum solutions to the Einstein field equations has been studied on such a basis, leading to additional geometric insights into this area of investigation[13].

There exists a natural correspondence between the formalism outlined in this article and the gravi-electric and gravi-magnetic field approach to rotation and related phenomena[6] [14]. It may be profitable to investigate this relation in greater detail. Astrophysical applications of the formal results may also lead to interesting results as has been the case with the latter formulation of rotational phenomena.

There has been considerable interest in the general relativistic versions of centrifugal and coriolis forces. A proper discussion of these forces involves several factors such as appropriate definition, interpretation, weeding out of spurious effects, induced dragging of inertial frames and, perhaps, Machian-type concept of inertial frames. 'Clean' model calculations involving exact solutions, e.g. a rotating cylindrical shell, can be performed in order to highlight these ideas[15]. On the other hand, for realistic astrophysical calculations specific methods like the one involving optical reference geometry[16] may have to be used. Such calculations indicate possible interesting effects such as the reversal of centrifugal force and its consequences[17]. The techniques described in this article may also be employed to elucidate the basic phenomena involved in this area of investigation.

Acknowledgements : It is a pleasure to thank B.R. Iyer for discussions and for his critical reading of this article. I would also like to thank R. Ramasubramaniyan for his help in the preparation of the manuscript.

References

[1] Honig E, Schucking E.L. and Vishveshwara C.V. (1974), *J.Math.Phys.*, **15**, 774.

[2] Vishveshwara C.V. (1968), *J.Math.Phys.*, **9**, 1319.

[3] Iyer B.R. and Vishveshwara C.V. (1988), *Class.Quantum.Grav.*, **5**, 961

[4] Rindler W. and Perlick V (1990, *General Relativity and Gravitation*, **22**, 1067

[5] Iyer B.R. and Vishveshwara C.V. (1992), (in preparation)

[6] See for instance, *Black Holes–The Membrane Paradigm*, Thorne K.S., Price R.H. and Macdonald D.A., eds., Yale University Press (1986)

[7] Greene R.D., Schucking E.L. and Vishveshwara C.V. (1975), *J.Math.Phys.*, **16**, 153.

[8] Carter B. (1968), *Phys.Rev.*, **174**, 1559.

[9] Penrose R. (1973), *Ann.N.Y. and Acad.Sci.*, **224**, 125.

[10] Faridi A.M. (1986), *Gen.Rel.Grav.*, **18**, 271; Samuel J. and Vishveshwara C.V., *The Killing Two-form and Particle Angular Momentum in Kerr Spacetime* (unpublished preprint)

[11] Carter B. (1987), *J.Math.Phys.*, **28**, 1535; (1987) *Gravitation in Astrophysics*, Carter B. and Hartle J. eds., Plenum.

[12] Letaw J.R. and Pfautsch J.D. (1982), *J.Math.Phys.*, **23**, 425; (1981), *Phys.Rev.D.*, **24**, 1491.

[13] Krisch J.P. (1988), *J.Math.Phys.*, **29**, 446.

[14] Jantzen R.T., Carini P. and Bini D. (1992), *Ann.Phys.*, **215**, 1.

[15] Cohen J.M., Sarill W.J. and Vishveshwara C.V. (1982), *Nature*, **298**, 829.

[16] Abramowicz M.A., Carter B. and Lasota J.P. (1988), *Gen.Rel.Grav.*, **20**, 1173.

[17] See for instance Abramowicz M.A. and Prasanna A.R. (1990), *Mon.Not.R.Astron.Soc.*, **245**, 720 and Abramowicz M.A. and Miller J.C.(1990), *Mon.Not.R.Astron.Soc.*, **245**, 729.

The First Law of Black Hole Mechanics

Robert M. Wald *

Abstract

A simple proof of a strengthened form of the first law of black hole mechanics is presented. The proof is based directly upon the Hamiltonian formulation of general relativity, and it shows that the the first law variational formula holds for arbitrary nonsingular, asymptotically flat perturbations of a stationary, axisymmetric black hole, not merely for perturbations to other stationary, axisymmetric black holes. As an application of this strengthened form of the first law, we prove that there cannot exist Einstein-Maxwell black holes whose ergoregion is disjoint from the horizon. This closes a gap in the black hole uniqueness theorems.

1. Derivation of the First Law

It was noted by Hilbert at the inception of general relativity that the Einstein field equations are derivable from an action principle,

$$S = \frac{1}{16\pi} \int R \sqrt{-g} \, d^4x \tag{1}$$

Thus, general relativity has a Lagrangian formulation. The corresponding Hamiltonian formulation was given many years later in a collaboration between Charles Misner, Richard Arnowitt, and Stanley Deser. The main results of this collaboration are summarized in [1].

The Hamiltonian formulation of general relativity is employed as a starting point in all attempts to formulate a quantum theory of gravity via the canonical approach. It plays a less essential role within the context of purely classical general relativity. However, even in that context, the Hamiltonian formulation of general relativity provides some penetrating insights into the structure of the theory. In this paper, I shall illustrate this point by showing how a strengthened form of the first law

---

*Enrico Fermi Institute and Department of Physics, University of Chicago, 5640 S. Ellis Avenue, Chicago, IL 60637, USA. This research was supported in part by the National Science Foundation under Grant No. PHY89-18388.

of black hole mechanics can be derived in a very simple and direct manner from the Hamiltonian formulation of general relativity. The results presented here were obtained in collaboration with D. Sudarsky and were first reported in [2]. I shall restrict attention here to Einstein-Maxwell theory–the more general case of Einstein-Yang-Mills theory was considered in [2]–but the analysis generalizes straightforwardly to allow other fields, provided only that a Hamiltonian formulation of the complete theory can be given.

In the Hamiltonian formulation of Einstein-Maxwell theory, a point in phase space corresponds to the specification of the fields $(h_{ab}, \pi^{ab}, A_a, E^a)$ on a three dimensional manifold Σ. Here h_{ab} is a Riemannian metric on Σ and A_a is the spatial part of the vector potential (i.e., the pull-back to Σ of the spacetime vector potential A_μ). The momentum canonically conjugate to h_{ab} is $\frac{1}{16\pi}\ \pi^{ab}$, where π^{ab} is related to the extrinsic curvature, K_{ab}, of Σ in the spacetime obtained by evolving this initial data by,

$$\pi^{ab} = \sqrt{h}(K^{ab} - h^{ab}K) \tag{2}$$

The momentum conjugate to A_a is $\frac{1}{4\pi}\ \sqrt{h}\ E^a$, where E^a is the electric field in the evolved spacetime.

Constraints are present in Einstein-Maxwell theory. The allowed initial data is restricted to the constraint submanifold in phase space defined by the vanishing at each point $x\epsilon\Sigma$ of the following quantities,

$$0 = \mathcal{C} = \frac{1}{4\pi}\sqrt{h}D_a E^a \tag{3}$$

$$0 = \mathcal{C}_0 = \frac{1}{16\pi}\sqrt{h}\ \{-R + 2E_a E^a + F_{ab}F^{ab} + \frac{1}{h}(\pi^{ab}\pi_{ab} - \frac{1}{2}\pi^2)\} \tag{4}$$

$$0 = \mathcal{C}_a = -\frac{1}{8\pi}\sqrt{h}\ \{D_b(\pi_a^b/\sqrt{h}) - 2F_{ab}E^b\} \tag{5}$$

where D_a is the derivative operator on Σ compatible with h_{ab}, R denotes the scalar curvature of h_{ab}, and $F_{ab} = 2D_{[a}A_{b]}$.

The ADM Hamiltonian, H, for Einstein-Maxwell theory has the "pure constraint" form,

$$H = \int_\Sigma (N^\mu \mathcal{C}_\mu + N^\mu A_\mu \mathcal{C}) \tag{6}$$

Here N^μ and A_0 are to be viewed as non-dynamical variables, which may be prescribed arbitrarily. In the spacetime obtained by solving Hamilton's equations, N^μ has the interpretation of being the time evolution vector field (i.e., its projection normal to Σ yields the lapse function, N, and its projection into Σ yields the shift vector, N^a), and A_0 has the interpretation of being the component of the vector potential normal to Σ. The "pure constraint" form of H is not special to Einstein-Maxwell theory; any Hamiltonian arising from a diffeomorphism invariant theory always takes such a form (see the appendix of [3]).

The derivation of the strengthened form of the first law of black hole mechanics is based upon the following three properties of the ADM Hamiltonian: (i) It vanishes identically on the constraint submanifold. Hence, its first order variation off of a solution vanishes whenever the varied initial data satisfies the linearized constraints. (ii) Its variation yields the Einstein-Maxwell equations. (iii) For suitable choices of time evolution vector field in an asymptotically flat spacetime, its variation is directly related to formulas for the variation of mass, charge, and angular momentum in the spacetime.

The first of these properties is manifest from eq. (6). The second property is just the statement of what we mean by H being a Hamiltonian for Einstein-Maxwell theory. More explicitly, let $(h_{ab}, \pi^{ab}, A_a, E^a)$ be initial data (satisfying the constaints) for an Einstein-Maxwell solution, and let $(\delta h_{ab}, \delta \pi^{ab}, \delta A_a, \delta E^a)$ be an arbitrary perturbation (not necessarily satisfying the linearized constraints) of compact support on Σ. By integrating by parts, we can express the variation of H in the form.

$$\delta H = \int_{\Sigma} [P^{ab}\delta h_{ab} + Q_{ab}\delta\pi^{ab} + R^a\delta A_a + S_a\delta(\sqrt{h}E^a)] \tag{7}$$

Then, the coefficient, P^{ab}, of δh_{ab} yields minus the "time derivative" (i.e., the Lie derivative with respect to N^μ) of the canonical momentum $\frac{1}{16\pi}\pi^{ab}$ in the solution to the Einstein-Maxwell equations arising from the initial data $(h_{ab}, \pi^{ab}, A_a, E^a)$. Similarly, the coefficient, $16\pi Q_{ab}$, of $\frac{1}{16\pi}\delta\pi^{ab}$ yields the time derivative of h_{ab}, etc. (Explicit formulas for P^{ab}, Q_{ab}, R^a, and S_a are given in eqs. (19)-(23) of [2] in the more general case of Einstein-Yang-Mills theory.)

The third property can be understood and derived from the following considerations [4]: Let $(h_{ab}, \pi^{ab}, A_a, E^a)$ be initial data (satisfying the constaints) for an asymptotically flat spacetime. Suppose, now, that we consider variations $(\delta h_{ab}, \delta\pi^{ab}, \delta A_a, \delta E^a)$ of this initial data which are merely asymptotically flat (rather than being of compact support). Then extra terms will appear in eq. (7) due to contributions from boundary terms at infinity which arise when one does the integrations by parts needed to put the volume terms in the form (7). In the case where N^μ asymptotically approaches a time transition (i.e., the lapse function N goes to 1 and the shift vector N^a goes to 0 at infinity), one obtains (assuming that no other boundaries are present on Σ – see below),

$$\delta H = \int_{\Sigma} [P^{ab}\delta h_{ab} + Q_{ab}\delta\pi^{ab} + R^a\delta A_a + S_a\delta(\sqrt{h}E^a)]$$
$$- \frac{1}{16\pi}\delta\oint_\infty dS^a[\partial^b h_{ab} - \partial_a h^b_b] - \frac{1}{4\pi}\delta\oint_\infty dS^a A_0 E_a \tag{8}$$

Equation (8) suggests that we modify the definition of the ADM Hamiltonian by addition of a surface term,

$$\tilde{H} \equiv H + \frac{1}{16\pi}\oint_\infty dS^a[\partial^b h_{ab} - \partial_a h^b_b] + \frac{1}{4\pi}\oint_\infty dS^a A_0 E_a \tag{9}$$

If we do so, then $\tilde{H}$ will act as a true Hamiltonian on phase space in the sense that its variation will be given by the right side of eq.(7) for all asymptotically flat perturbations. It is natural, then, to define the "canonical energy" E on the constraint submanifold of phase space to be the numerical value of this true Hamiltonian. Hence, we obtain,

$$\mathcal{E} = \frac{1}{16\pi} \oint_\infty dS^a [\partial^b h_{ab} - \partial_a h_b^b] + \frac{1}{4\pi} \oint_\infty dS^a A_0 E_a = m + VQ \tag{10}$$

where m is the ADM mass, V is the asymptotic value of A_0 at infinity, and Q is the electric charge. (In the case presently considered – where there are no "boundaries" aside from infinity – Q will vanish, but we keep this term in eq. (10) since it will be nonvanishing in the more general cases considered below.) In terms of the original ADM Hamiltonian H, we thereby obtain,

$$\delta H = \int_\Sigma [P^{ab}\delta H_{ab} + Q_{ab}\delta\pi^{ab} + R^a\delta A_a + S_a\delta(\sqrt{h}E^a)] - \delta m - V\delta Q \tag{11}$$

which yields the desired relationship between the variation of H and the variation of ADM mass, m, and charge, Q, in the case where N^μ approaches a time translation at infinity. In a similar manner, if N^μ asymptotically approaches a rotation at infinity, we obtain,

$$\delta H = \int_\Sigma [P'^{ab}\delta h_{ab} + Q'_{ab}\delta\pi^{ab} + R'^a\delta A_a + S'_a\delta(\sqrt{h}E^a)] + \delta J$$

$$\tag{12}$$

where J is the "canonical angular momentum" defined by [2] (see also [5]),

$$J = -\frac{1}{16\pi} \oint_\infty (2\phi_b\pi^{ab} + 4\phi^b A_b E^a)dS_a \tag{13}$$

where ϕ^a is an asymptotic rotational Killing field on Σ. (In eq. (12), I have inserted primes on the quantities P'^{ab}, etc. appearing in the volume integral to alert the reader to the fact that these quantities depend upon the choice of N^μ and, hence are different in eqs. (11) and (12), since different choices of N^μ have been made. The Hamiltonian functions appearing on the left sides of these equations also, of course, are different for the same reason, but since I prefer to use H to denote the Hamiltonian (6) for any choice of N^μ, I have not inserted a prime on H in eq. (12).) Equations (11) and (12) give explicit expression of property (iii) stated above.

The above formulas (11) and (12) are easily generalized to the case where Σ is a manifold with boundary, i.e., when, in addition to having an asymptotically flat "end", Σ also possesses a regular "interior boundary", S. In that case, the integrations by parts needed to put the "volume contribution" to the variation of H in the form (7) also give rise to surface terms from S. These additional surface terms are readily computed (see [2] for their explicit form).

The strengthened form of the first law of black hole mechanics follows directly from the above three properties of the Hamiltonian formulation of Einstein-Maxwell theory. Let $(M, g_{\mu\nu}, A_\mu)$ be a solution to the Einstein-Maxwell equations describing a stationary-axisymmetric black hole, whose event horizon is a bifurcate Killing horizon, with bifurcation surface S. Let t^μ and ϕ^μ denote the Killing fields on this spacetime which, respectively, asymptotically approach a time translation and rotation at infinity. We assume that a Maxwell gauge choice has been made so that A_μ is nonsingular everywhere outside the black hole and on the event horizon, and satisfies $\mathcal{L}_t A_\mu = \mathcal{L}_\phi A_\mu = 0$. (Note that the assumption that we can introduce a globally well defined vector potential restricts consideration to the case where the magnetic charge vanishes. However, there actually is no loss of generality in restricting attention to this case, since the magnetic charge always can be put to zero by means of a duality rotation.) Let

$$\chi^\mu = t^\mu + \Omega \phi^\mu \tag{14}$$

denote the linear combination of t^μ and ϕ^μ which vanishes on S. (Equation (14) defines the "angular velocity of the horizon", Ω.) Let Σ be an asymptotically flat hypersurface which terminates on the bifurcation surface S. Let $(h_{ab}, \pi^{ab}, A_a, E^a)$ denote the initial data which is induced on Σ. Finally, let H denote the ADM Hamiltonian asssociated with the time evolution vector field $N^\mu = \chi^\mu$.

Now, let $(\delta h_{ab}, \delta\pi^{ab}, \delta A_a, \delta E^a)$ denote any perturbation of the above initial data which is asymptotically flat, is nonsingular on Σ (including S), and which satisfies the linearized constraint equations. Then, by property (i) above, we have $\delta H = 0$ for this perturbation. However, by property (ii) above, together with the fact that χ^μ is a symmetry of the background solution

(i.e., $\mathcal{L}_\chi g_{\mu\nu} = \mathcal{L}_\chi A_\mu = 0$), it follows immediately that the "volume contribution" to δH vanishes. Thus it is clear from eqs. (11) and (12) that property (iii) will give rise to a formula relating the variations in mass, angular momentum, and charge associated with the perturbation (i.e., the "surface terms" from infinity) to a surface contribution from S (which was not included in eqs. (11) and (12) above). Since $N^\mu = \chi^\mu$ vanishes on S, the evaluation of this boundary term from S simplifies considerably. The final result thereby obtained is the following [2]: For any nonsingular, asymptotically flat perturbation of a stationary, axisymmetric black hole with bifurcate horizon, we have,

$$\delta m + V\delta Q - \Omega\delta J = \frac{1}{8\pi}\kappa\delta A \tag{15}$$

Here, κ denotes the surface gravity (see, e.g., [6]) of the horizon and

A denotes the area of S. (The term $\frac{1}{8\pi}\kappa\delta A$ is, of course, just the surface contribution from S.) Equation (15) expresses the first law of black hole mechanics.

The above derivation of eq. (15) is considerably simpler than the original derivation given in [7]. More significantly, the result obtained here is considerably stronger:

The derivation of [7] establishes that eq. (15) holds only for perturbations to other stationary, axisymmetric black holes. (An extension of the derivation of [7] to include a somewhat more general class of perturbations which are "$t = \phi$–symmetric" was given in [8].) The above derivation proves that eq. (15) holds for all nonsingular asymptotically flat perturbations which satisfy the linearized constraint equations. As we now shall show, this strengthened form of the first law will enable us to close a gap that had existed for many years in the proof of the black hole uniqueness theorems.

2. Application to the Black Hole Uniqueness Theorems

The conclusion that the charged Kerr solutions are the only stationary black hole solutions in Einstein-Maxwell theory rests on the combined work of many authors. One of the key steps in the argument leading to this conclusion is a theorem of Hawking [9], [10], which usually is quoted as asserting that a stationary black hole must either be static or axisymmetric. Under the assumption that the surface gravity, κ, is nonvanishing (corresponding to the case of a bifurcate Killing horizon – see [11]), Israel's theorem [12], [13] then proves that the only static black holes in Einstein-Maxwell theory are the Reissner-Nordstrom solutions (i.e., the charged Kerr solutions with vanishing angular momentum), whereas the combined work of Carter [14], Robinson [15], Mazur [16], and Bunting (see Carter [17]) establishes uniqueness of the charged Kerr solutions in the stationary, axisymmetric case.

However, Hawking's theorem actually states the following: First, the theorem asserts that the event horizon of a stationary black hole must be a Killing horizon, i.e., there must exist a Killing field χ^μ in the spacetime which is normal to the horizon. If χ^μ fails to coincide with the stationary Killing field t^μ, then it is shown that the spacetime must be axisymmetric as well as stationary. In that case it follows immediately that eq. (14) will hold with $\Omega \neq 0$ – i.e., the black hole will be "rotating" –and t^μ will be spacelike in a neighborhood of the horizon, so that the black hole will be enclosed by an "ergoregion." On the other hand, if t^μ coincides with χ^μ (so that the black hole is "non- rotating") AND if t^μ is globally timelike outside of the black hole (so that no "ergoregions" exist), then it is shown that the spacetime must be static. However, the case where t^μ coincides with χ^μ but fails to be globally timelike outside of the black hole is not ruled out by the theorem, although plausibility arguments against this possibility have been given [10]; see also [18]. Consequently, the standard black hole uniqueness theorems leave open the following loophole: In principle, there could exist additional stationary black hole solutions to the Einstein-Maxwell equations with bifurcate horizon which are neither static nor axisymmetric. Such black holes would have to be nonrotating (in the sense that t^μ coincides with χ^μ) and also would have to have a nontrivial ergoregion. Furthermore, since χ^μ automatically is timelike in a neighborhood of the horizon outside of the black hole (see [19]), this ergoregion would have to be disjoint from the horizon. I now shall show how this loophole can be closed by proving that any nonrotating black hole in Einstein-Maxwell theory whose ergoregion is disjoint from the horizon must be static–even if t^μ is not initially assumed to be globally timelike outside of the black hole. In particular, this gives a direct proof that there cannot exist black holes in Einstein-Maxwell theory whose ergoregion is disjoint from the horizon. The proof relies directly upon the strengthened form of first law of black hole mechanics obtained in the previous section, and thus provides an excellent example of the utility of this result.

Although the derivation of eq. (15) was given above for the case of a stationary, axisymmetric black hole, it is immediately clear that the derivation also applies for a black hole which is merely stationary (i.e., which possesses a Killing field t^μ which approaches a time translation near infinity) but is non-rotating in the sense that t^μ vanishes on S. In that case, we obtain,

$$\delta m + V \delta Q = \frac{1}{8\pi} \kappa \delta A \qquad (16)$$

i.e., eq. (15) holds with $\Omega = 0$. Hence, as an immediate corollary of our strengthened form of the first law of black hole mechanics, we obtain the following result: *For an arbitrary stationary, nonrotating Einstein-Maxwell black hole, any nonsingular, asymptotically flat perturbation of the initial data which satisfies the linearized constraint equations, preserves the charge, **Q**, of the black hole and preserves the area, **A**, of **S** cannot result in a first order change the ADM mass, **m**, of the spacetime.*

We now shall attempt to explicitly construct a perturbation which violates this corollary. As we shall see, this attempt will succeed unless the spacetime is static. Consequently, we shall conclude that every stationary, non-rotating Einstein-Maxwell black hole must be static.

The first (and, technically, most difficult) step in the argument is to prove that in the (unperturbed) stationary black hole spacetime, a maximal (i.e., vanishing trace extrinsic curvature) slice, Σ, always can be chosen which intersects the bifurcation surface, S, is asymptotically flat and is asymptotically orthogonal to t^μ at infinity. A proof that such a slice exists is given in [20], and we refer the reader to that reference for further details.

Now, let $(h_{ab}, \pi^{ab}, A_a, E^a)$ be the initial data which is induced on the maximal slice, Σ, of the previous paragraph for an (unperturbed) nonrotating Einstein-Maxwell black hole. Consider the following perturbation of this initial data:

$$\delta h_{ab} = 4\phi h_{ab} \qquad (17)$$

$$\delta \pi^{ab} = -4\phi \pi^{ab} - \pi^{ab} \qquad (18)$$

$$\delta A_a = -A_a \qquad (19)$$

$$\delta E^a = -6\phi E^a \qquad (20)$$

where ϕ is the solution to

$$D^a D_a \phi - \mu \phi = \rho \qquad (21)$$

on Σ determined by the boundary conditions $\phi \to 0$ at infinity and $\phi = 0$ on S, where

$$\mu = \frac{1}{h}\pi^{ab}\pi_{ab} + E^a E_a + \frac{1}{2}F_{ab}F^{ab} \qquad (22)$$

$$\rho = \frac{1}{4}[\frac{1}{h}\pi^{ab}\pi_{ab} + F_{ab}F^{ab}] \qquad (23)$$

Then it may be verified directly that this perturbation satisfies the linearized constraint equations and also satisfies $\delta Q = \delta A = 0$. However, it also can be proven [2] that this perturbation satisfies $\delta m < 0$ unless $\rho = 0$. Consequently, a contradiction with the first law of black hole mechanics will be obtained unless $\pi^{ab} = 0$ (and also $F_{ab} = 0$), i.e., the first law implies that the full extrinsic curvature of Σ must vanish. By isometry invariance, the one-parameter family of slices, Σ_t, obtained by "time translating" Σ along the orbits of t^μ also must have vanishing extrinsic curvature. However, it then follows that the projection of t^μ normal to these hypersurfaces (i.e., $t'^\mu = -(t^\nu n_\nu)n^\mu$, where n^μ is the unit normal field to Σ_t) must be a Killing field. (Indeed, since t'^μ approaches t^μ at infinity, we actually must have $t'^\mu = t^\mu$.) Hence, the nonrotating black hole possesses a hypersurface orthogonal, timelike Killing field which is everywhere timelike outside the black hole and approaches a time translation at infinity. Thus, the black hole is static, as we desired to show.

Interestingly, this proof does not generalize to Einstein-Yang-Mills case. Indeed, it is argued in [2] that nonrotating black holes which fail to be static will occur in Einstein-Yang-Mills theory, although such solutions, if they exist, should be unstable.

References

[1] R. Arnowitt, S. Deser, and C.W. Misner, "The Dynamics of General Relativity", in *Gravitation: An Introduction to Current Research*, ed. L. Witten (Wiley, New York, 1962).

[2] D. Sudarsky and R.M. Wald, Phys. Rev. D **46**, 1453 (1990).

[3] J. Lee and R.M. Wald, J. Math. Phys. **31**, 725 (1990).

[4] T. Regge and C. Teitelboim, Ann. Phys. **88**, 286 (1974).

[5] J.D. Brown, E.A. Martinez, and J.W. York, Jr., Phys. Rev. Lett. **66**, 2281 (1991).

[6] R.M. Wald, *General Relativity* (University of Chicago Press, Chicago, 1984).

[7] J.M. Bardeen, B. Carter, and S.W. Hawking, Commun. Math. Phys. **31**, 161 (1973).

[8] S.W. Hawking, Commun. Math. Phys. **33**, 323 (1973).

[9] S.W. Hawking, Commun. Math. Phys. **25**, 152 (1972).

[10] S.W. Hawking and G.F.R. Ellis, *The Large Scale Structure of Spacetime* (Cambridge University Press, Cambridge, 1973).

[11] I. Racz and R.M. Wald, Class. and Quant. Grav. **9**, 2643 (1992).

[12] W. Israel, Phys. Rev. **164**, 1776 (1967).

[13] W. Israel, Commun. Math. Phys. **8**, 245 (1968).

[14] B. Carter, in *Black Holes*, ed. by C. DeWitt and B.S. DeWitt (Gordon and Breach, New York, 1973).

[15] D.C. Robinson, Phys. Rev. Lett. **34**, 905 (1975).

[16] P.O. Mazur, J. Phys. **A15**, 3173 (1982).

[17] B. Carter, Commun. Math. Phys. **99**, 563 (1985).

[18] P. Hajicek, Phys. Rev. **D7**, 2311 (1973).

[19] B.S. Kay and R.M. Wald, Phys. Rep. **207**, 49 (1991).

[20] P. Chrusciel and R.M. Wald, "Maximal Hypersurfaces in Aymptotically Stationary Spacetimes", to be published. See also, R. M. Wald, "Maximal Slices in Stationary Spacetimes with Ergoregions" in Brill Festschrift (Vol. 2 of this Proceeding), ed. B. L. Hu and T. A. Jacobson

Gravitational Radiation Antenna Observations

J. Weber* and G. Wilmot *

Abstract

Gravitational radiation antennas have been operating since 1965. A large number of pulses have been observed, coincident on widely separated antennas. These data and the Supernova 1987A observations are reviewed. It is concluded that some of these pulses may have a gravitational radiation origin.

1. Introduction

The theory of elastic solid, and interferometer gravitational radiation antennas has been under development[2] at the University of Maryland since 1957. Aluminum bar systems have been operating continuously since 1965.[1,2]

It is very important to stress that the output of a gravitational antenna (and a neutrino detector as well) differs in fundamental ways from the output of an optical telescope.

When an optical telescope collects light from a star, it can be concluded that most of the light came from the star.

A gravitational antenna - bar or interferometer - has electrical output pulses. For a single antenna, there is no way to guarantee that observed pulses are not noise of internal origin or noise from local disturbances such as lightning. If statistically significant numbers of coincident pulses are observed on widely separated antennas, this is evidence that the pulses have a common origin. Directive information is useful but not conclusive, in identifying the source.

There is no way to be certain that such pulses are or are not due to gravitational radiation.

Therefore the statement which is frequently made that gravitational radiation has not yet been observed is meaningless.

---
*University of Maryland, College Park, Maryland 20742

2. Observations

Interaction with radiation may change the amplitude ρ and phase ϕ of the antenna voltage. Amplitude changes are observed by study of the envelope of the electronics amplifier output. A very stable quartz oscillator tuned to the bar normal mode frequency provides a reference for observing changes in antenna phase.

The instrumentation generates the two dynamical variables

$$x = \rho cos\phi \tag{1}$$

$$y = \rho sin\phi \tag{2}$$

the quantities

$$\left[\frac{d}{dt} \left(x^2 + y^2 \right) \right]^2 , \; algoithm \; A \tag{3}$$

$$\left(\frac{dx}{dt} \right)^2 + \left(\frac{dy}{dt} \right)^2 , \; algorithm \; B \tag{4}$$

have been recorded on tape, for widely separated antennas.

If both antenna outputs exceed a given threshold at the same time, that is defined as a coincidence. A certain number of coincidences will occur, due to chance and it is essential to measure these.

If the two antenna outputs are correlated, the correlations will disappear if data from one antenna are delayed sufficiently with respect to the other antenna. A computer inserts a series of time delays. An average of the numbers of coincidences for different time delays gives the chance coincidence rate, and the standard deviation. If the coincidence rate without time delay exceeds the chance coincidence rate by more than one standard deviation, this is regarded as evidence for correlation of the antenna outputs.

Figure 1 is a histogram for June 1-5, 1973 for bar antennas[6] at the University of Maryland and the Argonne National Laboratory near Chicago, Illinois. Algorithm A is employed. There are 2065 coincidences due to chance, and 2345 coincidences at zero time delay. The excess is 280 coincidences, 5.7 standard deviations.

Algorithms A and B were studied for the period May 1973 - November 1974. Algorithm A gave an excess at zero delay exceeding 3 standard deviations for the period May 20 - June 13, 1973, and a zero delay excess exceeding 4 standard deviations for the periods May 21 - June 25, 1974, and August 3 - October 17, 1974. Figure 2 is the histogram for the August 3 - October 17, 1974 period. Each period data are a continuous sequence of tapes, each with a zero delay excess at the same threshold. For the November 1973 - November 1974 period, algorithm B gave a 3.8 standard deviation zero delay excess only for the period June 19 - July 1, 1974.

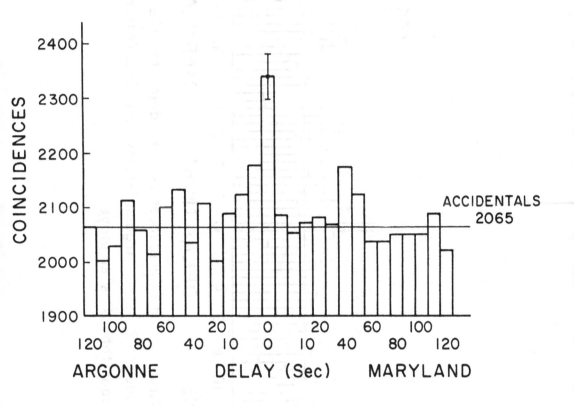

Figure 1. Histogram for June 1-5,1973 data

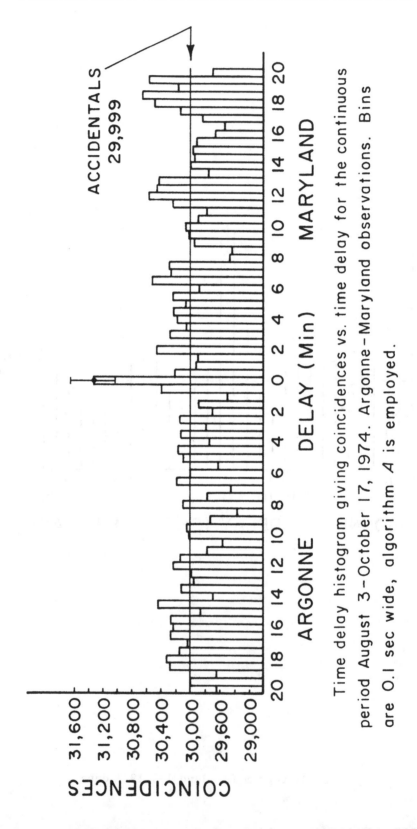

Figure 2. Histogram for a continuous 2½ month period in 1974

There was a data exchange with the University of Rome in 1978, employing algorithm B. Data from the Maryland 3100 kilogram antenna were compared with data from the 390 kilogram University of Rome antenna operating at 4.2 Kelvin for the June 5-13 period. The zero delay excess was 140 coincidences, 3.6 standard deviations.[5]

A gravimeter was emplaced on the lunar surface by the astronauts of the Apollo 17 lunar mission, to employ the moon as a gravitational radiation detector. Some of the data were analyzed and reported in the 1978 PH.D. thesis of R. L. Tobias (Engineering and Physical Sciences Library LD 3231 M 70 d). Correlations were found with the University of Maryland and Argonne National Laboratory gravitational radiation detectors. With a one second delay of the Argonne and Maryland data, coincidences with lunar surface accelerations 2.7 standard deviations above the mean occur for the Maryland data, and 1.5 standard deviations above the mean for the Argonne data. At 6.5 seconds delay of the Argonne and Maryland data, the correlations are 3.7 standard deviations above the mean for the Argonne data and 2.3 standard deviations above the mean for the Maryland data. These correlations can be understood in terms of a lunar model with regions of varying elastic properties.

3.   Supernova 1987A Data

Gravitational antennas at the Universities of Rome and Maryland recorded[4] data during the rapid evolutionary period of Supernova 1987A. Both antennas were operating at room temperature. The Rome antenna has a mass of 2300 kilograms and instrumented normal mode frequency 858 Hertz. The Maryland antenna has a mass of 3100 kilograms and a normal mode frequency at 1660 Hertz.

The neutrino detector at Mont Blanc observed a 5 neutrino burst beginning at 02-52-37 Universal time February 23, 1987. The Rome and Maryland gravitational antennas observed pulses 1.2 seconds earlier.

Professor G. Pizzella of the University of Rome performed the following correlation analyses for a two hour period centered on the 5 neutrino burst, and for other periods as well.

Let $E_M(t)$ be the output of the Maryland antenna, and let $E_R(t)$ be the output of the Rome antenna. $X(t))$ will be one of the four quantities $E_R(t)$; $E_M(t)$; $E_R(t) + E_M(t)$; $E_R(t) \cdot E_M(t)$, and let N_ν be the number of Mont Blanc neutrino detector events. Evaluation of the following quantity is performed:

$$C(\delta, \phi) = \frac{1}{N_\nu} \sum X\left[(t_i + \phi) + \delta\right] \qquad (5)$$

t_i is the time of the ith neutrino event, ϕ is a time in steps of 0.1 seconds for varying the delay between neutrino and gravitational antenna events with initial value 1.2 seconds. δ may be given a sequence of values. If there are correlations between the neutrino and gravitational wave detectors, the value of $C(\delta, \phi)$ for $\delta = 0$ will be the largest value.

There are 79 Mont Blanc neutrino detector events in the two hour period centered on the 5 neutrino burst. 2000 integral values of δ are employed, with ϕ fixed at 1.2 seconds. The value of $C(\delta, \phi)$ with $\delta = 0$ is found to be the largest. Computer analyses show that 12 large (about 100 Kelvin) pulses of the Rome-Maryland gravitational antennas account for most of the correlations, in this two hour period.

The central point is now moved away from the Mont Blanc 5 neutrino burst in steps of 30 minutes. Let n be the number of values of δ giving larger sums than the $\delta = 0$ value. Figure 3 is a curve of n as a function of the central point time. The open circles are for $E_R(t) \cdot E_M(t)$, and the black filled circles are for the sum $E_R(t) + E_M(t)$.

Figure 4 is a display of comparable data for the Mont Blanc, Kamioka, and Frejus elementary particle detectors.

Correlations of the two gravitational wave antennas are studied in the following way. For the central time 0245 Universal time, δ is kept fixed at zero, and ϕ of equation (5) is varied in steps of 0.1 seconds. Rome and Maryland gravitational antennas are treated separately. Figure 5 lower curve shows the Rome gravitational antenna correlation with the Mont Blanc neutrino detector, and the upper curve shows the Maryland gravitational antenna correlation with the Mont Blanc neutrino detector.

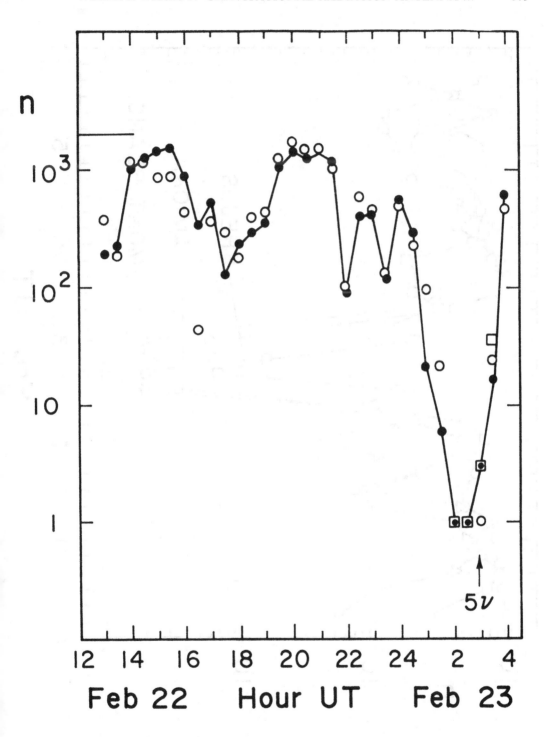

Figure 3. Supernova 1987 A correlations of combined Rome and Maryland gravitational antenna data with Mont Blanc neutrino detectors

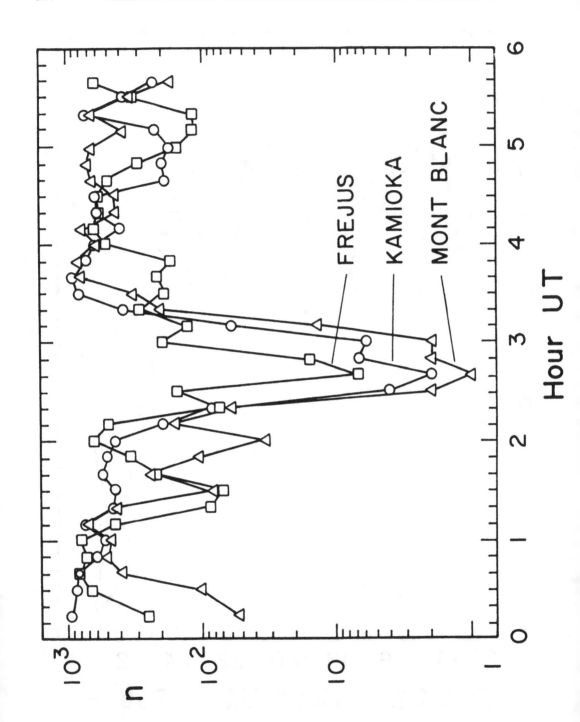

Figure 4. Supernova 1987 A correlations of combined Rome and Maryland gravitatio
antenna data with Mont Blanc,Kamioka,and Frejus observatories

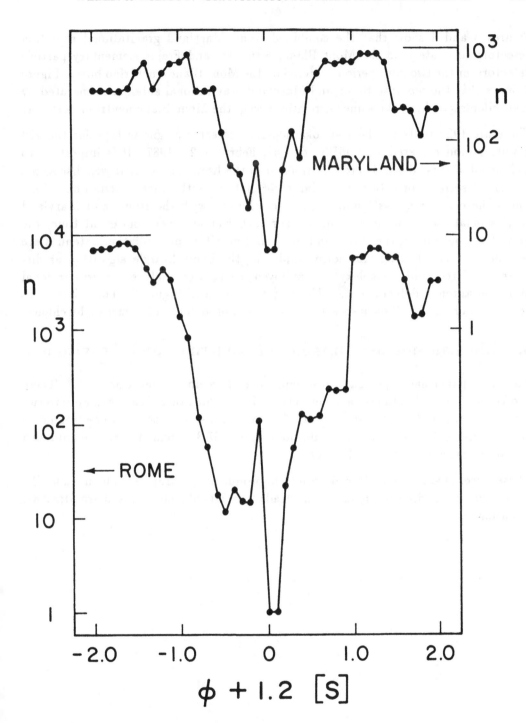

Figure 5. Supernova 1987 A correlations of Mont Blanc neutrino detectors with Rome
gravitational antenna (lower curve) and with Maryland gravitational
antenna (upper curve)

Figures 3 and 4 show that the combined Rome-Maryland gravitational radiation detectors correlate with the Mont Blanc, Kamioka, and Frejus elementary particle detectors for the two hour period centered on the Mont Blanc 5 neutrino burst. Figure 5 shows that the separate Rome and Maryland gravitational antennas, separated by 8000 kilometers, have the same correlations with the Mont Blanc neutrino detector.

There are 192 events on the Kamioka neutrino detector magnetic tape for the 120 minute period centered at 0245UT, day 54, February 23, 1987. It is important to include all of these data. There was an error in the Kamioka clock. It was discovered that a 7.5 second correction led to large correlations with other neutrino detectors. The same correction was found to give correlations with the Rome and Maryland gravitational radiation antenna data. For each Kamioka neutrino event time, the sum of the outputs of the Rome and Maryland gravitational radiation antennas are recorded. A fast Fourier transform employing the Lomb Scargle algorithm as discussed by Press and Teukolsky[7] is employed, to Fourier analyze the gravitational radiation antenna detector data. The result is shown in Figure 6. There is a large peak at a period of 67.3 seconds with a 1.5 percent probability of occurring by chance.

4.   Mechanisms for Emission of Neutrinos and Gravitons

An earlier paper had reported millisecond correlations in the neutrino data.[8] These and the 67.3 second data give a supernova model in which there is a break up into two very dense fragments orbiting each other with a 67.3 second period. The fragments have millisecond period internal oscillations. Every 67.3 seconds their close approach produces extremely large tidal forces.

These forces change the matter distribution, producing quadrupole moments and lifting the Pauli principle degeneracy. This leads to bursts of neutrinos and gravitational radiation.

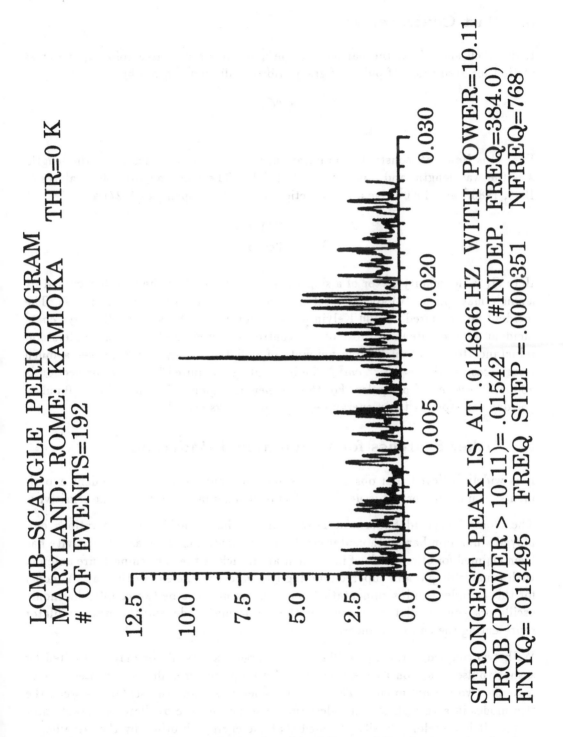

Figure 6. Fourier transform analysis of combined Rome and Maryland gravitational
antenna data with all Kamioka pulses during the two hour period centered
at 0245 UT day 54, February 23, 1987

5. Past Controversy

If the 1960 analysis of the bar as a single mass quadrupole is employed, the cross section for absorption of pulses of gravitational radiation is given by

$$\sigma = \frac{8\pi^3 G M L^2}{c^2 \lambda} \tag{6}$$

Here G is Newton's constant of gravitation, M is the reduced mass, L is the length, λ is the wavelength, and c is the speed of light. The many quadrupole analyses of 1984, 1986,[1] and 1990[3] give a cross section, derived in appendix (A.27) as

$$\sigma_N = \frac{2\pi^3 G M L^3 Q_1}{3 c^2 \lambda S_a}$$

Here Q_1 is the quality factor of a single mass plane and S_a is the length occupied by a single mass plane. (6) gives a total cross section $\sim 1.4 \times 10^{-24}$ cm^2 for pulses. The incident flux required for 100 Kelvin pulses according to (6) is about 10^{10} ergs cm^{-2} incident on the antenna. Most of the controversy originated from the understood reluctance to accept such large amounts of radiated energy. (A.27) gives a much larger cross section $\sim 10^{-18}$ cm^2. The incident flux required by this cross section is about 10^4 ergs cm^{-2} per pulse. For the wavelength region $\sim 10^7$ cms, about 10 pulses per day corresponds to an energy density 10^{-11} ergs cms^{-3}.

6. Other Searches for Gravitational Radiation

It is widely believed that observations should be confirmed by independent experiments. A most important issue is the effect of changes in apparatus on the sensitivity.

The Rome-Maryland Supernova 1987A, and earlier coincidence observations employed aluminum bars with center crystal instrumentation. Increased sensitivity has been claimed for other kinds of instrumentation such as the low temperature antenna with tunable diaphragm superconducting transducer at the end of the bar. A model for this cylinder end instrumentation is usually assumed to be two coupled harmonic oscillators—one oscillator representing the bar normal mode, and the other oscillator representing the end transducer.

With two coupled harmonic oscillators, the large mass oscillator may be excited by a short pulse. As soon as the exciting pulse drops to zero, all the accepted energy is in the two normal modes. The absorbed energy is found to oscillate between the two modes in a very short time—less than a second for the oscillator constants now in use. It is therefore usually assumed that the energy absorbed by the large bar is very quickly transferred to the small end transducer, giving it a large amplitude for accurate observation.

The quantum theory of the gravitational antenna implies that a short pulse of high frequency gravitational radiation may excite phonons in any mass elements. Unlike the two coupled harmonic oscillator model, when the excitation which produced the phonons has dropped to zero, the phonons remain, oscillating from end to end of the bar. Center crystal instrumentation senses the large number of repeated nearly identical pulses. The bar end transducer requires a long period to absorb a significant fraction of the bar energy. Clearly the two kinds of instrumentation have different total cross sections.

7.   Conclusion

For extended periods including the rapid evolutionary period of Supernova 1987A, statistically significant numbers of coincident pulses are observed on widely separated gravitational antennas.

It is not certain what fraction of these pulses are, or are not, due to gravitational radiation.

Pulse heights observed during the Supernova 1987A rapid evolutionary period are in good agreement with predictions based on the antenna theory cross sections published before the supernova.

These data are evidence that at least some of the observed pulses have a gravitational radiation origin.

A   Appendix

A1.   Antenna Cross Sections

General Relativity Theory employs curved spacetime Riemannian Geometry. The four dimensional squared interval between two closely spaced events is given by

$$- ds^2 = g_{\mu\nu} dx^\mu dx^\nu \tag{A.1}$$

Repeated indices are summed over the 4 coordinates. The raised index metric tensor is defined by

$$g_{\mu\alpha} g^{\nu\alpha} = \delta_\mu^\nu \tag{A.2}$$

The Kronecker tensor δ_μ^ν is one if $\mu = \nu$ and zero if $\mu \neq \nu$. Christoffel symbols are defined by

$$\Gamma_{\alpha\beta}^\mu = \frac{1}{2} g^{\mu\gamma} \left(\frac{\partial g_{\gamma\alpha}}{\partial x^\beta} + \frac{\partial g_{\gamma\beta}}{\partial x^\alpha} - \frac{\partial g_{\alpha\beta}}{\partial x^\gamma} \right) \tag{A.3}$$

The Riemann tensor is formed from the $\Gamma_{\alpha\beta}^\mu$ and derivatives as

$$R_{\alpha\beta\nu}^\mu = \frac{\partial \Gamma_{\alpha\nu}^\mu}{\partial x^\beta} - \frac{\partial \Gamma_{\alpha\beta}^\mu}{\partial x^\nu} + \Gamma_{\alpha\nu}^\sigma \Gamma_{\sigma\beta}^\mu - \Gamma_{\alpha\beta}^\sigma \Gamma_{\sigma\nu}^\mu \tag{A.4}$$

These equations are valid in arbitrary coordinates. However certain special coordinate systems are useful in achieving intuitive understanding of important defined quantities.

The Lorentz metric of Special Relativity theory has $g_{\mu\nu}$ diagonal with elements +1 and -1.

Such a metric may be employed everywhere if spacetime is flat. For curved spacetime, coordinates may be chosen which have the Lorentz metric in the vicinity of one point, called the pole, and vanishing first derivatives of the $g_{\mu\nu}$ at the pole. This is the normal coordinate system which we shall employ in what follows.

Plane gravitational waves which propagate in the x^1 direction may be described by the metric tensor components g_{22}, g_{33}, and g_{23}. There are two polarizations. One involves g_{22} and g_{33} satisfying

$$\frac{\partial^2 g_{22}}{\partial x^{o2}} + \frac{\partial^2 g_{33}}{\partial x^{o2}} = 0 \tag{A.5}$$

The second state of polarization is described by g_{23}.

The spacetime curvature is propagated by gravitational radiation. A quadrilateral made up of light rays in the $x^1 x^2$ plane would have the appearance shown in Figure

7. For the curved space the sum of the angles $\theta_1 + \theta_2 + \theta_3 + \theta_4$ differs from 2π. The "excess" is $\Delta\theta = \theta_1 + \theta_2 + \theta_3 + \theta_4 - 2\pi$ and related to the Riemann tensor by

$$R_{1212} = -\frac{1}{2}\frac{\partial^2 g_{22}}{\partial x^{1^2}} = g_{s_q \to 0}^{lim}\frac{\Delta\theta}{S_q} \tag{A.6}$$

In (A.6), g is the determinant of $g_{\mu\nu}$. S_q is the area of the quadrilateral.

The propagating gravitational wave will have $\Delta\theta$ positive, negative or zero depending on the instantaneous value of the Riemann tensor.

An early analysis considered a bar gravitational antenna as a single large mass quadrupole, with reduced mass m, as a classical harmonic oscillator driven by the Riemann tensor. The equation of motion is

$$c^2 m\frac{d^2\xi^\mu}{dx^{o^2}} + Dc\frac{d\xi^\mu}{dx^o} + k\xi^\mu = -mc^2 R^\mu_{oao}r^\alpha \tag{A.7}$$

A bar gravitational radiation antenna is a large number of atoms coupled by chemical forces and described by modern 20th century quantum theory and elementary particle physics. It is a many body problem.

The propagating gravitational wave Riemann tensor is a spin 2 rest mass zero graviton field with interaction Hamiltonian for a single small quadrupole

$$H' = -mc^2 R_{\ell ojo}r^\ell q^j \tag{A.8}$$

In (A.8) Latin indices are summed only over the three space coordinates. The single quadrupole has reduced mass m, mass separation r^ℓ and harmonic oscillator displacement operator q^j. Absorption cross sections are computed by writing the S matrix as

$$S = \frac{mc}{\hbar}\int < F \mid \left(R_{otoj} + R^\dagger_{otoj}\right) r^t \left(q^j + q^{jt}\right)\bar\phi_A\phi_A \mid 0 > d^4x \tag{A.9}$$

The second quantized Riemann tensor is the summation

$$R_{otoj} = \frac{1}{\sqrt{2\pi}}\sum_k R^k_{otoj}e^{\frac{i}{\hbar}(\bar p_k \cdot \bar r)}d_k \tag{A.10}$$

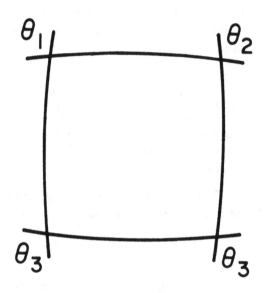

Figure 7. Light ray quadrilateral in curved spacetime

R_{otoj}^k is a normalization factor and the operator d_k is an annihilation operator for a graviton with wavenumber k. $\dot{\phi}_A$ is a creation operator for a mass quadrupole oscillator with harmonic oscillator coordinate q^j and ϕ_A is the annihilation operator.

$$\phi_A = \sum_S \sum_R \phi_{SR}(\bar{r} - \bar{r}_R)a_{SR} \tag{A.11}$$

In A(11) a_{SR} is an annihilation operator for the harmonic oscillator state with wavefunction ϕ_{SR}. If the one mass quadrupole undergoes a transition from the state with energy E_ν to the state with energy $E_{\nu\pm1}$, A(9) is evaluated, with time coordinate t as

$$S = \frac{mc^2}{\hbar\sqrt{2\pi}} \int \sum_k R_{otoj}^k r^t q_{\nu\pm1,\nu}^j e^{-\frac{i}{\hbar}(\bar{p}_k \cdot \bar{r}) + i(\omega\pm\omega_0)t} dt \tag{A.12}$$

In A(12) $q_{\nu\pm1}$ is the harmonic oscillator matrix element for the $\nu \to \nu \pm 1$ transition.

As noted earlier, the bar antenna is a many body problem. Gravitons which are absorbed are observed at the detection system with no way of determining which element of mass absorbed them. (A.12) must therefore be summed over all mass elements. Present bars are instrumented for the lowest frequency compressional mode for which the bar length is half an acoustic wavelength - very short compared with the wavelength of the gravitational radiation. Phase factors are close to unity. The N quadrupole sum is then

$$S_N = \frac{Nmc^2}{\hbar\sqrt{2\pi}} \int \sum_k R_{oloj}^k r^\ell q_{\nu\pm1,\nu}^j e^{i(\omega\pm\omega_0)t} dt \tag{A.13}$$

Integration of A(13) over time gives a delta function in frequency, and the result is

$$S_N = \frac{Nmc^2\sqrt{2\pi}}{\hbar} R_{oloj}(\omega_o) r^\ell q_{\nu\pm1,\nu}^j \tag{A.14}$$

In A(14) $R_{oloj}(\omega_o)$ is the Fourier transform of the Riemann tensor at angular frequency ω_o. The cross section for net absorption is given by

$$\sigma_N = \frac{2\pi N^2 m^2 c^4 V \mid R_{oloj}(\omega_o) r^\ell \mid^2 (q_{\nu+1,\nu}^{j2} - q_{\nu-1,\nu}^{j2})}{\hbar^2 \tau c} \tag{A.15}$$

in (A.15), τ is a very long time.

The harmonic oscillator squared matrix elements are

$$q_{\nu+1,\nu}^2 = \frac{(\nu + 1)\hbar}{2m\omega} \tag{A.16}$$

$$q_{\nu-1,\nu}^2 = \frac{\nu\hbar}{2m\omega} \tag{A.17}$$

The gravitational stress energy pseudotensor leads to

$$| R_{o\ell oj}(\omega_o) |^2 = \frac{4\pi\omega_o G U_{INCIDENT}}{Ac^7} \tag{A.18}$$

In (A.18) $U_{INCIDENT}$ is the incident gravitational radiation energy, A is the cross sectional area for the normalization volume V.

$$U_{INCIDENT} = \frac{c\tau\hbar\omega A}{V} \tag{A.19}$$

for N quadrupoles these equations give a cross section

$$\sigma_N = \frac{8\pi^3 G m r^2 N^2}{c^2 \lambda} \tag{A.20}$$

In (A.20) λ is the gravitational wavelength corresponding to the normal mode frequency ω_o.

The quadrupoles may be chosen in a number of ways. For one dimensional (longitudinal) waves at the lowest normal mode frequency with bar a half acoustic wavelength long, consider the acoustic impedance looking to the right. This is the complex conjugate of the acoustic impedance looking to the left. Therefore a plane of atoms will have a resonance frequency the same as the normal mode.

Consideration of the planar mass and spring constants leads to a resonant frequency which is the Debye frequency, orders higher than the normal mode frequency. However the lowest frequency normal mode is one in which planes are moving together coherently, and the relative displacement between planes is orders smaller than it would be if a single plane above were displaced.

For the coherent process considered here the atoms do move coherently and a resonant angular frequency of each plane is ω_o.

The mean square distance for the planes is

$$< r^2 > = \frac{2}{L} \int_0^{L/2} r^2 dr = \frac{L^2}{12} \tag{A.21}$$

The total mass M is Nm and (A.20) becomes

$$\sigma_N = \frac{2\pi^3 G M L^3}{3c^2 \lambda S_a} \tag{A.22}$$

In (A.22), S_a is the length occupied by one of the mass planes.

Professor Guiliano Preparata[3] has given a brilliant and clever discussion of the cross section. He considers individual atoms instead of planes.

Each quadrupole is then an atom driven with respect to the center of mass, with restoring forces associated with the Debye angular frequency ω_D. No significant Riemann tensor components are present at the Debye frequency ω_D. Nonetheless energy may be exchanged if each oscillator is imagined to go from a quantum state with energy E_α to the state

$$\psi = (1 - \mid a_{\alpha \pm 1} \mid^2)\psi_\alpha + a_{\alpha \pm 1}\psi_{\alpha \pm 1} \tag{A.23}$$

Energy conservation requires

$$\int \psi^* H \psi d^3 x = \alpha \hbar \omega_D \pm \hbar \omega_o \tag{A.24}$$

Evaluating (A.24) gives

$$\mid a_{\alpha \pm 1} \mid^2 = \frac{\omega_o}{\omega_D} \tag{A.25}$$

These relations plus the harmonic oscillator matrix elements give a cross section

$$\sigma_{NP} = \frac{2\pi^3 G M L^2}{3c^2 \lambda} \left(\frac{\omega_0}{\omega_D}\right)^2 N_{ATOMS} \tag{A.26}$$

For bar antennas now operating (A.26) and (A.20) are approximately equal.

Important corrections must be made for very large damping associated with single quadrupoles for a room temperature detector. These and possible restrictions on the Fourier components which may reach the detector lead to

$$\sigma_N = \frac{2\pi^3 G M L^3 Q_1}{3c^2 \lambda S_a} \tag{A.27}$$

$$\sigma_{NP} = \frac{2\pi^3 G M L^2 N_{ATOMS}}{c^2 \lambda} \left(\frac{\Delta\omega}{\omega_0}\right)\left(\frac{\omega_0}{\omega_D}\right)^2 \tag{A.28}$$

In (A.27), Q_1 is the quality factor of individual crystal planes, estimated[1] as 10^{-4} in 1986 for the room temperature antennas. For a high quality factor antenna such as a single crystal of silicon, $Q_1 \to 1$.

B Appendix

Data Analyses Checks

Prior to 1970, the outputs of two widely separated gravitational antennas were compared by transmission of data to a common collection point by telephone lines. An analogue coincidence detector marked pen and ink recorder charts when the two detector output exceeded a preset threshold within a specified short time. A human observer studied the charts.

Small numbers of coincident pulses were observed. In 1970, magnetic tape recorders were installed, and tapes searched for coincidences by computer.

There were many checks of the procedures. Pen and ink analogue coincidence experiments continued. These had two outputs, one with and the other without a time delay in one channel. Separate charts were provided for data with and without the time delay. These charts were marked in coded form which was changed at certain times. Mrs. Alessandra Exposito searched the charts for coincidences without knowledge of which charts had the time delay. Excess coincidences were found in the channel with no time delay during the same periods that the computer-magnetic tape system found significant numbers of coincidences.

For an extended period, the computer programmer was Mr. Michael Lee. As a further check, Dr. William Davis inserted artificial pulses into the gravitational antennas at times not known to Lee. Lee correctly identified times when artificial pulses were present and when they were absent.

Mr. Bruce Webster prepared copies of some old magnetic data tapes. Some were identical with originals, others had certain channels replaced by detector data from different periods of time. These were coded and not further identified until Lee analyzed them. The analysis of the exact copies by Lee agreed with his analyses originally done on those tapes. No significant zero delay excess was found on tapes with exchanged channels.

Very considerable controversy was generated as a result of disagreements among scientists concerning one tape, number 217, and discussed in the text for the period June 1-5, 1973. The histogram shown as Figure 1 was obtained by computer programming analyses, with checks independently performed by Dr. Paul C. Joss, Dr. S. Steppel, and Mr. Michael Lee.

The telephone circuit from one antenna to the second site transmitted the data in coded digital form. For an extended period the telephone circuit was removed, and data collected by separate magnetic tape units at each site. A significant zero delay excess was observed for this period, exceeding 4 standard deviations.

Bibliography

1. J. Weber, Phys. Rev. 117, 306 (1960), General Relativity and Gravitational Waves, Interscience-John Wiley, New York, London, 1961; Chapter 8 General Relativity and Gravitation, Volume 2, A. Held (Editor) Plenum Publishing Company (1980); Chapter 1, Volume 3, Sir Arthur Eddington Centenary Symposium, World Scientific 1986.

2. Moss, Miller, Forward Applied Optics 10, 2495 (1971).

3. G. Preparata, Modern Physics Letters A, Volume 1, 1990.

4. M. Aglietta et al., IL Nuovo Cimento, Volume 12C, N.1, 75, (1989); Volume 14C, N.2, 171, (1991); Volume 106B, N.11, 1257 (1991).

5. Ferrari, Pizzella, Lee, Weber, Phys. Rev. D25, 2471 (1982).

6. Gretz, Lee, Steppel, Weber, Phys. Rev. D14, 4, 893 (1976).

7. W. H. Press and S. A. Teukolsky, "Search Algorithm for Weak Periodic Signals in Unevenly Spaced Data," Computers in Physics, Nov/Dec 1988, 77-82.

8. P. Raychaudhuri and D. Saha, Modern Physics Letters A, Volume 1, 61, (1990).

The Back-Reaction Is Never Negligible: Entropy of Black Holes and Radiation

JAMES W. YORK, JR.

University of North Carolina, Chapel Hill

Even though the quantum stress-energy tensors of conformal scalar fields and electromagnetic fields renormalized on a Schwarzschild background violate the classical energy conditions by a wide margin, the corresponding equilibrium thermodynamical entropy ΔS by which these fields augment the usual black-hole entropy is found to be positive if the back-reaction is taken into account. This is the expected conclusion if the statistical interpretation of entropy is valid in the semi-classical approach to quantum field theory in curved spacetime. However, the calculated thermodynamical entropy ΔS would not turn out to be strictly positive if the back-reaction were ignored. Furthermore, the derivative of ΔS with respect to radius, at fixed black hole mass, is found to vanish at the horizon for *all* regular renormalized quantum stress-energy tensors. This physically natural property would also fail if the back-reaction were ignored. In this context, therefore, the back-reaction can never be regarded as "negligibly small".

1 DEDICATION

This paper is dedicated to Charles W. Misner on his sixtieth birthday, with best wishes and with sincere gratitude for his encouragement early in my career.

2 STRESS-ENERGY TENSORS

Stress-energy tensors renormalized on a Schwarzschild background have been calculated for conformal scalar fields and for $U(1)$ gauge fields (Maxwell fields), respectively, by Howard (1984) and by Jensen and Ottewill (1989). Both results can be written in the form

$$\langle T^\mu_\nu \rangle_{\text{renormalized}} = \langle T^\mu_\nu \rangle_{\text{analytic}} + \left[\frac{\hbar}{\pi^2 (4M)^4} \right] \Delta^\mu_\nu, \tag{1}$$

where the analytic piece, in the case of the conformal scalar field, was given by Page (1982), and Δ^μ_ν is obtained from numerical evaluation of a mode sum. The numerical piece is small compared to the analytic piece that will be used in the present work. This does not change any of the results qualitatively because both pieces obey all requisite regularity conditions. The analytic part will be denoted below simply by

T^μ_ν. The analytic piece, in both cases, contains the exact trace anomaly. Units are chosen in which $G = c = k_B = 1$, but $\hbar \neq 1$.

The stress-energy tensors satisfy $\hat{\nabla}_\mu T^\mu_\nu = 0$ on the Schwarzschild background with metric

$$\hat{g}_{\mu\nu} = \text{diag}\left[-\left(1 - \frac{2M}{r}\right), \left(1 - \frac{2M}{r}\right)^{-1}, r^2, r^2 \sin^2\theta\right]. \tag{2}$$

These tensors represent the stress-energy required to equilibrate the black hole with its Hawking radiation. Each satisfies the regularity condition $T^t_t = T^r_r$ at the horizon $r = 2M$ and each has the asymptotic form of a flat-space radiation stress-energy tensor at the uncorrected Hawking temperature of an ordinary Schwarzschild black hole, which is denoted here by $T_H = \hbar(8\pi M)^{-1}$. (The scalar field has only one helicity state, while the vector field has two, so an explicit $1/2$ is displayed explicitly below.)

For the conformal scalar field (Page 1982), we have

$$T^t_t = -\frac{1}{3}aT^4_H\left(\frac{1}{2}\right)(3 + 6w + 9w^2 + 12w^3 + 15w^4 + 18w^5 - 99w^6), \tag{3}$$

$$T^r_r = \frac{1}{3}aT^4_H\left(\frac{1}{2}\right)(1 + 2w + 3w^2 + 4w^3 + 5w^4 + 6w^5 + 15w^6), \tag{4}$$

$$T^\theta_\theta = T^\phi_\phi = \frac{1}{3}aT^4_H\left(\frac{1}{2}\right)(1 + 2w + 3w^2 + 4w^3 + 5w^4 + 6w^5 - 9w^6), \tag{5}$$

where $w = 2Mr^{-1}$, $3^{-1}aT^4_H = 3^{-1}[\pi^2(15\hbar^3)^{-1}][\hbar(8\pi M)^{-1}]^4 = \epsilon(48\pi KM^2)^{-1}$, $\epsilon = \hbar M^{-2}$, and $K = 3840\pi$. For the $U(1)$ vector field (Jensen and Ottewill 1989), we have

$$T^t_t = -\frac{1}{3}aT^4_H(3 + 6w + 9w^2 + 12w^3 - 315w^4 + 78w - 249w^6), \tag{6}$$

$$T^r_r = \frac{1}{3}aT^4_H(1 + 2w + 3w^2 - 76w^3 + 295w^4 - 54w^5 + 285w^6), \tag{7}$$

$$T^\theta_\theta = T^\phi_\phi = \frac{1}{3}aT^4_H(1 + 2w + 3w^2 + 44w^3 - 305w^4 + 66w^5 - 579w^6). \tag{8}$$

In both cases $T^r_r > 0$ and the energy density $-T^t_t$ is negative in the vicinity of the horizon, which violates the weak energy condition. For the scalar field this occurs from $r = 2M$ to $r \approx 2.34M$ and for the vector field from $r = 2M$ to $r \approx 5.14M$. Both tensors also violate the dominant energy condition in a region surrounding and bordering on the horizon.

3 EFFECT ON THE SCHWARZSCHILD SPACETIME

We will obtain the back-reaction from the semi-classical Einstein equation

$$G^\mu_\nu = 8\pi\langle T^\mu_\nu\rangle_{\text{analytic}} \equiv 8\pi T^\mu_\nu. \tag{9}$$

To be consistent with $\hat{\nabla}_\mu T^\mu_\nu = 0$, we take $g_{\mu\nu} = \hat{g}_{\mu\nu} + \delta g_{\mu\nu}$ and solve the linearized version of (9) for fractional corrections to the Schwarzschild metric through first order in $\epsilon = \hbar M^{-2}$. A suitable framework has been set up in detail by York (1985a). The method yields a metric of the form

$$ds^2 = -\left(1 - \frac{2m(r)}{r}\right)(1 + 2\epsilon\bar{p})dt^2 + \left(1 - \frac{2m(r)}{r}\right)^{-1} dr^2 + r^2 d\omega^2 \qquad (10)$$

with $d\omega^2$ the standard metric of a unit sphere. The mass is given by

$$m(r) = M(1 + \epsilon\mu(r)) = M + M_{\text{rad}}(r), \qquad (11)$$

where an additive constant of integration in the function $\mu(r)$ has been absorbed into the definition of the "dressed" or "renormalized" black-hole mass M in (11), which will not be distinguished notationally from the "bare" Schwarzschild mass because the latter has no further physical significance in this work. Thus, we have arranged that $\mu(r = 2M) = 0$ and the effective mass function of the radiation is

$$M_{\text{rad}}(r) = \int_{2M}^r (-T^t_t) 4\pi\tilde{r}^2 d\tilde{r} = \epsilon M\mu(r), \qquad (12)$$

which defines $\mu(r)$. For the scalar field, one finds (York 1985a)

$$K\mu_{\text{s}} = \frac{1}{3}w^{-3} + w^{-2} + 3w^{-1} - 4\ell n(w) - 5w - 3w^2 + 11w^3 - \frac{22}{3}. \qquad (13)$$

For the vector field, one finds (Hochberg and Kephart 1992),

$$K\mu_{\text{v}} = \frac{2}{3}w^{-3} + 2w^{-2} + 6w^{-1} - 8\ell n(w) + 210w - 26w^2 + \frac{166}{3}w^3 - 248. \qquad (14)$$

In both (13) and (14), note that the first term on the right multiplied by $\epsilon M K^{-1}$ corresponds to the naive flat-space value $aT^4_{\text{H}}V$ for the radiation energy.

The metric is completed by the determination of $\bar{p}$ which, like μ, can be found from an elementary integration. Defining $K\bar{p} = K\rho + k$, with k an integration constant, we have

$$\rho = \frac{1}{\epsilon}\int_{2M}^r (T^r_r - T^t_t)(\tilde{r} - 2M)^{-1}4\pi\tilde{r}^2 d\tilde{r}. \qquad (15)$$

For the scalar field, one finds (York 1985a)

$$K\rho_{\text{s}} = \frac{1}{3}w^{-2} + 2w^{-1} - 4\ell n(w) - \frac{20}{3}w - 5w^2 - \frac{14}{3}w^3 + \frac{42}{3}. \qquad (16)$$

Note that at the horizon $r = 2M$, or $w = 1$, we have $\rho(1) = 0$. The constant k for the scalar field is denoted by k_{s} and will be determined later from a boundary condition. Similarly, for the vector field, we find $K\bar{p}_v = K\rho_v + k_v$, where (Hochberg and Kephart 1992)

$$K\rho_v = \frac{2}{3}w^{-2} + 4w^{-1} - 8\ell n(w) + \frac{40}{3}w + 10w^2 + 4w^3 - 32. \tag{17}$$

Because both radiation tensors approach asymptotically constant values, it is clear that the system composed of black hole plus equilibrating radiation must be put in a finite "box". Otherwise, $\delta g_{\mu\nu}$ would not remain small compared to $g_{\mu\nu}$. Physically, this means that the radiation would collapse onto the black hole, producing a larger one. It is convenient to impose microcanonical boundary conditions (York 1985a). We fix the radius at some value $r = r_0$ where we imagine placing an ideal massless perfectly reflecting wall. Outside r_0 we have then an ordinary Schwarzschild spacetime with Arnowitt-Deser-Misner (ADM) mass $m(r_0)$. Continuity of the three metric induced on the world tube $r = r_0$ fixes the constant of integration k (*i.e.*, k_s or k_v) in $\bar{p}$ by means of

$$kK^{-1} = -\rho(r_0). \tag{18}$$

The finite discontinuities in the extrinsic curvature of $r = r_0$ show, from $T_r^r > 0$, that the wall is in tension, as expected (York 1985a). However, the wall plays no explicit role in what follows. The spacetime geometry, with back-reaction, is now completely determined.

4 TEMPERATURE

If we imagine releasing a small packet of energy from the box through a long thin radial tube, it will red-shift and attain the asymptotic temperature

$$T_\infty = \frac{\kappa_H \hbar}{2\pi} \tag{19}$$

where κ_H is the surface gravity of the event horizon. For an ordinary Schwarzschild black hole (ignoring the radiation), one has $\kappa_H = (4M)^{-1}$ and $T_\infty = T_H = \hbar(8\pi M)^{-1}$. However, the radiation's stress-energy changes the surface gravity κ_H of the horizon. A straightforward calculation (York 1985a) yields

$$\kappa_H = \frac{1}{4M}[1 + \epsilon(\bar{p} - \mu) + 8\pi r^2 T_t^t]|_{r=2M} . \tag{20}$$

With the stated microcanonical boundary condition, we can use (18) to obtain from (19) and (20)

$$T_\infty = \frac{\hbar}{8\pi M}[1 - \epsilon\rho(r_0) + \epsilon n K^{-1}], \tag{21}$$

where n has the value $n_s = 12$ for the scalar field and $n_v = 304$ for the vector field. The local temperature, blue-shifted from infinity back to r_0, is given by

$$T_{\text{loc}}(r_0) = T_\infty[-g_{tt}(r_0)]^{-1/2}. \tag{22}$$

This quantity can be shown to be independent of the choice of boundary conditions at r_0 (York 1985a). Thus we find

$$T_{\text{loc}}(r_0) = \frac{\hbar}{8\pi M}[1 - \epsilon\rho(r_0) + \epsilon n K^{-1}]\left[1 - \frac{2m(r_0)}{r_0}\right]^{-1/2}. \tag{23}$$

Either measure of the temperature, T_∞ or T_{loc}, can be used to calculate the entropy in conjunction with an appropriate measure of energy, as we shall see.

5 THERMODYNAMIC ENTROPY

Suppose the radius r_0 of the box is fixed. Then the first law of thermodynamics for slightly differing equilibrium states tells us that

$$dS = \beta_\infty dm = \beta dE \quad (dr_0 = 0), \tag{24}$$

where $\beta_\infty \equiv T_\infty^{-1}$, $\beta \equiv T_{\text{loc}}^{-1}$, and E is the quasi-local energy (York 1986, Brown *et al.* 1990, Brown and York 1992), which is given in the static spherical case treated here by

$$E(r_0) = r_0 - r_0[g^{rr}(r_0)]^{1/2}. \tag{25}$$

Choosing M and r_0 as independent variables, and fixing r_0, we can readily integrate (24) and obtain S up to a function of r_0 and a constant, as follows.

From (21) we have

$$\beta_\infty = \frac{8\pi M}{\hbar}[1 + \epsilon\rho - \epsilon n K^{-1}], \tag{26}$$

and from (11), holding r_0 fixed,

$$dm = [1 - \epsilon\mu + \epsilon M(\partial\mu/\partial M)]dM. \tag{27}$$

One can see by inspection that (24) is correct, so that using β and dE will give the same result for the entropy as using β_∞ and dm.

Observe that from fractional changes of $\mathcal{O}(\epsilon)$ in the metric, and thence in the surface gravity and the temperature, we are able to calculate from (24) departures of $\mathcal{O}(\epsilon^\circ) = \mathcal{O}(1)$ from the usual black hole entropy $S_{\text{BH}} = (4\pi M^2)\hbar^{-1} = 4\pi\epsilon^{-1}$. But all of these departures are clearly of the same order as the naive flat-space radiation entropy itself:

$$\frac{4}{3}aT_{\text{H}}^3 V_0 = \frac{4}{3}\left(\frac{\pi^2}{15\hbar^3}\right)\left(\frac{\hbar}{8\pi M}\right)^3\left(\frac{4}{3}\pi r_0^3\right) = \frac{8\pi}{K}\left(\frac{8}{9}w_0^{-3}\right) = \mathcal{O}(1). \tag{28}$$

The $\hbar$'s in (28) cancel out, leaving only a function of $w_0 = 2Mr_0^{-1}$. Combining (26) and (27) yields

$$dS = \frac{8\pi M}{\hbar}dM + 8\pi[w_0^{-1}(\rho - \mu) + (\partial\mu/\partial w_0 - nK^{-1}w_0^{-1})]dw_0 \tag{29}$$

with $dr_0 = 0$. Integration of (29) gives an expression of the form

$$S = \frac{4\pi M^2}{\hbar} + \Delta S(w_0) + f(r_0/\ell_P), \qquad (30)$$

where the first term is the usual Bekenstein-Hawking expression for black hole entropy S_{BH}, the second term is a function of w_0 determined up to an additive constant by the second term on the right of (27), and f is a thus-far undetermined dimensionless function of r_0 divided by the Planck length $\ell_P = \hbar^{1/2}$.

Let us first dispose of f. Quantum gravity would be required to supply such a term, if it were to exist in this approximation, because f, which depends on r_0, can only be made dimensionless by using the Planck length. But the Planck length, in turn, enters the present calculation only in conjunction with the third length scale in this problem, the gravitational radius $2M$. However, all dependence of S on $2M$ is in ΔS, which does not depend on $\hbar$! We conclude that $f = 0$ in the semi-classical theory. Actually, a formal proof showing $f = 0$ has been constructed by Comer and York (1991) using the relations between thermodynamical conjugacy and canonical conjugacy established in Brown et $al.$ (1990) that, in turn, are based on analysis of the action principle of general relativity.

In considering ΔS, which will be given explicitly below, we first note the very significant property that

$$\frac{\partial(\Delta S)}{\partial w_0} = 8\pi[w_0^{-1}(\rho - \mu) + \partial\mu/\partial w_0 - nK^{-1}w_0^{-1}] \qquad (31)$$

vanishes at the horizon $w_0 = 1$. Therefore, for fixed black hole mass M, there is initially no increase of entropy as the box wall is moved away from the horizon. This physically natural result follows from several general features of all renormalized stress-energy tensors on the Schwarzschild background and the back-reactions they induce, not just the two cases analysed here. First, both ρ and μ vanish at the horizon, as follows from (15) and the regularity condition that $T_t^t = T_r^r$ at the horizon (Page 1982). More precisely, we have that

$$\lim_{w \to 1}\left(\frac{T_t^t - T_r^r}{1 - w}\right) \text{ exists.} \qquad (32)$$

Second, the last two terms on the right of (31) add to zero at the horizon because the Hamiltonian constraint, the Einstein equation $G_t^t - 8\pi T_t^t = 0$, holds there. Furthermore, if the fractional effects of $\mathcal{O}(\epsilon)$ in the temperature induced by the back-reaction were neglected, the derivative (31) would not vanish at the horizon, as the reader can readily verify.

The vanishing slope of ΔS at the horizon and the presence of S_{BH} in expression (28), with M in S_{BH} having already been renormalized by the absorption of an undetermined integration constant arising from the back-reaction, motivate the choice of the additive constant remaining in ΔS, which can only be a pure number, to be such that $\Delta S = 0$ at $w_0 = 1$. With this choice we obtain for the conformal scalar field (York 1985b, Comer 1990)

$$\Delta S_{\mathrm{s}} = \frac{8\pi}{K}\left(\frac{1}{2}\right)\left(\frac{104}{9}w_0^3 - 8w_0^2 - \frac{40}{3}w_0 + \frac{32}{3}\ell n(w_0) + 8w_0^{-1} + \frac{8}{3}w_0^{-2} + \frac{8}{9}w_0^{-3} - \frac{16}{9}\right). \quad (33)$$

Similarly, for the electromagnetic or $U(1)$ gauge field, we find (Hochberg et al. 1992)

$$\Delta S_{\mathrm{v}} = \frac{8\pi}{K}\left(\frac{344}{9}w_0^3 - 8w_0^2 + \frac{40}{3}w_0 - 96\ell n(w_0) + 8w_0^{-1} + \frac{8}{3}w_0^{-2} + \frac{8}{9}w_0^{-3} - \frac{496}{9}\right). \quad (34)$$

In both expressions, the naive flat-space radiation entropy (28) appears as the next-to-last term on the right. Both ΔS_{s} and ΔS_{v} are positive for $w_0 \in (0,1)$ and vanish at $w_0 = 1$. The reader can verify, by omitting the back-reaction terms in the inverse temperature in (24), that not only is the vanishing slope of ΔS at $w_0 = 1$ lost, but also that the value of the resulting "ΔS", normalized as above, is no longer positive. In this fundamental sense, the back-reaction is never negligible!

ACKNOWLEDGMENTS
I thank G. L. Comer, D. Hochberg, and T. W. Kephart for helpful discussions. This research was supported by National Science Foundation grants PHY–8407492 and PHY–8908741.

REFERENCES

Brown, J. D., G. L. Comer, E. A. Martinez, J. Melmed, B. F. Whiting, and J. W. York (1990). Thermodynamic ensembles and gravitation. *Classical and Quantum Gravity*, **7**, 1433–1444.

Brown, J. D. and J. W. York (1992). Quasi-local energy in general relativity. In *Mathematical Aspects of Classical Field Theory*, Eds M. J. Gotay, J. E. Marsden, and V. E. Moncrief. American Mathematical Society, Providence.

Comer, G. L. (1990). The thermodynamic stability of systems containing black holes. University of North Carolina doctoral dissertation. Unpublished.

Comer, G. L. and J. W. York (1991). Integrability of the entropy. Unpublished.

Hochberg, D. and T. W. Kephart (1992). Down the rabbit hole with gauge bosons. To be published.

Hochberg, D., T. W. Kephart and J. W. York (1992). Positivity of entropy in the semi-classical theory of black holes and radiation. To be published.

Howard, K. W. (1984). Vacuum $\langle T^\mu{}_\nu \rangle$ in Schwarzschild spacetime. *Physical Review*, **D30**, 2532–2547.

Jensen, B. P. and A. Ottewill (1989). Renormalized electromagnetic stress tensor in Schwarzschild spacetime. *Physical Review*, **D39**, 1130–1138.

Page, D. N. (1982). Thermal stress tensors in static Einstein spaces. *Physical Review*, **D25**, 1499–1509.

York, J. W. (1985a). Black hole in thermal equilibrium with a scalar field: The back reaction. *Physical Review*, **D31**, 775–784.

York, J. W. (1985b). Entropy of a conformal scalar field and a black hole. Unpublished.

York, J. W. (1986). Black-hole thermodynamics and the Euclidean Einstein action. *Physical Review*, **D33**, 2092–2099.

Toward a Thesis Topic

J. A. Wheeler

Of all obstacles to understanding the foundations of physics, it is difficult to point to one more challenging than the question,"How Come the Quantum?" unless it be the twin question, "How Come Existence?" Stuck, but studying every available clue, (Box 1), from the papers of Bohr, Einstein, Planck and Schrödinger to the thoughts of the presocratic philosophers, (Box 2), I remember one of the great messages I have received from sixty-five years of research: Why does a university have students? To teach the professsors! Not least in convincing me of that lesson is the wealth of learning that I owe to Charles W. Misner, graduate student at Princeton University from 1953 to 1957.

Already from the time Misner dropped into my office to talk about a conceivable thesis topic, I gained a vivid impression of what it was to see his active mind at work comparing researchable issues in elementary particle physics and in general relativity. "What is timely and tractable?" That is the proper criterion of choice, according to John R. Pierce, that great guide of productive research at Bell Telephone Laboratories and animating spirit of the travelling-wave tube and the Tel Star satellite.

Charles Misner, so far as I could see, used the same criterion in making his decision. It led to a Ph.D. thesis and a 1957 paper in the *Reviews of Modern Physics*, entitled "Feynman quantization of general relativity,"[1] forerunner to the great and influential 1962 paper of R. Arnowitt, S. Deser and Misner on the "Dynamics of General Relativity."[2] That Misner, a top-brain among the graduate students, opted for gravity physics rather than particle physics may have had its part in the decision about this time by a handful of other very able students also to go into gravity physics: Dieter Brill, Jacob Bekenstein, Kip Thorne, Demetrios Christodoulou, John Klauder and Arthur Komar, David Sharp, Robert Euwema, John Fletcher, Kent Harrison, and others.

How would I react if another student came in the door just now asking to discuss possibilities for a thesis topic, someone reputed like Misner to be top ranking among the graduate students? Oh, what a joy that would be! How else can one make headway with a deep problem except by discussing it with a student? And how else can one be taught or set straight in one's thinking except by talking out an idea with someone who satisfies James Bryant Conant's criterion for a student: One with an uncommitted mind?

Any discussion will depend on the promise and challenge of the problem. "Daunting" might have been the word for the question that was put to the real Misner: How to quantize general relativity.? His response to that challenge appears partly in his own writings[1] and partly in papers[2,3] for which he supplied much of the necessary mathematical instruction and inspiration.[4]

Since 1988 I have been bewitched with two questions of a still more difficult cast: "How come the quantum" and "How come existence." If another Misner came in the door at this moment, he might reduce me to sanity with a few well-chosen comments. Or he might ask me to discuss some of the wonderful foundation stones of cosmology and quantum theory as we see them today (Box 1).

Box 1.   Some Key Features of Physics as a Description of Nature

GENERAL RELATIVITY IN BRIEF: Spacetime tells mass how to move, and mass tells spacetime how to move.[5]

ALBERT EINSTEIN: ". . .time and space are modes by which we think and not conditions in which we live."[6]

TODAY'S PHOTON-TO-PARTICLE RATIO, 10^{10}. [7]

NIELS BOHR, COMPLEMENTARITY: "Since, in the observation of these [atomic] phenomena, we cannot neglect the interaction between the object and the instrument of observation, the question of the possibilities of observation again comes to the foreground. **Complementarity: "Any given application of classical concepts precludes the simultaneous use of other classical concepts which in a different connection are equally necessary for the elucidation of the phenomena.**[8]

THE AHARONOV-BOHM EFFECT.[9]

QUANTUM-GRAVITY ANALOGS OF THE AHARONOV-BOHM EFFECT.[10]

IT FROM BIT: Do all things physical submit to an information-theoretic description, in line with the thesis "*it from bit*," that every *it*, every particle, every field of force, even the spacetime continuum itself, derives its way of action and its very existence entirely, even if in some contexts indirectly, from the detector-elicited answers to yes or no questions, binary choices, *bits*? [11]

WILLARD VAN ORMAN QUINE: "Just as the introduction of the irrational numbers...is a convenient myth [which] simplifies the laws of arithmetic...so physical objects are postulated entities which round out and simplify our account of the flux of exis-tence...The conceptual scheme of physical objects is a convenient myth, simpler than the literal truth and yet containing that literal truth as a scattered part." [12]

BLACK HOLE'S HORIZON AREA, expressed in units of the Bekenstein[13]-Hawking[14] multiple, $4 \log_e 2$ ($\hbar$ G/c^3) $\sim 10^{-66}$ cm^2, of the Planck area, measures in BITS how much information it takes to describe, entity by entity and quantum state by quantum state, what went in to the making of that black hole, that IT, an example of the theme of IT FROM BIT.[15]

R. A. FISHER'S proof that the mathematical description of the distinguishability of populations requires REAL PROBABILITY AMPLITUDES in a REAL HILBERT SPACE.[16]

MAX BORN: Translation of the quantum mechanics of Schrödinger and Heisenberg into the language of complex vectors and complex operators in a complex Hilbert space.[17]

STÜCKELBERG and SAXON: Clarification why complementarity demands this COMPLEX feature for the mathematical representation of quantum mechanics.[18,19]

ENVIRONMENT'S influence in extinguishing interference phenomena.[20,21]

RESUSCITATION OF AN OBLITERATED INTERFERENCE PATTERN.[22,23]

It would be an even greater challenge and joy to ferret out and to discuss every clue and every writer in every domain of thought that promises significant insight on "How come existence" and "How come the quantum". Wild goose chase? Find out which are wild geese and which is the Merlin with a revelation to disclose! And argue that philosophy is too important to be left to the philosophers. Search out potential clues in the fragments that come down to us from the writings of the presocratic philosophers and some of the later thinkers (Box 2).

To get on with the double issue, "How come existence?" and "How come the quantum?" and to try to make something of such scattered hints as are collected in Boxes 1 and 2, give a course in lectures at some long-established university on "Foundation Problems of Quantum Physics."

Invite the new Charles Misner to come along and pursue his thesis research, each of us serving as sounding board for the other, as I was privileged to have the company of the real Charles Misner in my lectures on Gravitational Physics at the University of Leyden in the 1956 Spring semester. We had a definite question to guide our dicussions: Is it possible to push field theory to the limit? Accept as motto, "Mass without mass" and "Charge without charge"? And go on to describe all of classical physics in terms of curved empty space and nothing more? Take seriously the proposal of Rainich that the electromagnetic field, through its energy-momentum tensor, makes a footprint on the geometry of space so characteristic that from it one can read back to all he needs to know about the electromagnetic field. Take equally seriously the idea of Hermann Weyl that space may be multiply connected and that through what Misner and I came to call a "wormhole" can thread electric lines of force, or, as Weyl put it, "One cannot say, here *is* charge, but only that this closed surface cutting through the field includes charge."[34]

REFERENCES

1.   Misner, C. W., "Feynman quantization of general relativity," *Rev. Mod. Phys.* **29**, 497-509 (1957).

2.   Arnowitt, R., S. Deser and C. W. Misner, "The dynamics of general relativity" in L. Witten, ed., *Gravitation: An Introduction to Current Research,* Wiley, New York, 227-265 (1962).

3.   Misner, C. W., K. S. Thorne and J. A. Wheeler, *Gravitation*, hereafter referred to as MTW, W. H. Freeman and Co., San Francisco, now in New York, 1222 pp. (1973).

4.   Misner, C. W. and J. A. Wheeler, ""Conservation laws and the boundary of a boundary," in Shelest 1972, pp. 338-351; and J. A. Wheeler, *A Journey Into Gravity and Spacetime*, Scientific American Library, W. H. Freeman and Co., New York, Chap. 7 (1990).

5.   MTW, p. 5.

6.   Einstein, A.: as quoted by A. Forsee in *Albert Einstein Theoretical Physicist*, Macmillan, New York, p. 81 (1963).

7.   Olive, Schramm, Steigman and Walker, *Phys. Lett. B*, **236**, p. 454 (1990).

Box 2. A few fragments from the pre-Socratic philosophers and later thinkers suggestive of a link to "How come the quantum" and "How come existence"

PYTHAGORAS OF SAMOS (b. ca. 560 B.C.-d. ca. 480 B.C.): "All is number."[24]

HERACLITUS (c. 540-475 B.C.): "All things are one; hot and cold, good and evil, night and day, etc. are the same in the sense that they are inseparable halves of one and the same thing. In Nature, the sole actuality is change. This rhythm of events and order in change is the reason, *logos*, of the universe. Knowledge consists in comprehending the all-pervading harmony as embodied in the manifold of perception."[25]

"Virtue consists in the subordination of the individual to the laws of this harmony as the universal reason wherein alone true freedom is to be found. The law of things is a law of Reason Universal; but most men live as though they had a wisdom of their own."

"It is not possible to step twice into the same river."

PARMENIDES OF ELEA (c. 515 B.C.-450 B.C.). "What is, is identical with the thought that recognizes it."[26]

"By assuming that there is no *tertium quid* between being and absolute non-being, the goddess [through whom Parmenides speaks] construes a tight and cogent reasoning which shows that that which is (being) must be ungenerated, imperishable, homogenous, changeless, immovable, complete, and unique. These characteristics are meant to emphasize from the negative side the unique and unalterable existence of being, for it is implied that if being did not have any one of these characteristics, one would have to admit the existence of something different from being; and such an admission, given the original assumption, would be tantamount to accepting the existence of not-being."

"...For what birth will you seek for it? How and whenst did it grow? I shall not allow you to say nor to think from not being: For... what need would have driven it later rather than earlier, beginning from the nothing, to grow? Thus it must either be completely or not at all..."

"...So it is all continuous: for what is draws near to what is."[27]

ANAXAGORAS OF CLAZOMENAE (c. 500-428 B.C.) "'By Zeus, gentlemen of the jury, it is because he says that the sun is a stone, the moon earth.' Do you imagine, friend Meletus, that you are accusing Anaxagoras, and do you despise the jury and think them so illiterate that they do not not know that the rolls of Anaxagoras of Clazomenae are packed with such theories?"[28]

EMPEDOCLES OF ACRAGAS (c. 495-435 B.C.) "Hear first the four roots of all things; shining Zeus [fire], life-bringing Hera [air], Aidoneus [i.e. Hades, earth], and Nestis [water]..."[29]

"And these things never cease their continual interchange, now through Love all coming together into one, now again each carried apart by the hatred of Strife. So insofar as they have learned to grow one from many, and again as the one grows apart grow many, thus far do they come into being and have no stable life; but insofar as they never cease their continual interchange, thus far they exist always changeless in the cycle."[30]

PHILOLAUS OF CROTON (b.c. 470 B.C.) "And indeed all the things that are known have number; for it is not possible for anything to be thought of or known without this."[31]

MARTIN HEIDEGGER (1889-1976) : "Without the word, no thing may be."[32]

GEORGE BERKELEY's (1685-1753) doctrine: "To be is to be perceived" (*esse est percipi*)[33]

8.   Bohr, N., *Atomic Theory and the Description of Nature,* Cambridge University Press (1934).

9.   Aharonov, Y. and D. Bohm, *Phys. Rev.,* **115**:485-491 (1959).

10. Anandan, J. and Y. Aharonov, "Geometric quantum phase and angles," *Phys. Rev. D. 38,* 1863-1870 (1988), includes references to the literature of the subject; and J. Anandan: "Comment on geometric phase for classical field theories,"*Phys. Rev. Lett. ,***60,** 2555 (1988).

11. Wheeler, J. A., "Information, Physics, Quantum: The Search for Links" in *Proc. 3rd Int. Symp. Foundations of Quantum Mechanics, Tokyo, 1989,* pp. 354-368.

12. Quine, W. V. O., "On what there is" in *From a Logical Point of View, 2nd ed.,* Harvard University Press, Cambridge, Massachusetts, p.18 (1980).

13. Bekenstein, J. D., *Nuovo Cimento Lett.* **4**:737-740 (1972); Bekenstein, J. D. *Phy. Rev.* **D, 8**:3292-3300 (1973).

14. Hawking, S. W., *Commun. Math. Phys.* **43**:199-220 (1975).

15. See Ref. 11, p. 355.

16. Fisher, R. A., "On the dominance ratio," *Proc. Roy. Soc. Edin,* **42:** 321-341 (1922); and R. A. Fisher, *Statistical Methods and Statistical Inferenence,* Hafner, New York, 8-17 (1956).

17. History summarized in M. Born, "Die statistische Deutung der Quantenmechanik," *Les Prix Nobel en 1954* (Stockholm, 1955), pp. 79-90; *Nobel Lectures--Physics (1942-1962)* , Elsevier, Amsterdam, London, New York, 1964, pp. 256-267.

18. Stueckelberg, E. C. G., "Quantum theory in real Hilbert space," *Helv. Phys. Acta,* **33,** 727-752 (1960).

19. D. S. Saxon, *Elementary Quantum Mechanics*, Holden, San Francisco (1964).

20. Zeh, H. D., "On the Interpretation of Measurement in Quantum Theory." See footnote reprinted in *Quantum Theory and Measurement,* ed. J. A. Wheeler and W. H. Zurek, Princeton University Press, Princeton, NJ, p. 342 (1983).

21. Zurek, W. H. , 1982, "Environment-induced superselection rules,"*Phys. Rev. D. 26,* **8,** pp. 1862-1880 (Oct. 1982).

22-23. Kwiat, P. G., A. M. Steinberg and R. Y. Chiao, "Observation of a 'quantum eraser': A revival of coherence in a two-photon interference experiment," *Phys. Rev. Letts,* **45,** 11, p. 7729 (June 1992); and "Dispersion cancellation and high-resolution time measurements in a fourth-order optical interferometer." *Phys. Rev. Letts.,* **45,** 9, p. 6659 (May 1992); and "Dispersion cancellation in a measurement of the single-photon propagation velocity in glass," *Phys. Rev. Letts.,* **68,** p. 2421 (April 1992).

24. **Pythagoras.** From article in *Encyclopaedia Britannica,* hereafter referred to as EB, Chicago, IL, **18:**803 (1959).

25. **Heraclitus.** EB, Chicago, IL, **11:**455, (1979).

26-27. **Parmenides of Elea.** EB, Chicago, IL, **17:**327 (1959).

28. **Anaxagoras of Clazomenae.** *The Presocratic Philosophers,* 2nd Edition, hereafter referred to as TPP, eds. G. S. Kirk, J. E. Raven and M. Schofield, Cambridge University Press, Cambridge, U.K., Chap. XII, (1990).

29-30. **Empedocles of Acragas**. TPP, Chap. X, (1990).

31. **Philolaus of Croton**. TPP, Chap. XI, (1990).

32. Heidegger, M, *On the Way to Language*, ed. and trans. P. D. Hertz and J. Stambaugh, New York. Trans. of "Die Sprache." 1950: "Language." In *Poetry, Language, Thought.* Trans. and intr. A. Hofstädter. New York: Harper and Row, 1975, 189-210. lst ed. 1971. "Sprache-Language." Trans. T. J. Sheehan. *Philosophy Today* (Celina, Ohio), Vol. 20, 1976, 291 ff.

33. EB, **3:**439 (1959) .

34. Weyl, H., *Was ist Materie*, Springer, Berlin, p. 57 (1924).

Charles Misner:
A Celebration of Memories

In the Introduction to this volume we began with an appreciation of Charles Misner as a scholar and an educator, and we felt that it would be fitting to end it with a selection of reminiscences that would allow us to show something of Charlie Misner the man and the teacher. Few of the readers of this book in the year that it is published will not know the dry facts of his life, but since all books are at least a reach for immortality, we should consider the reader who may see it long after all of us are dust and give him or her the framework of a life on which so many memories rest.

Charles was born on June 13, 1932, attended Notre Dame University from 1948 to 1952, and received his Ph.D. from Princeton, where his advisor was John Wheeler, in 1957. He married Susanne Kemp in 1959 and has four children. From 1956 to 1963 he was an Instructor and then an Assistant Professor at Princeton. Since 1963 he has been on the faculty of the University of Maryland. He has been a visiting faculty member in universities and institutes throughout the United States and Europe. He is a fellow of the American Physical Society, the Royal Astronomical Society and the American Association for the Advancement of Science, and a member of the International Society on General Relativity and Gravitation, the Association of Mathematical Physicists, the American Mathematical Society, the International Astronomical Union, the Philosophy of Science Association and the Federation of American Scientists. This does not exhaust the varied activities that Charlie has engaged in, but we will mention these below when we talk about specific facets of his life that have been the origin of so many memories.

As part of the sixtieth birthday party that was organized for Charlie during GR13 held in Cordoba, Argentina in June of 1992, many people who had known him were asked to contribute to a scrapbook of tributes and memories, and the scrapbook was presented to him there. This scrapbook is a fine source of reminiscences, and many of the memories we will give here are taken from it.

However, it is best not just to list the reminiscences that all of us have in random order, but to try to organize them into memories of the many faces that Charlie has shown to his many colleagues, friends and students.

The first of these is that of an active researcher in general relativity. In the Introduction this aspect of his life was covered from the point of view of his many invaluable contributions to the field. Here we would like to pursue our goal of presenting Charlie Misner the man by recording memories of him as a physicist. Yet it becomes difficult to separate Charlie the researcher from what perhaps is the most prominent facet of what he is—a teacher. Perhaps this is not surprising since the best physicists are often the best teachers—the ability to communicate the excitement of the field is important to both pursuits. Thus we will begin with reminiscences that can best be classified as memories of Charlie Misner the physicist and allow them to slide slowly into memories of Charlie Misner the teacher.

Thomas B. Day, President of San Diego State University, writes:

> It has been a long time and a curiously short time since the winter of '51/52 in Notre Dame when our lives began to be so intertwined. Going five days a week to Alex[Petrauskas]'s Physics 200/201, doing test problems in fluid flow around journal bearings. You reading Bourbaki in uncut French original, and I struggling to understand the proof of Cauchy's theorem from first principles. From department beer-cooler guarded with combination lock set variously to different 4-digit physical and mathematical constants, to physics parties and dances where we both met Anne. Both graduating together, but not really.
>
> Going off to mesons and wormholes, respectively, then back together again, during the College Park boom times. Our families growing in parallel, sharing the Church, and children's joys and pains. The veering off again to current pursuits.
>
> Through the 40 years I've watched your career with friendship and admiration, seeing a mathematical power and physical intuition of rare strength and depth. As my scientific life has faded now to University President and Vice Chairman of the National Science Board, I realize all the more the importance of lifelong continuous scientific contributions like yours.
>
> I'm proud to have been your colleague, pleased to be your friend. Anne and I send you and your family greetings on this slightly time-retarded celebration day in space-displaced Argentina.

Two people who are in the unique position of remembering Charlie from the time he was a young student and saw him grow into the man he is today are John Toll, Chancellor Emeritus of the University of Maryland, and his thesis advisor, John Wheeler. The former says:

Our friendship goes back many years, since you were recognized as one of the best graduate students in physics that had come to Princeton for many years. Based on the comments of my many friends of Princeton, I determined at that time that the University of Maryland should do its best to attract you to its faculty. Indeed, one of the great steps in making the University of Maryland physics faculty one of the best in the nation was your decision to come to College Park and to devote your career to the University of Maryland.

You made the University of Maryland a great center for important research in general relativity and related fields. You have been an outstanding researcher who had the courage to tackle particularly difficult problems. You were also a gifted teacher. Your contributions to teaching included such heroic efforts as your magnificent text with Thorne and Wheeler, used with great joy throughout the world. You have also shown national leadership in the development of computers for effective teaching in physics.

One quality that makes you such an outstanding professor is your patience. I have marvelled at your willingness to take time to explain to me and others things that are immediately obvious to you but require for us some struggle to comprehend. Many undergraduates and graduate students, as well as grateful colleagues on the faculty, have come to appreciate your lucidity in presentation and your kind help to others.

John Wheeler has a more personal view:

Welcome home from Cordoba with happy memories of many a warm birthday greeting from one or other colleague there to celebrate your birthday. I do hope that meeting helped you to realize what an honored and respected place you hold in our profession. Like me, our friends know no one more authoritative to turn to when up against a problem of mathematical physics in the realm of general relativity that presents difficult points of principle. I know that I feel at a loss compared to the old days when I run up against one or another deep point. If only several duplicate copies of you existed, so I could always find "one of you" nearby to consult! Anyway, many more happy years of creativity to you.

It is interesting to see that from this point on memories of Charlie the teacher begin to dominate what everyone has to say. Ralph Baierlein, Professor of Physics at Wesleyan University, is perhaps one of the earliest of Charlie's students, and he remembers:

Only a short time ago—or so it seems—I sat next to your desk in a large Prince-ton room that you shared as an office with at least three other faculty members. Those were the ADM days, but the most valuable thing that I learned as your thesis student had nothing to do with ADM or any other formalism. It was this: If a problem is worth working on, then you can make progress by turning it this way, then that way, next flipping it over, and so on. You have to worry it but—sooner or later—some route that you try will produce an insight or some other bit of progress. If the first approach doesn't work, don't give up. Rather, look for another way to get at it.

Since the rest of these reminiscences will be about Charlie's career as a teacher, here is probably the best point to talk about this facet of his life. For many years he supervised a number of graduate students who have since gone on to become productive physicists. He then turned his talents toward physics instruction at more elementary levels. He has been a consultant to the AIP/AAPT "Active Physics" high school physics project, a member of the Curriculum Coordination Group of the National Science Teachers Association SS&C project, a member of the steering committee of the APS/AAUP Introductory University Physics Project, a member of the advisory committee for the Conference on Computers in Physics Instruction, and a coauthor of *Spreadsheet Physics*, which uses computers to teach university-level physics.

It is impossible to try to find what are probably myriad memories of undergraduate physics students about this person that they have never met, so we have to be content with the recollections of his graduate students. So many of them remember how his tremendous insight and careful, kind way of explaining physics started them on the road to research and still guide them not only in how they approach a problem, but also in their understanding of what a physics problem really is. Also, Charlie is one of the many inheritors of the style of Niels Bohr which makes for the best of teachers. This style is the opposite of that of some famous physicists who seem to feel, as some nineteenth-century economists felt about wealth, that there is a limited amount of academic glory, and that it can only be had by taking it from someone else. The style of Bohr seems much closer to the idea that wealth or academic glory is something that is infinitely creatable and that giving others more credit than is their due reflects well not only on those who receive the credit, but on those who bestow it.

After this description of Charlie's methods as a teacher, it comes as a surprise that many of his students (not only those whose recollections will be given here) felt in awe of him and often found it difficult to speak to him on an informal basis. Chris Stephens says:

Meeting Charlie for the first time as a beginning graduate student and a prospective disciple of the Maryland relativity group is quite a daunting experience. Not, as we all know, because there is anything daunting about Charlie's personality (in a profession full to the brim with giant egos it is difficult to think of a more amiable and unassuming man) but because of his universal reputation as a physicist. The nature of the man is to set you at ease—the nature of the intellect is the opposite.

I well remember giving my first seminar at Maryland, at the time a compulsory rite of passage into the relativity group. It was on gravity on the lattice. I had been preparing the seminar for some weeks and was making a reasonable, if probably somewhat staccato attempt at fielding questions. What was memorable was that if I got into trouble on a question Charlie would answer it at once as if he had written the papers himself. This was to be a regular occurrence. Any feelings of inadequacy on my part eventually disappeared after I saw some others, far more eminent than myself, who were subject to the same phenomenon.

To my knowledge I was Charlie's last relativity student—I hope that was no indication of any trauma I might have caused. On my last trip to Maryland I was much heartened to see that Charlie was getting back into relativity research. I am sure there are few people whose papers will be more eagerly anticipated.

Jim Nester also begins his reminiscences with almost the same feeling:

I remember how much I was in awe of you when we first met. I knew that you were a leader in investigating relativity and gravitation. Later, I learned (from your writings, for you never taught me these things directly or suggested investigating them) that you were a pioneer in applying modern differential geometry (especially differential forms) to gravitational theory and in developing the canonical Hamiltonian formulation, identifying expressions for conserved quantities, especially mass-energy. You have inspired me to apply these techniques and investigate these topics.

The best teachers teach by example. Certainly that was true in my case. Your treatment seemed optimal to encourage me to bring out my best. First, you employed me in checking equations for MTW. Then you suggested a certain direction ("Maybe you would like to look into Trautman's torsion theory?") and just monitored my (slow) progress. I recall my days as your student at Maryland, and remember how I didn't want to leave that situation.

Your insight was amazing. I remember when you were leafing through the long awaited first draft of my thesis. You stopped on one page, pointed to one of the many equations and said "Something is wrong here." After much checking I found that you were right. But I still don't know how you did it.

You were an excellent teacher. You perceived an individual's strengths and weaknesses and advised them accordingly. When Jim Isenberg and I were just beginning to write one of our joint papers, you suggested that Jim should write the equations and I should write the text, that way the paper would be much shorter.

However, the feeling of awe never seemed to have dulled Charlie's effect as a teacher on all of his students and all of them remember him both fondly and gratefully.

It was with great pleasure that we all have noted that after many years in which Charlie has dedicated himself to expanding his role as teacher, he has begun to return to active research. We hope our field will continue to benefit for many years from his new insights as it has so significantly benefited in the past.

In order not to end on too serious a note, our last "reminiscence" is a delightful birthday greeting by Harry Zapolsky. He writes:

In the attempt to draft a birthday greeting, my own words have failed me, so the following is based on entries which I found in the index of Misner, Thorne, and Wheeler (aka "The Telephone Directory").

When I learned that you were about to attain the grand age of 60, you could have bowled me over with a rotating steel beam.[1] Mere chronological age, as a carrier of information about time,[2] defined so motion looks simple,[3] is really quite subtle and sneaky. As you have pointed out, Earth years are not always a good measure of physically elapsed time. After[4] all, you still have your hair on[5], advanced potential[6], a past history not much affected by k[7], and absence of frame dragging[8], and you continue to display every sign of intellectual expansion forever vs. recontraction[9]. In a comparison of ages deduced by various methods[10], I suggest that, in your case, we treat the humble year as an imaginary coordinate, not to be used[11], and thought of as improperly posed data[12].

*On an occasion like this you deserve a celebration with your colleagues. For such an event, there are at least three needed[13], but I know that you will have many more. I would love to join you in Cordoba, where the Big Dipper[14] cannot be seen in the night sky, but time constraints prevent me from doing so. My geodesic can't change from timelike to spacelike[15], and even the less rapid mode of traveling by **tunnel**[16] is currently unavailable to me. I will miss seeing many old friends with whom I have translated from one (GRG) convention to another[17] over the years. If I could be there, I would testify how, as friend, mentor, and colleague, you have been a buoyant force[18] in my life, and I would regale the audience with humorous tales due to my interaction with[19] you. I imagine that many others will do the same, and I will miss a great chance to see Misner exploited in the simplest form[20,21].*

So Charlie, have a ball, and stay away from the machine with slots[22]! I propose a toast with three cheers and 60 Bell bongs[23], along with my wishes for the very best in the good years to come.

¹ Gravitational waves, sources of, rotating steel beam, 879-980.

² Three geometry, as carrier of information about time, 488, 533.

³ Time, defined so motion looks simple, 23-29.

⁴ After, undefined term in quantum geometrodynamics, 1183.

⁵ Black hole, "hair on", 43, 863, 876.

⁶ Advanced potential, 121.

⁷ Cosmology, history of the universe according to the "standard big bang model", past history not much affected by k (by geometry of hypersurfaces), 742f, 763.

⁸ Spinning body, spin precessions, frame dragging, 1119f.

⁹ Cosmology, history of the universe according to the "standard big-bang model", expansion forever vs. recontraction, 747, 771, 774.

¹⁰ Ibid.; observational probes of standard model, comparison of ages deduced by various methods, 797f.

¹¹ Time, imaginary coordinate for, not to be used, 51.

¹² Initial-value data for geometrodynamics, improperly posed data, 534-535.

¹³ Test particles, three needed to explore Lorentz force, 72.

¹⁴ Big Dipper, shape unaffected by velocity of observer, 1160-1164.

¹⁵ Geodesics, can't change from timelike to null or spacelike *en route*, 321.

¹⁶ Earth, particle oscillating in a hole bored through, 39.

¹⁷ Parametrized post-Newtonian formalism, parameter, translated from one convention to another, 1093.

[18] Buoyant force, 606.

[19] Gravitational waves, propagation through curved spacetime, tails due to interaction with background curvature, 957.

[20] Variational principles for geometrodynamics, Arnowitt, Deser, Misner, exploited, 526.

[21] Variational principles for geometrodynamics, Arnowitt, Deser, Misner, in simplest form, 521.

[22] Covariant derivative, as a machine with slots, 253ff.

[23] Bell bongs, 55f, 60, 99, 202, 231.

Michael P. Ryan, Jr.

Curriculum Vitae

CHARLES W. MISNER

University of Maryland

EDUCATION

B.S., 1952, University of Notre Dame
M.A., 1954, Princeton University
Ph.D., 1957, Princeton University

UNIVERSITY POSITIONS

1956–59 Instructor, Physics Department, Princeton University
1959–63 Assistant Professor, Physics, Princeton University
1963–66 Associate Professor, Department of Physics and Astronomy, University of Maryland
1966– Professor, Physics, University of Maryland, College Park

VISITOR

1983 (January) Cracow, Poland: Pontifical Academy of Cracow
1980–81 Institute for Theoretical Physics, U. Cal. Santa Barbara
1977 (May) Center for Astrophysics, Cambridge, Mass.
1976 (June) D.A.M.T.P. and Caius College, Cambridge, U.K.
1973 (spring term) All Souls College, Oxford
1972 (fall term) California Institute of Technology
1971 (June) Inst. of Physical Problems, Academy of Sciences, USSR, Moscow
1969 (fall term) Visiting Professor, Princeton University
1976 (summer) Niels Bohr Institute, Copenhagen, Denmark
1966–67 Department of Applied Mathematics and Theoretical Physics, Cambridge University, England
1960 (spring term) Department of Physics, Brandeis University, Waltham, Massachusetts

1959 (spring and summer) Institute for Theoretical Physics, Copenhagen, Denmark

1956 (spring and summer) Institute for Theoretical Physics, Leiden, the Netherlands

HONORS AND AWARDS

Computers in Physics (AIP magazine/journal) annual educational software contest winner in 1991

Visiting Scientist, Inst. Theor. Phys., Santa Barbara, 1980–81

Visiting Fellow, All Souls College, Oxford, 1973

Guggenheim Fellowship, 1972–73

Washington Academy of Sciences Award for Scientific Achievement, 1967

National Science Foundation Senior Postdoctoral Fellowship 1966–67

Gravity Research Foundation, Third Award for an Essay on Gravity, 1967

Maryland Academy of Sciences Award: Maryland's Outstanding Young Scientist of 1965

Science Centennial Award, University of Notre Dame, 1965

Alfred P. Sloan Research Fellowship 1958–62

Junior Fellowship, Harvard Society of Fellows (appointment declined, 1957)

PROFESSIONAL ACTIVITIES

American Physical Society fellow

International Society on General Relativity and Gravitation member

Association of Mathematical Physicists member

American Mathematical Society member

Royal Astronomical Society fellow

International Astronomical Union member

Philosophy of Science Association member

American Assoc. for the Advancement of Science fellow

Federation of American Scientists member

Consultant to the AIP/AAPT "Active Physics" high school physics project, 1991–present

Associate Editor for Education, *Computers in Physics*, 1992–present

Member of the AAPT Resource Letters Editorial Board, 1991–94

Member of the Curriculum Coordination Group of the National Science Teachers Association SS&C project, 1991

Member of the steering committee of the APS/AAUP Introductory University Physics Project, 1987–present

Member of the advisory committee for the Conference on Computers in Physics
    Instruction, N.C. State University 1–5 August 1988
Consultant to the Committee on Human Values of the National Conference of
    Catholic Bishops, 1987–91
Member of Editorial Board, *Phys. Rev. D*, 1978 and 1979
NSF Advisory Committee for Research 1973–77
Member of NASA Sub-Panel on Relativity and Gravitation 1975–76
Member of NAS Astronomy and Physics Survey Panel: Relativistic Astrophysics
    1971–74

INVITED LECTURER

Invited lecturer and workshop leader, Davidson College Conference on
    Computational Physics in the Undergraduate Curriculum, Davidson,
    North Carolina, October 1991
Invited lecturer and workshop leader, Dickinson College Summer Seminar on
    Workshop Physics, Dickinson College, Carlisle PA, June 1991
Invited participant, Denver Workshop on the Role of the Research Physicist in
    Undergraduate Curriculum Development, 16–18 November 1990
Bertha Halley Ross lecturer in mathematics and science at Ohio State University,
    August 1990
International Symposium on the Evaluation of Physics Education, Helsinki, 25–29
    June 1990
Joint AAPT-APS-AAAS meeting in San Francisco, 19 January 1989, AAPT
    session of invited papers in Computational Physics
Conference on Computers in Physics Instruction, North Carolina State University,
    1–5 August 1988
Study Week at Vatican Observatory, Castel Gandolfo, on "Our Knowledge of God
    and Nature: Physics, Philosophy and Theology", 21–25 September 1987
IBM Forum for the Physical Sciences, Tucson, Arizona, 1–3 November 1987
IBM ACIS University Conference, Boston, Massachusetts 27–30 June 1987
11th University Study Conference, IBM Academic Information Systems, Fort
    Lauderdale, Florida, November 1986
10th University Study Conference, IBM Academic Information Systems, Santa
    Clara California, November 1985
IBM University Advanced Education Projects Conference, Alexandria, Virginia,
    June 1985
Conference on Foundation Problems of Physics, Austin, Texas, February 1984
Conference on Evolution and Creation, University of Notre Dame, March 1983

Annual Meeting of the American Catholic Philosophical Association, St. Louis, Missouri, April 1981

Einstein Centennial Symposium, Institute for Advanced Study, Princeton, March 1979

Symposium on the Impact of Modern Scientific Ideas on Society (Einstein Centenary/UNESCO), Ulm, September 1978

Berkeley Symposium on Relativity and Cosmology in honor of Abraham H. Taub, August 1978

Alfred Schild Memorial Lectures, Austin, Texas, 1978

AAAS 1978 Annual Meeting, Symposium on the Role of Models in Scientific Inquiry

Chandrasekhar Symposium on Astrophysics and Relativity, University of Chicago, 1975

International School E. Majorana, Erice, Sicily, March 1975

International Congress of Mathematicians, Vancouver, 1974

Philosophy of Science Association 1972 Biennial Meeting, Symposium: Space, Time, and Matter

Brandeis Summer Institute for Theoretical Physics, 1968

Les Houches Summer School of Theoretical Physics, 1963

BIOGRAPHICAL LISTINGS

Who's Who in America

World Who's Who

Dictionary of International Biography

American Men of Science

see also A. Lightman and R. Brawer, *Origins—the Lives and Worlds of Modern Cosmologists*, Harvard University Press, 1990, pp. 232–249

Physics Ph.D. Theses
Supervised by Charles W. Misner

University of Maryland
(except as noted)

1   Carl H. Brans 1961 (Princeton University, Prof. R. H. Dicke co-supervisor)
       Mach's Principle and a Varying Gravitational Constant

2   Ralph F. Baierlein 1962 (Princeton University) *On the Existence and
       Properties of Spacelike Hypersurfaces in Quantized General Relativity*

3   Walter C. Hernandez Jr. 1966 *Relativistic Descriptions of Material Sources
       for Gravitational Fields, including Elastic Materials as Sources of
       Asymmetric Static Fields*

4   Richard A. Isaacson 1967 *Gravitational Radiation in the Limit of High
       Frequency*

5   Richard A. Matzner 1967 *Scattering of Massless Scalar Waves by a
       Schwarzschild "Singularity"*

6   C. V. Vishveshwara 1968 *The Stability of the Schwarzschild Metric*

7   Samaresh C. Maitra 1968 *A Stationary and Complete, but Unstable,
       Inhomogeneous Cosmological Solution of Einstein's Equations*

8   Michael P. Ryan Jr. 1970 *Qualitative Cosmology: Diagrammatic Solutions for
       Bianchi Type IX Universes with Expansion, Rotation, and Shear*

9   D. M. Chitre 1972 *Investigation of Vanishing of a Horizon for Bianchi
       Type IX (the Mixmaster) Universe*

10   Leslie G. Fishbone 1972 *The Relativistic Roche Problem: Bodies in Equatorial, Circular Orbit around Kerr Black Holes*

11   Beverly K. Berger 1972 *A Cosmological Model Illustrating Particle Creation through Graviton Production*

12   Reinhard A. Breuer 1973 (Würzburg University, Prof. Rolf Ebert co-supervisor) *Polarization of Gravitational Synchrotron Radiation*

13   Paul L. Chrzanowski 1973 (Prof. Dieter Brill co-supervisor) *Gravitational Synchrotron Radiation*

14   James M. Nester 1977 *Canonical Formalism and the ECSK Theory*

15   James A. Isenberg 1979 *The Construction of Spacetimes from Initial Data*

16   William A. Hiscock 1979 *Black Holes and Cosmic Censorship*

17   Terrence J. Honan 1986 *The Geometry of Lattice Field Theory*

18   Mark D. Somers 1986 *An Investigation of the Strong Equivalence Principle through the use of Freely Falling Geocentric Coordinates*

19   Christopher R. Stephens 1986 *On Some Aspects of the Relationship Between Quantum Physics, Gravity and Thermodynamics*

List of Publications

CHARLES W. MISNER

BOOKS AND THESIS

A   *Feynman Quantization of General Relativity*, Charles W. Misner (Ph.D. thesis, Princeton University, June 1957).

B   *Gravitation*, Charles W. Misner, Kip S. Thorne and John Archibald Wheeler (W. H. Freeman and Co., San Francisco 1973) 1279 pp.

B′ *Gravitatsiya*, (3 Vols.) Charles W. Misner, Kip S. Thorne and John Archibald Wheeler, translated by A. G. Polnarev (Pub. Co. 'Mir', Moscow 1977).

C   *Spreadsheet Physics*, Charles W. Misner and Patrick J. Cooney (Addison-Wesley Pub. Co., Reading, Mass., 1991) xviii + 228 pp.

JOURNAL ARTICLES

1=A   "Feynman Quantization of General Relativity", Charles W. Misner, *Reviews of Modern Physics* **29**, 497–509 (1957).

2   "Geometrodynamics", Charles W. Misner and J. A. Wheeler, *Annals of Physics* **2**, 525–603 (1957).

3   "Some New Conservation Laws", David Finkelstein and Charles W. Misner, *Annals of Physics* **6**, 230–243 (1959).

4   "Active Gravitational Mass", Charles W. Misner and Peter Putnam, *Physical Review* **116**, 1045–1046 (1959).

5   "Dynamical Structure and Definition of Energy in General Relativity", R. Arnowitt, S. Deser, and C. W. Misner, *Phys. Rev.* **116**, 1322–1330 (1959).

6 "Canonical Variables, Expression for Energy, and the Criteria for Radiation in General Relativity", R. Arnowitt, S. Deser, and C. W. Misner, *Il Nuovo Cimento* (X) **15**, 487–491 (1959).

7 "Canonical Variables for General Relativity", R. Arnowitt, S. Deser, and C. W. Misner, *Physical Review* **117**, 1595–1602 (1960).

8 "Finite Self-energy of Classical Point Particles", R. Arnowitt, S. Deser, and C. W. Misner, *Physical Review Letters* **4**, 375–377 (1960).

9 "Energy and the Criteria for Radiation in General Relativity", R. Arnowitt, S. Deser, and C. W. Misner, *Physical Review* **118**, 1100–1104 (1960).

10 "Wormhole Initial Conditions", Charles W. Misner, *Physical Review* **118**, 1110–1111 (1960).

11 "Note on the Positive-definiteness of the Energy of the Gravitational Field", R. Arnowitt et al., *Annals of Physics* **11**, 116–121 (1960).

12 "Consistency of the Canonical Reduction of General Relativity", R. Arnowitt et al., *Journal of Mathematical Physics* **1**, 434–439 (1960).

13 "Gravitational-Electromagnetic Coupling and the Classical Self-energy Problem", R. Arnowitt et al., *Physical Review* **120**, 313–320 (1960).

14 "Interior Schwarzschild Solutions and Interpretation of Source Terms", R. Arnowitt et al., *Physical Review* **120**, 321–324 (1960).

15 "Heisenberg Representation in Classical General Relativity", R. Arnowitt, S. Deser, and C. W. Misner, *Il Nuovo Cimento* (X) **19**, 668–681 (1961).

16 "The Wave Zone in General Relativity", R. Arnowitt, S. Deser, and C. W. Misner, *Physical Review* **121**, 1556–1566 (1961).

17 "Coordinate Invariance and Energy Expressions in General Relativity", R. Arnowitt et al., *Physical Review* **122**, 997–1006 (1961).

18 "Gravitational Field Energy and g_{00}", Charles W. Misner, *Physical Review* **130**, 1590–1594 (1963).

19 "Fermi Normal Coordinates and Some Basic Concepts in Differential Geometry", F. K. Manasse and C. W. Misner, *J. math. Phys.* **4**, 735–745 (1963).

20 "The Flatter Regions of Newman, Unti, and Tamburino's Generalized Schwarzschild Space", Charles W. Misner, *J. math. Phys.* **4**, 924–937 (1963).

21 "The Method of Images in Geometrostatics", Charles W. Misner, *Annals of Physics* **24**, 102–117 (1963).

22 "High-density Behavior and Dynamical Stability of Neutron Star Models", C. W. Misner and H. S. Zapolsky, *Physical Review Letters* **12**, 635–637 (1964).

23 "Relativistic Equations for Adiabatic, Spherically Symmetric Gravitational Collapse" Charles W. Misner, and David H. Sharp, *Physical Review* **136**, B571–B576 (1964).

24 "Relativistic Equations for Spherical Gravitational Collapse with Escaping Neutrinos", Charles W. Misner, *Physical Review* **137**, B1360–B1364 (1965).

25 "Vaidya's Radiating Schwarzschild Metric", R. W. Lindquist, R. A. Schwartz and C. W. Misner, *Physical Review* **137**, B1364–B1368 (1965).

26 "Spherical Gravitational Collapse with Energy Transport by Radiative Diffusion", C. W. Misner and D. H. Sharp, *Physics Letters* **15**, 279–281 (1965).

27 "Minimum Size of Dense Source Distributions in General Relativity", R. Arnowitt et al., *Annals of Physics* **33**, 88–107 (1965).

28 "Observer Time as a Coordinate in Relativistic Spherical Hydrodynamics", Walter C. Hernandez, Jr., and Charles W. Misner, *Astrophysical Journal* **143**, 452–464 (1966).

29 "Gravitational Field Equations for Sources with Axial Symmetry and Angular Momentum", Richard A. Matzner and Charles W. Misner, *Physical Review* **154**, 1229–1232 (1967).

30 "Transport Processes in the Primordial Fireball", Charles W. Misner, *Nature* **214**, 40–41 (1967).

31 "Neutrino Viscosity and the Isotropy of Primordial Blackbody Radiation", Charles W. Misner, *Physical Review Letters* **19**, 533–535 (1967).

32 "The Isotropy of the Universe", C. W. Misner, *Astrophysical Journal* **151**, 431–457 (1968).

33 "A Singularity-free Empty Universe", C. W. Misner and A. H. Taub, *Zh. Eksp. Teor. Fiz.* **55**, 233–255 (1968); English original in *Soviet Physics—JETP* **28**, 122–133 (1969).

34 "Mix-master Universe", Charles W. Misner, *Physical Review Letters* **22**, 1071–1074 (1969).

35 "Cohesive Force of Metals", G. H. Wannier, C. Misner, and G. Schay, Jr., *Physical Review* **185**, 983–984 (1969).

36 "Quantum Cosmology. I", Charles W. Misner, *Physical Review* **186**, 1319–1327 (1969).

37 "Absolute Zero of Time", Charles W. Misner, *Physical Review* **186**, 1328–1333 (1969).

38 "Dissipative Effects in the Expansion of the Universe. I", R. A. Matzner and C. W. Misner, *Astrophysical Journal* **171**, 415–432 (1972).

39 "Interpretation of Gravitational-Wave Observations", C. W. Misner, *Physical Review Letters* **28**, 994–997 (1972).

40 "Gravitational Synchrotron Radiation in the Schwarzschild Geometry", C. W. Misner, R. A. Breuer, D. R. Brill, P. L. Chrzanowski, H. G. Hughes III, and C. M. Pereira, *Physical Review Letters* **28**, 998–1001 (1972).

41 "A Minisuperspace Example: The Gowdy T^3 Cosmology", Charles W. Misner, *Physical Review D* **8**, 3271–3285 (1973).

42 "Geodesic Synchrotron Radiation", R. A. Breuer, P. L. Chrzanowski, H. G. Hughes III, and C. W. Misner, *Physical Review D* **8**, 4510–4524 (1973).

43 "Geodesic Synchrotron Radiation in the Kerr Geometry by the Method of Asymptotically Factorized Green's Functions", P. L. Chrzanowski and C. W. Misner, *Physical Review D* **10**, 1701–1721 (1974).

44 "Harmonic Maps as Models for Physical Theories", Charles W. Misner, *Physical Review D* **18**, 4510–4524 (1978).

45 "Relativistic Effects on an Earth-Orbiting Satellite in the Barycenter Coordinate System", C. F. Martin, M. H. Torrence, and C. W. Misner, *Journal of Geophysical Research* **90**, 9403–9410 (1985).

46 "Problem: Zero-gravity Pendulum", C. W. Misner, *American Journal of Physics* **55**, 657–668 (1987).

ARTICLES FROM BOOKS/CONFERENCES

1[B] "Remarks on Unquantized General Relativity", Charles W. Misner in *Conference on the Role of Gravitation in Physics*, Chapel Hill Conference 1957, (Wright Air Development Center Technical Report 57-216, ASTIA Document No. AD 118180, 1957) pp 18–29, 134.

2=2[B] "Classical Physics as Geometry", C. W. Misner and J. A. Wheeler in *Geometrodynamics*, ed. J. A. Wheeler (Academic Press, N.Y. 1962), [reprinted from *Annals of Physics* **2**, 525–603 (1957)].

3[B] "Further Results in Topological Relativity", David Finkelstein and Charles W. Misner in *Les Théories Relativistes de la Gravitation*, Royaumont Conference 1959, (Centre National de la Recherche Scientifique, Paris 1962).

4[B] "The Dynamics of General Relativity", R. Arnowitt, S. Deser, and C. W. Misner, Chapter 7 in *Gravitation: An Introduction to Current Research*, L. Witten, ed., (J. Wiley, New York 1962) pp 227–265.

5[B] "Canonical Analysis of General Relativity", R. Arnowitt, S. Deser, and C. W. Misner in *Recent Developments in General Relativity* dedicated to Leopold Infeld (Pergamon Press, New York, 1962) pp 127–136.

6[B] "Mass as a Form of Vacuum", C. W. Misner in *The Concept of Matter*, Ernan McMullin, ed., (University of Notre Dame Press 1963), pp 596–608.

7[B] "Waves, Newtonian Fields, and Coordinate Functions", C. W. Misner (reporting work done in collaboration with R. A. Arnowitt and S. Deser) in *Proceedings on Theory of Gravitation*, Conference in Warszawa and Jabłonna, 24–31 July 1962, (PWN-Warsaw, and Gauthier-Villars, Paris 1964), pp 189–205.

8[B] "Differential Geometry and Differential Topology", C. W. Misner in *Relativity Groups and Topology, Les Houches 1963*, C. DeWitt and B. DeWitt, eds., (Gordon and Breach, New York, 1964) pp 881–929.

9[B] "Taub-Nut Space as a Counterexample to Almost Anything", C. W. Misner in *Relativity Theory and Astrophysics 1*, Lectures in Applied Mathematics, Vol. 8, edited by J. Ehlers, (American Mathematical Society, Providence, R.I. 1967), pp 160–169.

10[B] "Observer-time as a Coordinate in Relativistic Spherical Hydrodynamics", C. W. Misner in *Relativity Theory and Astrophysics 3*, Lectures in Applied Mathematics, Vol. 10, edited by J. Ehlers, American Mathematical Society, Providence R.I. 1967), pp 117–128.

11[B] "Relativistic Fluids in Cosmology", C. W. Misner in *Fluids et champ gravitationnel en relativité générale*, Colloques international du CNRS No. 170 (Éditions du Centre National de la recherche scientifique, Paris 1969) pp 155–157.

11[B'] "Relativistic Fluids in Cosmology", C. W. Misner in DeWitt and Wheeler, eds., *Battelle Recontres 1967*, (Benjamin, N.Y.), pp 117–120, reprinted from CNRS conference.

12[B] "Gravitational Collapse", C. W. Misner in *Astrophysics and General Relativity, 1968 Brandeis Summer Institute*, edited by M. Chretein, S. Deser, and J. Goldstein (Gordon and Breach, New York 1969), Vol. 1, pp 113–215.

13[B] "The Equations of Relativistic Spherical Hydrodynamics", Charles W. Misner and David H. Sharp in *Quasars and High-Energy Astronomy*, ed. K. N. Douglas, et. al. (New York: Gordon and Breach 1969) pp 393–395.

14[B] "Energy Transport by Radiative Diffusion in Relativistic Spherical Hydrodynamics", Charles W. Misner and David H. Sharp in *Quasars and High-Energy Astronomy*, ed. K. N. Douglas et. al. (New York: Gordon and Breach, 1969) pp 397–400.

15[B] "Gravitational Forces Accompanying Bursts of Radiation" R. W. Lindquist, R. A. Schwartz and C. W. Misner in *Quasars and High-Energy Astronomy*, ed. K. N. Douglas, et. al. (New York: Gordon and Breach 1969)

16[B]   "Classical and Quantum Dynamics of a Closed Universe", C. W. Misner in
        Carmeli, Fickler, and Witten, editors, *Relativity* (Plenum Pub. Co., San
        Francisco 1970), pp 55–79.

17[B]   "Minisuperspace", C. W. Misner in J. Klauder, ed., *Magic Without Magic—
        J. A. Wheeler 60th Anniversary Volume* (W. H. Freeman and Co., San
        Francisco 1972) pp 441–473.

18[B]   "Conservation Laws and the Boundary of a Boundary", Charles W. Misner and
        John A. Wheeler in V. P. Shelest, ed., *Gravitatsiya: Problem i Perspektivi:
        pamyati, Alekseya Zinovievicha Petrova posvashaetsys* (Naukova Dumka,
        Kiev 1972) pp 338–351.

19[B]   "Some Topics for Philosophical Inquiry Concerning the Theories of Mathe-
        matical Geometrodynamics and of Physical Geometrodynamics", Charles
        W. Misner in *PSA 1972: Proceedings of the 1972 Biennial Meeting of the
        Philsophy of Science Association*, Kenneth F. Schaffner and Robert S. Co-
        hen, eds. (D. Reidel Pub. Co., Dordrecht 1974), pp 7–29.

20[B]   "Radiation from Highly Relativistic Geodesics", C. W. Misner in the proceed-
        ings of Colloque International C.N.R.S. No. 220 (held at Institute Henri
        Poincaré, Paris, June 1973): *Ondes et Radiations Gravitationelles* (CNRS,
        Paris 1974), pp 145–160.

21[B]   "Mechanisms for the Emission and Absorption of Gravitational Radiation", C.
        W. Misner in the Proceedings of IAU Symposium No. 64 (held September
        1973 in Warsaw, Poland): *Gravitational Radiation and Collapse*, C. DeWitt-
        Morette, ed., (Reidel, Dordrecht 1974), pp 3–15.

22[B]   "Quantum Descriptions of Singularities Leading to Pair Creation", C. W. Mis-
        ner in the Proceedings of I.A.U. Symposium No. 64 held September 1973
        in Cracow, Poland: *Confrontation of Cosmological Theories with Observa-
        tional Data*, M. S. Longair, ed., (Reidel, Dordrecht 1974), pp 319–327.

23[B]   "Cosmology and Theology", Charles W. Misner in *Cosmology, History, and
        Theology*, Wolfgang Yourgrau and Allen D. Breck, eds. (Plenum Press, New
        York 1977) pp 75–100; proceeding of the Third International Colloquium,
        University of Denver, November 1974.

24^B "Values and Arguments in Homogeneous Spaces", Charles W. Misner in *Essays in General Relativity*, Frank J. Tipler, ed. (Academic Press, New York 1980), pp 221–231; A. H. Taub Conference at Berkeley, August 1978.

25^B "Symmetry Paradoxes and Other Cosmological Comments", Charles W. Misner in *Some Strangeness in the Proportion—A Centennial Symposium to Celebrate the Achievements of Albert Einstein*, H. Woolf, ed. (Addison-Wesley Pub. Co., Reading, Pa. 1980) pp 405–415; Einstein Centennial Celebration, Institute for Advanced Study, Princeton, New Jersey, March 1979.

26^B "The Immaterial Constituents of Physical Objects", Charles W. Misner in *The Impact of Modern Scientific Ideas on Society*, C. M Kinnon, A. N. Kholodilin and J.G. Richardson, eds. (D. Reidel Pub. Co., Dordrecht 1981) pp 129–135; UNESCO Conference/Einstein Centennial, Munich and Ulm, September 1978.

26$^{B'}$ "Niematerialne Składowe Obiektów Fizycznych", Charles W. Misner in *Filozofować w Kontekście Nauki*, M. Heller, A. Michalik, and J. Życiński, eds. (Polskie Towarzystwo Teologiczne, Kraków 1987) pp 164–169; translated from UNESCO Conference, Munich and Ulm, September 1978 by M. Głódź.

27^B "Infinity in Physics and Cosmology", Charles W. Misner in *Infinity*, D. O. Dahlstrom, D. T. Ozar, and L. Sweeney, eds. (Proceedings of the American Catholic Philosophical Association, Vol. 55, Washington D. C. 1981) pp 59–72; 55th Annual Meeting ACPA, St. Louis, Missouri, April 1981.

28^B "Non-linear Model Field Theories", Charles W. Misner in *Spacetime and Geometry: The Alfred Schild Lectures*, Lawrence C. Shepley and Richard Matzner, eds. (University of Texas Press, Austin 1982) pp 82–101.

29^B "Spreadsheets in Research and Instruction", C. W. Misner in E. F. Redish and J. S. Risley, editors, *Proceedings of the Conference on Computers in Physics Instruction*, (Addison-Wesley Pub. Co., Reading, Mass., 1990) pp 382–398.

30^B "Quality of Physics Teaching Through Building Models and Advancing Research Skills", C. W. Misner, invited talk, pp 3–8 in *Proceedings of the International Symposium on the Evaluation of Physics Education—Criteria, Methods and Implications*, (25–29 June 1990) M. Ahtee, V. Meisalo, H. Saarikko eds., (Dept. of Teacher Education, Univ. of Helsinki, 1991).